EPILEPSY AND THE CORPUS CALLOSUM 2

ADVANCES IN BEHAVIORAL BIOLOGY

Recent Volumes in This Series

Volume 33 LECITHIN: Technological, Biological, and Therapeutic Aspects
Edited by Israel Hanin and G. Brian Ansell

Volume 34 ALTERATIONS IN THE NEURONAL CYTOSKELETIN IN ALZHEIMER'S DISEASE
Edited by George P. Perry

Volume 35 MECHANISMS OF CEREBRAL HYPOXIA AND STROKE
Edited by George Somjen

Volume 36 NOVEL APPROACHES TO THE TREATMENT OF ALZHEIMER'S DISEASE
Edited by Edwin M. Meyer, James W. Simpkins, and Jyunji Yamamoto

Volume 37 KINDLING 4
Edited by Juhn A. Wada

Volume 38A BASIC, CLINICAL AND THERAPEUTIC ASPECTS OF ALZHEIMER'S AND PARKINSON'S DISEASES
Volume 1
Edited by Toshiharu Nagatsu, Abraham Fisher, and Mitsuo Yoshida

Volume 38B BASIC, CLINICAL AND THERAPEUTIC ASPECTS OF ALZHEIMER'S AND PARKINSON'S DISEASES
Volume 2
Edited by Toshiharu Nagatsu, Abraham Fisher, and Mitsuo Yoshida

Volume 39 THE BASAL GANGLIA III
Edited by Giorgio Bernardi, Malcolm B. Carpenter, Gaetano Di Chiara, Micaela Morelli, and Paolo Stanzione

Volume 40 TREATMENT OF DEMENTIAS: A New Generation of Progress
Edited by Edwin M. Meyer, James W. Simpkins, Jyunji Yamamoto, and Fulton T. Crews

Volume 41 THE BASAL GANGLIA IV: New Ideas and Data on Structure and Function
Edited by Gérard Percheron, John S. McKenzie, and Jean Féger

Volume 42 CALLOSAL AGENESIS: A Natural Split Brain?
Edited by Maryse Lassonde and Malcolm A. Jeeves

Volume 43 NEUROTRANSMITTERS IN THE HUMAN BRAIN
Edited by David J. Tracey, George Paxinos, and Jonathan Stone

Volume 44 ALZHEIMER'S AND PARKINSON'S DISEASES: Recent Developments
Edited by Israel Hanin, Mitsuo Yoshida, and Abraham Fisher

Volume 45 EPILEPSY AND THE CORPUS CALLOSUM 2
Edited by Alexander G. Reeves and David W. Roberts

A Continuation Order Plan is available for this series. A continuation order will bring delivery of each new volume immediately upon publication. Volumes are billed only upon actual shipment. For further information please contact the publisher.

EPILEPSY AND THE CORPUS CALLOSUM 2

Edited by

Alexander G. Reeves and David W. Roberts

Dartmouth–Hitchcock Medical Center
Lebanon, New Hampshire

SPRINGER SCIENCE+BUSINESS MEDIA, LLC

The Library of Congress cataloged the first volume of this title as follows:

Main entry under title:

Epilepsy and the corpus callosum.

Based on the proceedings of the First Dartmouth Conference on the Corpus Callosum and Epilepsy, held July 1982 in Hanover, N.H.
Includes bibliographies and index.
1. Corpus callosum—Surgery—Congresses. 2. Epilepsy—Surgery—Congresses. I. Reeves, Alexander G. II. Dartmouth Conference on the Corpus Callosum and Epilepsy (1st: 1982: Hanover, N.H.)[DNLM: 1. Corpus Callosum—congresses 2. Epilepsy—congresses. WL 385 E64185]

RD594.E65 1985 616.8'53 85-3465

Based on proceedings of the Second International Conference on Epilepsy and the Corpus Callosum, held August 14–18, 1991, at Dartmouth-Hitchcock Medical Center in Hanover, New Hampshire

ISBN 978-1-4899-1429-3 ISBN 978-1-4899-1427-9 (eBook)
DOI 10.1007/978-1-4899-1427-9

Originally published by Plenum Press, New York in 1995
Softcover reprint of the hardcover 1st edition 1995

10 9 8 7 6 5 4 3 2 1

CONTRIBUTORS

FREDERICK ANDERMANN, *Montreal Neurological Institute and Hospital, McGill University, Montreal PQ H3A 2B4, Canada*

K. BAYNES, *Center for Neuroscience, University of California at Davis, Davis, California*

JOSEPH E. BOGEN, *Department of Neurological Surgery, University of Southern California, Los Angeles, California 90033*

A. BOUTHILLIER, *Service de Neurochirurgie, Hôpital Notre-Dame, Université de Montréal, Montréal, Québec, Canada*

G. BOUVIER, *Service de Neurochirurgie, Hôpital Notre-Dame, Université de Montréal, Montréal, Québec, Canada*

ARTHUR CUKIERT, *Department of Neurology and Neurosurgery, University Sao Paulo, Brazil*

COLETTE DEHAY, *Institut National de la Santé et de la Recherche Médicale, Bron, France*

R. FENDRICH, *Center for Neuroscience, University of California at Davis, Davis, California*

F. FUKUDA, *Department of Neurosurgery, Asahikawa Medical College, Nishikagura, Asahikawa 07811, Japan*

JOHN R. GATES, *The Minnesota Epilepsy Group, P.A., St. Paul, Minnesota 55102*

M.S. GAZZANIGA, *Center for Neuroscience, University of California at Davis, Davis, California*

GUY GEOFFROY, *Département de Neurologie et Neurochirurgie, Hôpital Sainte-Justine de Montréal, Montréal, Québec, Canada*

N. GIARD, *Service de Neurologie, Hôpital Notre-Dame, Université de Montréal, Montréal, Québec, Canada*

MITCHELL GLICKSTEIN, *Department of Anatomy and Developmental Biology, Neuroscience and Behaviour Group, University College London, London, WC1E 6BT, England*

GARY GRONICH, *Division of Functional Neurosurgery, University of Sao Paulo, School of Medicine, Sao Paulo, Brazil*

JOHN H. KAAS, *Department of Psychology, Vanderbilt University, Nashville, Tennessee 37240*

HENRY KENNEDY, *Institut National de la Santé et de la Recherche Médicale, Bron, France*

HERBERT P. KILLACKEY, *Department of Psychobiology, University of California, Irvine, California 92717*

MARYSE LASSONDE, *Groupe de Recherche en Neuropsychologie Expérimentale, Départment de Psyhchologie, Université de Montréal, Montréal, Québec, Canada*

G.R. MANGUN, *Center for Neuroscience, University of California at Davis, Davis, California*

C. MARESCAUX, *Clinique Neurologique, CHU, 67091 Strasbourg Cedex, France*

RAUL MARINO, JR., *Division of Functional Neurosurgery, University of Sao Paulo, School of Medicine, Sao Paulo, Brazil*

DAN C. MCINTYRE, *Department of Psychology, Carleton University, Ottawa, Ontario, K1S 5B6, Canada*

CHRISTIAN MENINI, *Institute A. Fessard, CNRS, 91190, Gif-sur-Yvette, France*

CLAUDE MERCIER, *Départment de Neurologie et Neurochirurgie, Hôpital Sainte-Justine de Montréal, Montréal, Québec*

FRANK MORRELL, *Department of Neurological Sciences, Rush Medical Center, Chicago, Illinois 60612*

R. NAQUET, *Institute A. Fessard, CNRS, 91190, Gif-sur-Yvette, France*

RICHARD E. NORDGREN, *Section of Neurology, Dartmouth-Hitchcock Medical Center, Lebanon, New Hampshire 03756*

ANDRE OLIVIER, *Montreal Neurological Institute and Hospital, McGill University, Montreal, Québec, PQ H3A 2B4, Canada*

E.A. PHELPS, *Department of Psychology, Yale University, New Haven, Connecticut 06510*

S.J. PURVES, *Sioux City Neurology, Sioux City, Iowa 51101*

ALEXANDER G. REEVES, *Section of Neurology, Dartmouth-Hitchcock Medical Center, Lebanon, New Hampshire 03756*

P.A. REUTER-LORENZ, *Department of Psychology, University of Michigan, Ann Arbor, Michigan*

ROBERT RHOADES, *Department of Anatomy, Medical College of Ohio, Toledo, Ohio 43699*

GAIL RISSE, *Department of Neurology, Hennepin County Medical Center, Minneapolis, Minnesota*

FRANK RITTER, *The Minnesota Epilepsy Group, PA, St. Paul, Minnesota 55415*

DAVID W. ROBERTS, *Section of Neurosurgery, Dartmouth-Hitchcock Medical Center, Lebanon, New Hampshire 03756*

I. ROULEAU, *Service de Neurologie, Hôpital Notre-Dame, Université du Québec à Montréal, Montréal, Québec, Canada*

J.M. SAINT-HILAIRE, *Service de Neurologie, Hôpital Notre-Dame, Université de Montréal, Montréal, Québec, Canada*

HANNELORE SAUERWEIN, *Groupe de Recherche en Neuropsychologie Expérimentale, Départment de Pshchologie, Université de Montreal, Montreal, Quebec*

MICHAEL L. SCHWARTZ, *Section of Neurobiology, Yale University School of Medicine, New Haven, Connecticut 06437*

DENNIS D. SPENCER, *Department of Neurosurgery, Yale University School of Medicine, New Haven, Connecitcut 06510*

SUSAN S. SPENCER, *Department of Neurology, Yale University School of Medicine, New Haven, Connecitcut 06510*

C. SILVA-BARRAT, *Institute A. Fessard, Gif-sur-Yvette, France*

MICHAEL SMITH, *Department of Neurological Sciences, Rush Medical Center, Chicago, Illinois 60612*

M. J. TRAMO, *Department of Neurobiology, Harvard Medical School, Neurology Service, Massachusetts General Hospital, Boston, Massachusetts 02115*

M. VERGNES, *INSERM, Centre de Neurochimie du CNRS, 67084 Strasbourg Cedex, France*

JEAN-GUY VILLEMURE, *Montreal Neurological Institute and Hospital, Montreal PQ H3A 2B4, Canada*

JUHN A. WADA, *Division of Neurosciences and Neurology, University of British Columbia, Vancouver, British Columbia V6T 2A1, Canada*

WALTER WHISLER, *Department of Neurosurgery, Rush Medical Center, Chicago, Illinois 60612*

PETER D. WILLIAMSON, *Section of Neurology, Dartmouth-Hitchcock Medical Center, Lebanon, New Hampshire 03756*

CHARLES L. WILSON, *Department of Neurology and Brain Research Institute, Reed Neurological Research Center, UCLA School of Medicine, Los Angeles, California 90024*

W. B. WOODHURST, *Department of Neurosurgery, Vancouver General Hospital, Vancouver, British Columbia V5Z 4E5, Canada*

PREFACE

In August of 1991, a second Dartmouth International Workshop on the corpus callosum was convened to share and discuss the progress that had been made over the decade that had passed following the first workshop. A nucleus of basic and clinical scientists came together to discuss their work and the work of others in a field that has been broadened clinically by the addition of many new centers throughout the world that are now performing corpus callosotomy for intractable epilepsy.

This text was stimulated by the participants' presentations and associated fertile discussions. It is compiled from the conference and subsequent studies. It reflects, both at the basic and clinical level, an important and expanding field of neural science endeavor.

In keeping with the present and rapidly expanding field of outcomes assessment, callosotomy is again evaluated in light of a further decade of surgery and follow-up. Callosotomy continues to be a useful, palliative procedure and the indications for its use have been better established.

The basic science section is a supplement to the first edition and elaborates progress in both new data and ideas. The section on experimental epilepsy models adds further support to the clinical rationale for callosotomy. Perhaps of greater importance is the contribution of experimental models to our understanding of the propagation of seizure activity.

The section on the neuropsychology of the split brain patient demonstrates the continuing major contributions to the understanding of brain and behavior that pour forth from this cornucopia.

Thanks are due to Juhn Wada, Herbert Killackey, Susan Spencer, John Gates, Dennis Spencer, Peter Williamson, and Mark Tramo for ably chairing the various sessions and organizing subsequent studies and contributions to the text.

Joan Sargent is again to be acknowledged for her extraordinary effort in getting this manuscript to press. It would not have happened without her.

Alexander G. Reeves
David W. Roberts

Lebanon, New Hampshire

CONTENTS

EPILEPSY AND THE CORPUS CALLOSUM 2

I

CORPUS CALLOSUM ORGANIZATION AND DEVELOPMENT

CORPUS CALLOSUM: HISTORY

Mitchell Glickstein, Ph.D.

University College London
Gower St.
London WC1E 6BT
England

The corpus callosum was recognized in the very first anatomical descriptions as an identifiable part of the brain. In this article I will trace some of the anatomical, clinical and experimental evidence that led to current knowledge about its structure and functions. As in all other brain pathways, the corpus callosum can best be understood in terms of its anatomical connections, and the functions of the structure which it interconnects. The corpus callosum links the two hemispheres of the cerebral cortex. As knowledge about the functions of the cortex increased, understanding of the role of the corpus callosum grew.

Study of the corpus callosum illustrates an important principle about the development of neuroscience; the relationship between anatomical observation, animal experiment, and clinical description. Much of our understanding of the structure and function of the brain is based on research with animals. Callosal damage in people is often associated with lesions of other brain structures, hence the identification of which symptoms are due specifically to loss of the callosum was not always clear. Animal research helped to clarify the role of the corpus callosum in interhemispheric integration.

Up to sixty years ago the prevalent view among physicians was that damage to the corpus callosum causes vague and rather non-specific symptoms. As late as 1929 Ironside and Guttmacher describing the "syndrome of the corpus callosum" wrote:

> *The mental changes which are so commonly present are often the first symptom. Apathy, drowsiness, and defect of memory, are met with more commonly than in growth elsewhere, and occur with such regularity in the corpus callosum syndrome as to be characteristic.*

Epilepsy and the Corpus Callosum II, Edited by A.G. Reeves
and D.W. Roberts, Plenum Press, New York, 1995

More recent evidence suggest that many of the symptoms which were ascribed to callosal damage were probably due to general impairment of brain function or to lesions of other, nearby brain structures. (In animals section of the corpus callosum of itself causes few obvious deficits.) With special testing, its unique and important role in brain function began to emerge. Although there were often misinterpretations or false conclusions, the history makes it clear that there has been a steady overall increase in understanding of the functions of the brain in general and the corpus callosum in particular.

The earliest anatomists recognized that the corpus callosum is different from the cerebral cortex, but their descriptions were vague in detail. In medieval times, interest in brain anatomy was often focused on the ventricles, since they were the presumed site of psychological function. Grey and white matter were distinguished, but possible differences in their functions were not emphasized.

Renaissance anatomists began to observe the brain more closely and to draw its structures more accurately. Vesalius, for example, in 1555 illustrated clearly the ventral surface of the brain, the origin of the cranial nerves, and the position of the cerebellum. Vesalius also described the gross position and appearance of the corpus callosum.

> *It comes into view of those dissecting when they manually separate the right side of the brain slightly from the left, for with the brain so separated, that which unites its parts there, the corpus callosum, is observed gleaming white, long, and arched. . .*

By the 18th century, the focus of anatomical studies had shifted from the ventricles to the substance of brain. With the new emphasis on brain substance, closer attention was paid to distinctions between white and grey matter in general, and to the corpus callosum in particular.

The French anatomist Viq D'Azyr, described the corpus callosum in 1786 and ventured an interpretation about its functions:

> *It seems to me that the commissures are intended to establish sympathetic communications between the different parts of the brain, just as the nerves do between the different organs and the brain itself, a consideration which can apply to all kinds of connections observed between the different parts of the brain. These communications can in general be divided into two classes: the first run from one hemisphere to the other, the second between different regions of the same hemisphere; by means of these structures impressions can be readily transmitted from the base of the brain to its center and towards its sides. I include in the first class the corpus callosum, appropriately named by some anatomists the great commissure of the brain, . . .*

The brain is soft and difficult to dissect in its natural state. Progress in understanding its structure required improved techniques for hardening. Some anatomists boiled the tissue in water or oil. Some froze the brain. Johann Christian Reil hardened the brain in alcohol and dissected it from below. With this better method, he could see its structure more clearly;

The corpus callosum,. . . consists of layers and transverse fibres in its center; however, when it spreads laterally and towards the peduncles of the cerebrum, it increases in bulk so that one can visualize it as if composed of spheres which come together at their points from both sides in the center of the corpus callosum. As the corpus callosum radiates like a cone, so it also extends in its anterior and posterior extremities and pushes into the apices of the cerebral lobes in front of the cerebrum.

The gross anatomy of the corpus callosum was clear. Little further progress could be made in relating its structure to its function without an understanding of the functions of the cerebral cortex.

The most famous proponent of the idea of cortical localization in the early part of the nineteenth century was Franz Gall. Gall believed that regional differences in the cerebral cortex of people might affect the shape of their skull. If one part of the cortex had a specialized function, and if that function were well developed in a given individual, then the skull underlying that region of the brain would be protuberant. Gall's speculations were based on the weakest of evidence. His enthusiastic disciples, nevertheless, went on to found a movement which was first called cranioscopy, and after phrenology. Phrenology purported to analyze human character and individual differences on the basis of bumps on the skull. Although Gall was a bad physiologist he was a good anatomist. His description of the structure of the corpus callosum was clear and lucid. His speculations about its possible function are of interest, but still limited.

All parts of the cerebrum are connected with analogous parts of the other hemisphere by a similar mechanism and are thus united for mutually influencing and the attainment of a common end. Some of these parts being much larger and more distinct than in the other systems, they have not escaped the attention of the most ancient anatomists. Galen described the <u>corpus callosum</u>.

Two great principles emerged and developed in the last half of the nineteenth century; the neuron doctrine and cortical localization. These conceptual advances deepened the level of understanding of the structure and function of the brain, and helped to define more accurately the role of the corpus callosum. The neuron doctrine is the idea that the basic elements of the nervous system are individual nerve cells. The processes of nerve cells may touch one another but they do not fuse. Cortical localization is the idea that different parts of the cerebral cortex do different things.

The importance of these advances for understanding brain function can be seen by considering a specific example. If a patient were to complain about troubles with his vision, an intelligent physician would think about the <u>locus</u> of those troubles. Is the problem in the eye or in the brain? Clearly diagnosis and treatment would be grossly different in the two cases. It is apparent from comparing textbooks which were written in the middle of the nineteenth century with those from the end of that century that advances in anatomy and physiology made a major difference in interpretation of neurological symptoms.

Romberg writing in 1850 about damage to the visual pathways wrote:

> *The individual affected with it sees objects indistinctly, even when they are lit up by a bright light; they appear surrounded by a fog or mist, and no effort nor the employment of artificial means increases the distinctness. The outlines of objects appear not only indistinct but also broken, and thus disfigured. The flame of the candle appears rent; while reading, the patient misses single syllables, words, and lines, and he is forced to follow them by moving his eye, his head or his entire body. At times the upper or lower, the right or left half, and circumference or center of the object only, is seen; at others the loss of vision is still more partial, and is confined to different spots of small extent, and with differently shaped outlines. Instances also occur in which the object is only seen, when it bears a definite relation to the eye, and it vanishes on the slightest movement of the eye or head. The faculty of distinguishing colour is often lost.*

Compare Romberg's text with that of Gowers written forty-two years later:

> *Each optic tract contains the fibres from the same-named half of each retina, e.e., from the temporal half of the retina on the same side, and the nasal half of the retina on the other side, and hence disease of the tract causes loss of vision in the opposite half of each field, the temporal half of the one, and nasal half of the other. This is termed <u>homonymous hemianopia, or lateral hemianopia</u>.*
> *The division between the two halves of the retina (from which fibres pass to the respective tracts) is generally in the vertical line of the macula lutes, . . .*
> *We have seen that the visual path passes, on each side, by the optic tract to the optic thalamus, and thence by the optic radiation to the cortex of the occipital lobe. This constitutes a half-vision center. . .*
> *. . . but in the half-vision centers there is no power of supplemental action, and no compensation for loss is possible. The effects of destruction of the path or center (when the results of the slighter damage have passed off) are permanent.*

If a medical student at University College were to submit Romberg's description of the visual pathways, he would fail our neuroanatomy course. The text confuses visual sensitivity with visual acuity. It fails to distinguish generalized visual impairment from scotomas. It says nothing about the symmetry of loss within the visual field of the two eyes. It fails to distinguish color vision deficits from other visual symptoms. Gowers' text, written forty-two years later, while still limited in scope, could be used to teach basic outline of the anatomy of the visual pathways today. The description of the consequence of lesions are all crisp and clear. What was the evidence that led to this advance?

The first solid support for the idea of cortical localization came from the report of Pierre Broca (1861) who found that a lesion of the left side of the frontal lobe produced a profound speech deficit. In Britain, John Hughlings-Jackson (1870) described the epileptic seizures which bear his name. Although Jackson himself was at first somewhat vague on the point, his findings have since been interpreted in terms of current views of motor control. Since neighboring parts of the body are successively affected in a jacksonian fit, perhaps adjacent parts of the body are controlled by neighboring structures in the brain.

The first experimental support for the idea of functional localization came from the physiological studies of Fritsch and Hitzig (1870). They demonstrated that electrical stimulation of a particular area of the frontal cortex of a dog produces movement on the opposite side of the body. Some fifty years earlier Magendie had shown that cutting the ventral spinal roots abolishes movement in the denervated region of the body. Fritsch and Hitzig raised the possibility that their electrical stimulation might merely have spread current to a distant spinal root, thus producing movement by activating motor nerves directly. But why, they asked, if that were the explanation, should the movement be elicited on the opposite side of the body since motor nerves on the same side are closer to the stimulating electrode? Fritsch and Hitzig also described the effects of small lesions of frontal cortex on movement. If electrical stimulation of a region of the frontal cortex produced movement, say of a forepaw, then ablation of that region of cortex led to the animal becoming clumsy in the use of that paw.

Fritsch and Hitzig's discovery led to a search for cortical localization of other functions, especially vision. Within ten years Hermann Munk (1881) studied the symptoms which were caused by lesions of the occipital lobe of a monkey and identified correctly the locus of the primary visual cortex in the occipital lobe. Unilateral lesions, he reported, produce hemianopia; bilateral lesions produce blindness.

Shortly after Munk's discovery of the role of the occipital lobe in vision, his work was extended to man by Salomon Henschen. Henschen reviewed a large number of cases to show that if the cortex on the banks and depths of the calcarine fissure is unilaterally damaged, patients become hemianopic. If the lesions spare this region of cortex, they do not.

By 1900 there was general agreement that motor cortex was in the frontal lobe and the visual cortex in the occipital lobe. The idea of cortical localization was established and has remained a fundamental principle of cortical organization.

In addition to evidence for cortical localization, there was increasing evidence in support of the neuron doctrine. Cajal (1909), the great proponent of the neuron doctrine, described and interpreted the connections of the corpus callosum. He wrote:

> *The bilateral location of perceptual centers, and the unilateral location of centers of memory, in my view, accounts for the existence of the corpus callosum. It suggests the necessity for two kinds of association fibers; or at least two kinds of collaterals: One direct which carries the homotopic part of the image to the memory center; the other commissural or callosal which carry the part of the image from the opposite hemisphere to the memory center.*

The two principles: cortical localization and the neuron doctrine, allowed the interpretation of symptoms in humans which are caused by interruption of interhemispheric pathways. Callosal lesions were interpreted in terms of connections between functional areas on the two sides of the brain.

Karl Hugo Liepmann was a physician who worked in Berlin in the early part of the twentieth century. His most famous case was that of the Regierungrast; the government counselor who sustained a large brain lesion which included the corpus callosum. Liepmann (1900) interpreted his patients symptoms in terms of damage to specific structures within the brain. In his description of this and of a later case with a more restricted midline lesion (Liepmann and Maas 1907), Liepmann attributed apraxia of the left hand to a lesion of the corpus callosum. Liepmann wrote:

> *However, closer examination has shown that generally motor apraxia in the right half of the body is accompanied by a lesser degree of apraxia (dyspraxia) in the left half (so-called sympathetic dyspraxia). In fact, in many cases where the right extremities are paralyzed, so that the apraxia is not observed, the unparalyzed upper left member displays dyspraxic phenomena. Numerous persons paralyzed on the right side exhibited dyspraxic disorders in the left arm and hand upon closer examination, although they were generally fit for gross object manipulation. These dextroplegics could not, for example, wave, shake, thumb the nose, toss a kiss, imitate demonstrated motions, or trace the movements of objects with their left hands. The decision as to whether these wrong reactions of the left hand are motor or ideational is extremely difficult, . . . In a fraction of the cases, the disorder in the left hand is more severe up to the inability to perform gross object manipulations.*

Liepmann ascribed such apraxias of the left hand to a broken connection between an area in the left hemisphere which had learned these semi-skilled movements and the right motor cortex which no longer could receive instructions from the center in the left hemisphere.

Although Liepmann's observations suggest that lesions of the corpus callosum might produce similar symptoms in experimental animals, the first experiments yielded contradictory results. Some authors reported diffuse symptoms to be caused by cutting the corpus callosum, some found none. An early negative report was that of Alexander von Koranyi who wrote in 1890:

> *The conclusion which can be drawn from these observations is that cutting the corpus callosum, if it can be accomplished without damaging the hemispheres, produces no obvious disturbances. If the hemispheres are damaged then there may be disturbances in vision, sensation or movement, . . .*

Similarly, Hartmann and Trendelenberg (1927) trained monkeys in tasks which require bi-manual coordination. They also found no obvious effects; the tasks could be performed accurately despite callosal section.

It is now clear that the experiments of Von Koranyi and of Hartmann and Trendelenberg failed to test adequately the role of the callosum, since the behavior which they studied does not require it. The earliest direct evidence of a role for the corpus callosum in interhemispheric integration came from work in Pavlov's lab. Anrep (1923), a student of Pavlov, conditioned dogs to salivate to cutaneous stimulation. He brushed a point on the skin of a dog (the conditioned stimulus) and followed it by presenting food (the unconditioned stimulus). Anrep tested for generalization of this conditioned response by brushing other points on the body surface. If a nearby point on the skin was brushed, the trained animal would also salivate, but not as much as it had done to the original locus. If a remote point was stimulated the animal might not salivate at all. Anrep also tested the effect of applying the conditioned stimulus to the symmetrically opposite side of the body. He found that the animal would salivate an equal amount when tested on the opposite side. If the conditioned response was extinguished by brushing one side without following the cutaneous stimulus with food, the response was extinguished on the other side.

Bykoff (1925), another student of Pavlov's, followed and extended Anrep's work by cutting the corpus callosum. Callosum-sectioned dogs, he reported could be trained to respond to cutaneous stimuli applied to either side, but the conditioned response failed to transfer between the two sides.

The psychologist Karl Lashley had a strong interest in interhemispheric transfer. Several of his students attempted to study interhemispheric transfer of visual learning by training an animal on a visual discrimination task using one eye and testing for transfer to the other. Some tested transfer in animals in which the optic chiasm had been divided in the midline, and found a high level of interocular transfer. Some sectioned the corpus callosum leaving the chiasm intact. The visual discrimination also transferred between the two eyes.

In their textbook Morgan and Stellar (1955) summarized these experiments:

> *We end up, then, without a critical experiment, i.e., one that cuts all crossings of the visual system. We cannot therefore draw any firm conclusions. We know only that neither the optic chiasm nor corpus callosum is solely necessary for interocular transfer.*

The proper tests for the role of the corpus callosum in interhemispheric transfer would be to study an animal in which visual input is restricted to one hemisphere by sectioning the optic chiasm in the midline and callosal fibres are cut. This experiment as first done by Ronald Myers (1956), then a graduate student working with Roger Sperry at the University of Chicago and later at the California Institute of Technology. Myers showed that cats in which the optic chiasm and corpus callosum had been sectioned in the midline readily learned a visual discrimination task using one or the other eye. When tested on the untrained eye, however, they failed to show any knowledge of the problem.

Working with Roger Sperry, I replicated this essential finding, testing monkeys for intermanual transfer of tactile learning (Glickstein and Sperry 1960). Since the great majority of cutaneous fibres are crossed it was not necessary to divide the sensory tracts. Monkeys were trained to reach out from a box and palpate two objects without seeing them. The animals readily learned to do so using either hand. Once trained the monkey was tested on the other hand. Normal monkeys show a high level of transfer between the hands. In monkeys in which the callosum had been cut, there was no transfer. Opposite habits could be learned easily and maintained on the two sides.

Many other studies support the general conclusion that when sensory information is restricted to one hemisphere and the corpus callosum and commissures are cut, there is no transfer of learning between the hemispheres. Other findings emerged from these experiments which remain rather puzzling. Downer (1959) reported that in a monkey in which the corpus callosum and optic chiasm had been sectioned in the midline the animal seemed to grope blindly when forced to use an ipsilateral eye-hand combination. This report made sense. Having cut the optic chiasm and corpus callosum there seemed to be no way for fibers from the visual cortex to connect to motor areas in the opposite hemisphere.

But although monkeys prefer to use the hand contralateral to the open eye, they can quickly learn to use any combination of hand and eye. With post-operative training monkeys are capable of skilled use of eye-hand combinations in the absence of corticocortical connections (Myers et al. 1962; Glickstein and Sperry 1963). Sperry summarized this rather puzzling state of affairs (1961) and concluded that subcortical routes must be able to serve functional connection between visual and motor areas of the brain.

It now seems likely that the route Sperry suggested is by way of the cerebellum (Glickstein 1990). Cells in the cerebral cortex project to the cerebellum by way of the pontine nuclei. Cells in the pontine nuclei in turn project by way of the middle cerebellar peduncle to both sides of the cerebellar cortex where they terminate as mossy fibres. There is abundant transmission of visual information along the corticoponto-cerebellar pathways. As Voneida (1963) suggested, bilateral pontocerebellar projections would account for the residual high level visuo-motor control found in split-brain animals.

The history of studies of the structure and function of the corpus callosum demonstrates the importance of interaction between anatomical, physiological and behavioral evidence. Confirmation of the neuron doctrine, and of the principle of cortical localization were critical advances. The sparing of sensory guided movements in split-brain animals in which the sensory input is restricted to one hemisphere and movements are controlled from the other, emphasizes the role that subcortical pathways must play in the sensory guidance of movement.

ACKNOWLEDGMENTS

I thank Dr. Edwin Clarke for permission to use translations from his book with C.D. O'Malley, "The Human Brain and Spinal Cord," Berkeley University of California Press, 1968. I also thank Erlbaum Associates for permission to quote from a translation of Liepmann's work in J.W. Brown, "Agnosia and Apraxia," 1988.

REFERENCES

Anrep, G.V., 1923, The irradiation of conditioned reflexes, *Proc. Roy. Sic. B.* 94:404-426.

Broca, P.P., 1861, Remarque sur la siège de la faculté du langage articulé suive d'une observation d'aphémie (perte de la parole), *Bull. Soc. Anat. Paris* 36:330-357.

Brown, J.W., 1988, "Agnosia and Apraxia," Plenum Press, New York.

Bykoff, K., (Bykov), 1925, Versuche an hunden mit durchschneiden des corpus callosum, *Zentrlbl. f. d. ges. Neurol. u. Psychiat.* 39/40:199.

Cajal, S. Ramon y., 1909-11, "Histologie du Systême Nerveux," C.S.I.C., Madrid (Translated into French by L. Azoulay, reprinted 1952).

Clarke, E. and O'Malley, C.D., 1968, "The Human Brain and Spinal Cord," University of California Press, Berkeley and Los Angeles.

Downer, Jh de C., 1959, Changes in visually guided behaviour following midsagittal division of optic chiasm and corpus callosum in monkey (Macaen Mulatta), *Brain* 82:251-259.

Fritsch, G. and Hitzig, E., 1870, Über die electrische erregbarkeit des grosshirns, *Anat., Physiol., und Wissenschaftl. Mediz. Leipzig* 300-332.

Gall, F.J. and Spurzhein, J.C., 1810-1819, Anatomie et physiologie du systême nerveux en général, et du cerveau en particulier, avec des observations sur la possibilité de reconnoitre plusieurs dispositions intellectuelles et morales de l'homme et des animaux, par la configuation de leur têtes, *4 Vols. and atlas Vol. 1 and 2 and atlas Paris: F. Schoell, Vol. 3 Paris: Libraire Greque-Latine-Allemande, 1818; Vol. 4 Paris, N. Maze 1819.* Translated in Clarke, E. and O'Malley, C.D., 1968, "The Human Brain and Spinal Cord," Berkeley U. of California, pp. 599-600.

Glickstein, M., 1990, Brain pathways in the visual guidance of movement, *in:* Brain Circuits and Functions of the Mind," Cambridge University Press, 157-167.

Glickstein, M. and Sperry, R.W., 1960, Intermanual somesthetic transfer in split-brain rhesus monkeys, *J. Comp. Physiol. Psychol.* 53:322-327.

Glickstein, M. and Sperry, R.W., 1963, Visuo-motor coordination in monkeys after optic tract section and commissurotomy, *Fed. Proc.* 22:456.

Gowers, W.R., 1892, Diseases of the nervous system, *in:* "Brain and General Function Diseases," 2nd Ed., Vol. II, J. and A. Churchill, London.

Hurtmann, jun. Fr. and Trendelenburg, W., 1927, Zur frage der bewegungsstörungen nach balkendurchtrennung an der katze und am affen, *Ztbl. f. d. ges. Neurol. u. Psychiat.* 47:149.

Henschen, S.Z., 1890, "Klinische und Anatomische Beiträge zur Pathologie des Gehirns, Part 1," Almquist and Wiksell, Upsala.

Ironside, R. and Guttmacher, M., 1929, The corpus callosum and its tumours, *Brain* 52:442-483.

Jackson, J. Hughlings, 1931, A study of convulsions, *reprinted in:* "Selected Writings of John Hughlings," J. Taylor, ed., Hodder & Stoughton, London, pp. 18-36.

Koranyi, A. von, 1890, Über die folge der durchschneidung des hirnbalkens, *Arch. F. Physiol.* 47:35-42.

Liepmann, H., 1900, "Das Krankheitsbild der Apraxie (Motorischen Asymbolie)," Karger, Berlin.

Liepmann, H. and Maas, O., 1907, Klinische-anatomischer beitrag zur lehre von der bedeutung der linken hemisphäre und des Balken für das handeln, *Berl. Klin. Wochenschr.* 44:757-758.

Morgan, C.T. and Stellar, E., 1955, "Physiological Psychology, 2nd Ed.," McGraw-Hill, New York.

Munk, H., 1881, "Über die functionen der grosshirnrinde," A. Hirschwald, Berlin, pp. 28-53, *English translation in:* von Bonin, G., 1960, *The Cerebral Cortex* 4:97-117.

Myers, E.E., 1956, Function of corpus callosum in interocular transfer, *Brain* 79:358-363.

Myeers, R.E., Sperry, R.W., and McCurdy, N.M., 1962, Neural mechanisms in visual guidance of limb movements, *Arch. Neurol.* 7:195-202.

Reil, J.C., 1795, Über den bau des hirns und der nerven, *Neues J. Physiol. Leipzip,* 1:96-114, *translation in:* ""The Human Brain and Spinal Cord," E. Clance and C.D. O'Malley, 1968, Univ. of California, Berkeley, p. 595.

Romberg, H., 1850, "Diseases of the Nervous System, Vol. 1," Edw. H. Sieveking, Transl. and ed., 1853, the Sydenham Society, London.
Sperry, R.W., 1961, Cerebral organisation and behaviour, *Science* 133:1749-1757.
Vesalius, A., 1543, "De Humani Corporis Fabrica Libri, Septem, Basel, *oponimus translation in:* ""The Human Brain and Spinal Cord," E. Clance and C.D. O'Malley, 1968, Univ. of California, Berkeley, p. 579.
Vicq d'Azytz, F., 1786, "Traité d'anatomie et de physiologie," Didot l'Aine, Paris, *translation in:* ""The Human Brain and Spinal Cord," E. Clance and C.D. O'Malley, 1968, Univ. of California, Berkeley, p. 592
Voneida, T.J., 1963, Performance of a visual conditioned response in split-brain cats, *Exp. Neurol.* 8:493-504.

DEVELOPMENT OF THE CORPUS CALLOSUM

Herbert P. Killackey, M.D.

Department of Psychobiology
University of California
Irvine, California 92717

During the course of development, the areal distribution of neurons which project across the corpus callosum undergoes significant changes. In the mature animal, the neurons which contribute axons to the corpus callosum are not uniformly distributed throughout the neocortex. Rather, within any given cortical area they are distributed in a differential fashion which varies from cortical area to cortical area. Generally speaking, the distribution of callosal projection neurons is more restricted within primary sensory areas than it is in other cortical areas. This discrete adult distribution of callosal projection neurons emerges out of a more widespread distribution in the developing animal. This phenomenon was initially reported by G. Innocenti and co-workers who described the initial widespread distribution of callosal projection neurons in the visual cortex of the neonatal cat. The term they used to refer to this initial widespread distribution was "exuberance." This phenomenon has since been reported to occur in the neocortex of a wide variety of mammalian species and in a second major class of neocortical projection neurons (cortico-subcortical) suggesting that it is a general feature of neocortical ontogeny. Two major questions concerning this phenomenon are: 1) what is the mechanism by which it occurs? and 2) what is its adaptive significance? Before turning to these questions, however, it is necessary to describe the phenomenon.

The changes in distribution of callosal projection neurons within the somatosensory cortex of the rat and rhesus monkey provide excellent examples. The pattern of callosal projection neurons in the somatosensory cortex of the adult rat is punctate and discontinuous. The overall pattern of callosal projection neurons and their terminations is clearly complementary to the distribution of thalamic afferents from the ventral posterior nucleus. Indeed, the distribution of callosal projection neurons and their termination clearly surrounds and interdigitates with the morphological map of the body surface which composes the primary somatosensory cortex of the rat. However, at birth the cells of origin of

Epilepsy and the Corpus Callosum II, Edited by A.G. Reeves
and D.W. Roberts, Plenum Press, New York, 1995

the immature callosal projection are continuously distributed in the deeper cortical layers below the cortical plate and callosal afferents are evenly distributed in the underlying white matter. The discontinuous adult pattern is established during the early postnatal period and is accompanied by the specific in growth of callosal afferents into the regions where they terminate in the adult.

This same phenomena occurs in the somatosensory cortex of the rhesus monkey during the last trimester of fetal life. In the adult rhesus monkey there is a gradient of callosal projections across the postcentral gyrus. They are most dense caudally in area 2, less dense in area 1, and least dense rostrally in area 3b. Within a given area, callosal projections are least dense in the hand and foot representations (regions which may be presumed to receive the densest thalamocortical projections) and denser in the regions representing the limbs, trunk and much of the head. At embryonic day 108 (approximately two months before birth) callosal projection neurons are located in the appropriate cortical layers but they are distributed in a uniform continuous band across the postcentral gyrus. The mature discontinuous distribution pattern of callosal projection neurons in the rhesus monkey is discernible by embryonic day 133, approximately one month before birth.

Since its initial description, this same phenomenon has been described in a number of cortical areas of a number of diverse species including mouse, rat, hamster, rabbit, opossum, cat, ferret, cynomolgus and rhesus monkey. The one apparent exception to the generality of this phenomenon is the striate cortex (area 17) of the monkey. This cortical area which is largely free of callosal projection neurons in the adult monkey also does not seem to contain callosal projection during development although "exuberant" changes in the distribution of callosal projection neurons can be found in the adjoining area 18 in this species. The early "exuberant" distribution of callosal projection neurons is often interpreted as evidence that the developing brain is more diffusely organized than the mature brain. This is not necessarily the case. It should be emphasized that while the areal distribution of callosal projection neurons is more widespread in the immature animal several other aspects of this projection are quite specific and discrete. The neuronal cell bodies are located in the appropriate laminar position. The position of the fibers within a pathway such as the corpus callosum is appropriate for their cortical area of origin and they appear to end in the appropriate position in the other hemisphere before they begin their ingrowth into the cortical layers.

The advent of fluorescent dyes as retrograde tracers brought two novel features to anatomical studies which allowed further investigation of the morphological events underlying "exuberant" change. First, these dyes remain in neurons for a relatively long period of time with no apparent damage to the neuron containing the dye. This feature allows the projection neurons to be labelled at an early time when the distribution is still widespread and then to be assayed at a later time after distributional changes have occurred. The second feature is that distinct dyes can be injected into different targets and both be easily discriminated in either the same or different neuron(s).

This major ontogenetic change in the distribution of callosal projection neurons is achieved in the same manner in the somatosensory cortex of the rat and the rhesus monkey, and presumably, in other cortical areas in other species as well. Callosal projection neurons initially send processes to at least two targets; one process via the corpus callosum to the other hemisphere and a second process to a target in the ipsilateral hemisphere. If a callosal process is located in an inappropriate position (with respect to the mature discrete distribution) it is eliminated by mechanisms which culminate in the glial phagocytosis of the callosal process and the projection neuron maintains an ipsilateral projection.

This raises the question of the mechanisms underlying "exuberant" distributional changes in cortical projection neurons. At present, all that can be noted is that the factors affecting such changes appear to be associated with the cortical area in which the projection neurons are located and that, in the case of callosal projection neurons, the periphery acting through the thalamus appears to play some role in sculpting their distribution.

In terms of adaptive significance, "exuberance" may be regarded as a variant of neuronal death which is a major morphogenetic feature of the development of most parts of the nervous system. Cell death does occur in the neocortex but it does not appear to be a particularly major phenomenon in this structure as compared to other portions of the central and peripheral nervous system. Indeed, it seems intuitively counterproductive for major cell loss to occur during the development of a structure in which there appears to be evolutionary pressure to increase the size of that structure as is the case for the neocortex. During development, cortical projection neurons face the dual task of coming into register with the source of afferent input into the cortical area in which they reside and with their target. Cortical "exuberance" appears to be part of a unique solution to this problem. It is a mechanism which allows the neocortex to solve the same problems which confront other neuronal structures without major cell loss.

REFERENCES

1. L.M. Chalupa and H.P. Killackey, Process elimination underlies ontogenetic change in the distribution of callosal projection neurons in the postcentral gyrus of the fetal rhesus monkey, *Proc. Nat. Acad. Sci.* 86:1076-1079 (1989).
2. G.M. Innocenti, General organization of callosal connections in the cerebral cortex, *in:* "Cerebral Cortex, Vol. 5," E.G. Jones and A. Peters, eds., Plenum Press, New York (1986).
3. G.O. Ivy and H.P. Killackey, Ontogenetic changes in the projections of neocortical neurons, *J. Neurosci.* 2:735-743 (1982).
4. H.P. Killackey, Neocortical expansion: an attempt towards relating phylogeny and ontogeny,*J. Cog. Neurosci.* 2:1-17 (1990).
5. H.P. Killackey and L.M. Chalupa, Ontogenetic change in the distribution of callosal projection neurons in the postcentral gyrus of the fetal rhesus monkey, *J. Comp. Neurol.* 244:331-348 (1986).
6. K. Koralek and H.P. Killackey, Callosal projections in rat somatosensory cortex are altered by early removal of afferent input, *Proc. Nat. Acad. Sci.* 87:1396-1400 (1990).

THE ORGANIZATION OF CALLOSAL CONNECTIONS IN PRIMATES

Jon H. Kaas, Ph.D.

Department of Psychology
Vanderbilt University
Nashville, TN 37240

INTRODUCTION

Information is sent from one cerebral hemisphere to the other almost exclusively via the corpus callosum and the anterior commissure, with the corpus callosum forming the major pathway. The regions of cortex subserved by these two pathways vary across mammalian species, so that all of the interhemispheric connections in monotremes and marsupials are via the anterior commissure, and there is no corpus callosum (Ebner 1967; Heath and Jones 1971; Granger et al. 1985; Cusick and Kaas 1986). In eutherian (placental) mammals, however, most of cortex is subserved by the corpus callosum, and only temporal cortex is interconnected by axons coursing through the anterior commissure (Ebner and Myers 1965). In primates, only a limited region near the pole of the temporal lobe is subserved by the anterior commissure (Jouandet and Gazzaniga 1979; Pandya, et al. 1975; Zeki 1973). Thus, most of cortex depends on the corpus callosum for the transfer of information.

A major problem in describing the corpus callosum connections in primates is that primates vary greatly in most features, and callosal connections have been studied in only a few primate species. The primate order contains 15 families, about 60 genera, and over 200 species. Primates range in body size from the 100 gram mouse lemur to the 200 kg gorilla, and the surface area of cortex varies from about 10 cm^2 in mouse lemurs to around 2,500 cm^2 in humans (Hofman 1985). The neocortex of all mammals is divided into a number of functionally distinct areas, but only a few areas have been identified with great certainty, and it is not yet clear how many areas exist in any primate. However, comparative studies indicate that all mammals, and hence all primates, have several areas in common, the number of areas has increased independently in several

Epilepsy and the Corpus Callosum II, Edited by A.G. Reeves
and D.W. Roberts, Plenum Press, New York, 1995

lines of evolution, and the number of areas is probably considerably greater in humans than in mouse lemurs (Kaas 1982, 1987, 1989a). The most primitive of mammals probably have on the order of 10-15 cortical areas and humans may have 100 or more (see Kaas 1989b). A review of callosal connections in primates cannot fully address this diversity at this time, and a description of connections, area by area, is not possible. Furthermore, direct information about the arrangement of callosal connections in humans is limited, and human brains are clearly the most unusual. Nevertheless, comparative studies and reviews (e.g., Cusick and Kaas 1986) reveal general features that apply across mammalian and primate species, and the features most likely pertain to humans as well. The connection patterns indicate that all cortical areas transfer information from one hemisphere to the other, connections provide both feedforward and feedback inputs in hierarchical processing systems, and these connections function by activating neurons and contributing to their excitatory receptive field centers, or more often modulate activation by contributing to broad antagonistic surrounds (see Allman et al. 1986).

CALLOSAL CONNECTIONS ARE BROADLY, BUT UNEVENLY, DISTRIBUTED

There have been several studies of the total pattern of callosal connections for large regions of cortex in primates. These include the pioneering report on the commissural connections in macaque monkeys (Fig. 1; Myers 1962; also Myers 1965) as well as more recent descriptions of interhemispheric connections in macaques (e.g., Jones and Powell 1969; Karol and Pandya 1971; Pandya et al. 1971; Van Essen et al. 1982; Killackey et al. 1983), owl monkeys (Cusick et al. 1984; Killacky et al. 1983; Gould et al. 1986; Newsome and Allman, 1980), squirrel monkeys (Gould et al. 1987); marmosets (Cusick et al. 1984) and prosimian galagos (Newsome and Allman 1980; Cusick et al. 1984), as well as tree shrews (Cusick et al. 1985), which are close relatives of primates. Direct information on callosal connections in humans is also available, but limited (Clark and Miklossy 1990; Shoumura et al. 1975). Results from all these studies indicate that callosal connections are broadly distributed and involve all or nearly all subdivisions of cortex. Thus, all or nearly all cortical areas are involved in the interhemispheric transfer of information.

In addition, large portions of most areas are involved, so that roughly 70-80% of cortex is callosally connected (e.g., Van Essen and Zeki, 1978). Yet, callosal connections are markedly uneven in distribution, being very dense in some locations and very sparse or absent in others. This unevenness raises the obvious but largely unanswered question: What is the significance of the uneven distribution? Of course, there are two long prevailing concepts relating to this unevenness (Ebner and Myers 1965). One is that primary visual cortex and a number of other visual areas are restricted to representing the contralateral visual hemifield, and thus callosal connections are dense along the borders of such areas because they connect neurons representing identical or nearly identical portions of visual space (e.g., Choudhury et al. 1965; Zeki and Sandman 1976; Lepore et al. 1986). They thereby subserve "local circuit" functions that

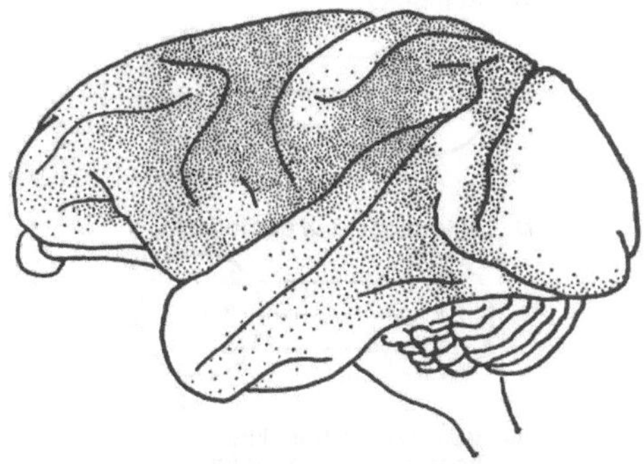

Figure 1. The first comprehensive portrayal of the distribution of callosal terminations in the cerebral cortex of a primate (*Macaca mulatta*). A lateral view of the cerebral hemisphere with dots indicating the distribution of commissural fibers (from Myers, 1965). More recent studies are consistent with this overview in that most regions of cortex are callosally connected, and local distributions tend to be uneven. More is now known about how connection patterns relate to proposed subdivisions of cortex (areas), the nature of connections for given locations in cortex, and laminar patterns of cells of origin and terminations (see text).

mediated in other parts of these areas by short intrinsic connections. In addition, since callosal connections within the borders of such areas would form visuotopic mismatches in visual inputs, callosal connections are sparse or absent (see Kennedy et al. 1991). Similar lines of reasoning lead to the conclusion that for the representations of the contralateral half of the body in somatosensory cortex, regions within these representations that relate to the body midline are densely connected via the callosum, while the representations of body parts away from the body midline are not (see Manzoni et al. 1989). These generalizations are useful, but they do not fully account for the variations in density of callosal connections. Callosal connections are also concentrated in locations in visual and somatosensory fields other than at the representations of the vertical midline of the visual field or the midline of the body.

HOMOTOPIC, HETEROTOPIC, AND HETEROAREAL CONNECTIONS

Callosal connections are of three basic types (Fig. 2; Weller and Kaas 1981). One type can be termed *homotopic* in that regions of one cortical area connect with regions of the corresponding cortical area in the other hemisphere that represent the same receptor surface. The classical example of this type of connection is along the outer border of primary visual cortex (V-I) or the inner border of the second visual area, V-II, where the vertical midline of the visual field (zero vertical meridian corresponding to the line of decussation of the retina) is represented. Receptive fields for the neurons along the border of V-I with V-II in one hemisphere overlap those for neurons in the other hemisphere (e.g., Choudhury et al. 1965), and matched locations along the border are interconnected (Sesma et al. 1984; Pritzel et al. 1988). Such interconnections are likely to contribute to the excitatory receptive fields of neurons along the border, and perhaps facilitate binocular activation (e.g., Payne et al. 1984).

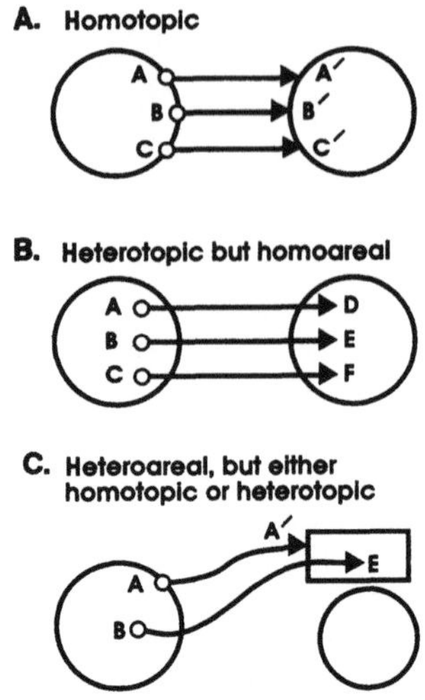

Figure 2. A schematic portrayal of types of callosal connections. A. Homotopic connections are between areas representing the right or left body surface or visual hemifield. Locations along the borders of such areas can be somatotopically matched for locations along the body midline or visuotopically matched for locations along zero vertical meridian through the point of fixation. Such homotopic connections may be functionally similar to short-range, lateral connections within areas. A long-standing view is that homotopic callosal connections subserve "midline fusion" and "ensure the unitory perception of visual and somatosensory space" (see Manzoni et al., 1989). B. Heterotopic callosal connections are between mismatched locations in sensory representations. Locations A, B, and C, for example, could be all 20° from the midline in the right visual hemifield, but have connections with locations D, E, and F some 20° from the midline in the left visual hemifield. Such connections are visuotopically mismatched, but they are often matched for mirror-image locations (upper field to upper field). Similarly, connections between the representations of body parts such as the hand in the two hemispheres are somatopically mismatched, but matched for mirror image locations. Heterotopic connections can be between pairs of areas (homoareal) or between different areas (heteroareal). C. Cortical areas typically connect with their counterpart in the opposite hemisphere, and several other areas as well. The callosal connections between different areas (VI with VII, for example) are heteroareal. If they are matched for visual or somatosensory space, they are also homotopic; if they are mismatched they are also heterotopic.

A concentration of visual homotopic connections characterizes the V-I/V-II border of all mammals (e.g., Ebner and Myers 1965; see Cusick and Kaas 1986). Because of this conspicuous concentration, and the logical premise that homotopic callosal connections related to the zero vertical meridian should characterize all visual areas with topographic representations of the visual hemifield, various investigators have attempted to use total callosal connection patterns to identify extrastriate visual areas with boundaries representing the zero vertical meridian (e.g., Zeki and Sandeman 1976; Van Essen et al. 1982; Clarke and Miklossy 1990). However, callosal connection patterns appear to have limited usefulness in locating such borders, largely because other types of callosal connections can be quite dense and they confound simple interpretations. In preparations of flattened cortex from New World monkeys, cut parallel to the surface so that callosal patterns can be viewed over large regions of cortex in single sections (e.g., Cusick et al. 1984), only the V-I/V-II border is obviously delimited. Even the middle temporal visual area, MT, another orderly representation of visual field (easily identified by myeloarchitecture), has only a slight increase in overall callosal connections along the outer border, and these connections are too broad and uneven to locate the border with any certainty.

In somatosensory cortex homotopic connections would be limited to portions of representations devoted to the body midline. In anterior parietal cortex, each of four architectonic fields, areas 3a, 3b, 1 and 2, represent the contralateral body (see Kaas 1983) and it has been proposed that borders between these areas are marked by dense callosal connections largely related to proximal body parts and the body midline (e.g., Shanks et al. 1975). This conclusion has not been supported by further studies of callosal connections in primates (Killacky et al. 1983), but some callosal axons do have receptive fields along the body midline (e.g., Lepore et al. 1986; Guillemot et al. 1987; Manzoni et al. 1989) and some callosal connections are homotopic, possibly contributing to the excitatory portions of receptive fields (Conti et al. 1986; Manzoni et al. 1989). In addition, the midline representations of the back, belly, and upper face do form part of the boundaries of at least area 3b and 1 (see Sur et al. 1982; Kaas 1983; Conti et al. 1986), but these parts of the body do not have high receptor densities and are not represented in large portions of cortex (they have a low "magnification factor"). Thus, even dense connections related to the body midline would indicate only small portions of the boundaries of areas 3b and 1. In addition, non-homotopic connections are so prevalent in somatosensory cortex (e.g., Gould and Kaas 1981; Krubitzer and Kaas 1990c) that even this limited information is blurred.

Auditory cortex differs from visual and somatosensory cortex in that auditory areas that are tonotopically organized form tonotopically matched pairs. In monkeys, three auditory areas, A-I, R (for rostral), and RT (for rostrotemporal) contain orderly representations of tones, so that in surface views of the brain, isofrequency bands or contours course across one dimension of each area while the other dimension forms an orderly progression, from high to low, of frequency contours (see Morel and Kaas 1992). Each isofrequency band is preferentially connected with a band matched for the same frequency in the same area of the opposite hemisphere (e.g., Luethke et al. 1989). Thus, for these auditory areas, homotopic or frequency-matched callosal connections prevail. Nevertheless, total patterns of callosal connections reveal little about the boundaries or organizations of auditory areas, since homotopic matches occur throughout auditory areas.

The representations of body movements in subdivisions of motor cortex are complex and fragmented in somatopic organization (see Gould et al. 1986). Thus, it is difficult to denote areas of even primary motor cortex (M-I) as devoted to the body midline. Nevertheless, it is possible to delimit portions of M-I related to movements of the proximal body or distal body. Callosal connections are dense in parts of M-I related to the shoulder and trunk, and sparse for parts related to the hand (see Gould et al. 1986). Thus, callosal connections of M-I favor portions commonly involved in bilaterally synergistic movements, but these connections are not strictly homotopic.

Many callosal connections are *heterotopic* in that they are between mismatched portions of representations of sensory surfaces. These connections are typically between mirror-image positions in pairs of sensory representations (Fig. 2). For example, the middle temporal visual area (MT) of primates systematically represents the contralateral visual hemifield (in a manner similar to that in V-I). Connections between MT of one hemisphere with MT of the other hemisphere are most dense between symmetrical positions. However, except for the outer border of MT, devoted to the zero vertical meridian, matched locations in the two fields correspond to different, mirror-image locations in the right and left visual hemifields. Thus, these callosal connections can not contribute to the excitatory receptive field centers of MT neurons, which are on the order of 5-20° in diameter and all located in the contralateral hemifield (e.g., Felleman et al. 1984). Instead, they contribute to the large antagonistic surrounds of these receptive fields, since the surround includes large portions of the ipsilateral as well as the contralateral visual hemifield (see Allman et al. 1986). More specifically, neurons in MT are best activated by stimuli within the receptive field center that are moving in a particular direction; stimuli moving in the same direction in the large surround decrease the response, while stimuli moving in the opposite direction in the surround enhance the response. Neurons at many levels of the visual system are characterized by antagonistic surrounds, and in extrastriate cortex, heterotopic callosal connections may commonly contribute to large, bilateral surrounds (see Land et al. 1983). Topographic mismatches in callosal inputs also occur in the somatosensory system, where the hand region of S-II, for example, receives dense inputs from hand representations in the other hemisphere (Manzoni et al. 1984; 1989; Krubitzer and Kaas 1990a). Such connections could contribute to antagonistic "surrounds" so that stimuli on both hands decrease the response. Similarly, some "off-focus" callosal connections between the primary auditory areas that result in tonotopic mismatches may contribute to antagonistic surrounds.

Heterotopic callosal connections may also mediate large, bilateral, excitatory receptive fields. Receptive fields in parts of inferior temporal cortex of monkeys, for example, have large receptive fields that include parts of both the ipsilateral and contralateral visual hemifields. The extensions of these receptive fields into the ipsilateral hemifield depend on callosal inputs (Rocha-Miranda et al. 1975). Likewise, the receptive fields in the S-II region of monkeys that include both forepaws (Robinson and Burton 1980) undoubtedly depend on the merging of ipsilateral and callosal inputs.

A third type of interhemispheric connection is between one area in one hemisphere and different areas in the other hemisphere. MT connects callosally, not only with MT, but with two or three adjoining visual areas as well (see Krubitzer and Kaas 1990a). Likewise, V-II has connections with three other visual areas in the opposite hemisphere: V-I, dorsolateral visual cortex, and MT (Cusick and Kaas 1988; Kennedy et al. 1986; Kennedy and Dehay 1988). S-I (area 3b) has callosal connections with S-II and the parietal ventral area (PV), in addition to 3b, and S-II has callosal connections with area 3b, area 1, PV, M-I and rostral parietal cortex (Krubitzer and Kaas 1990c). Connections of A-I include A-I and adjoining auditory fields of the opposite hemisphere (Luethke et al. 1989). Thus, the common observation is that each area interconnects with several areas in the other hemisphere. Such mismatched callosal connections between different areas are *heteroareal* (Fig. 2). Of course, heteroareal connections might be matched (homotopic) or mismatched (heterotopic) for receptor surface information. Most of the heteroareal connections, however, appear to also be heterotopic.

Heteroareal callosal connections can be of feedforward, intermediate or feedback types. Visual, somatosensory, and auditory areas are arranged in orderly processing hierarchies. Even though each field contains multiple inputs, it is possible to place areas at different levels in the system, so that MT is higher than V-I, and the middle superior temporal (MST) area is higher than MT (see Kaas 1989b). Rockland and Pandya (1979) and Maunsell and Van Essen (1983) have noted that ipsilateral connections between areas differ in laminar patterns according to their respective levels, and thus laminar patterns of connections provide information about processing levels. Feedforward connections typically originate almost exclusively from neurons in supragranular cortical layers, and terminate largely in layer 4 of target areas. Feedback connections usually originate at least in part from infragranular layers, and terminate above and below layer 4. Intermediate connections terminate within and outside layer 4. Total patterns of callosal inputs tend to look intermediate (see Killackey et al. 1983) because feedforward and feedback inputs are mixed. However, callosal projections from individual areas tend to be of the feedforward type, if to higher areas, and of the feedback type, if to lower areas (see Kennedy et al. 1986, 1991; Kennedy and Dehay 1988). Callosal connections between matched areas tend to be of the feedback type, and this is consistent with the more common role of contributing to the surround rather than the center of receptive fields.

SPECIES VARY IN CALLOSAL PATTERNS

Mammals appear to vary in number of cortical areas as well as brain size (Kaas 1987). Thus, mammals with more areas have callosal connections that mammals with fewer areas lack. However, the extent of the variation in cortical organization across the different lines of primate evolution is not yet clear. There are basic similarities in how visual cortex is subdivided across primates. For example, all primates have a middle temporal visual area (see Krubitzer and Kaas 1990a), a field not clearly present in non-primates (see Sesma et al. 1984). Nevertheless, there are clear differences in cortical organization across primate taxa. Most notably, prosimian primates lack the systematic representations of the body surface in areas 1 and 2 that characterize the brains of monkeys and other higher primates (Kaas 1983). There are likely other major differences in brain organization as well, and these differences would be reflected in patterns of callosal connections, but comparative studies have not yet addressed this issue of variability related to differences in numbers of cortical fields.

Another type of variability is in the callosal connections for cortical areas common to all primates. Area 17 or V-I is perhaps the most easily identified of all cortical areas, and it is architectonically obvious especially in primates. Remarkably, the callosal connections of area 17 are quite variable across species. A widespread opinion is that area 17 of primates and even most mammals is devoid of callosal connections. To the contrary, area 17 of generalized mammals (with small brains and few cortical areas) have callosally projecting neurons and terminations that are broadly distributed across area 17, although connections are most dense along the lateral border (see Cusick and Kaas 1986). Even tree shrews, close relatives of primates, have widespread interhemispheric connections within area 17 (Pritzel et al. 1988; Cusick et al. 1985). Prosimian primates, such as galagos, have a more restricted pattern with nearly all callosally projecting neurons and terminations within 2-3 mm of the border of area 17, but since area 17 is small in these primates, perhaps as much as a third of area 17 is callosally connected (Cusick et al. 1984). Old World monkeys (Kennedy and Dehay, 1988) and presumably apes and humans have a much more restricted pattern, with most of area 17 devoid of callosal connections. Yet, connections within area 17, along the border, are obvious and they extend as far as 2.5 mm from the border in macaques.

The primary somatosensory area, S-I or area 3b (see Kaas 1983), also varies in callosal connections across species. In most mammals, the representation of the forepaw in S-I is characterized by a considerable population of callosally connecting neurons, although they are less dense than in other parts of S-I (Weller et al. 1987; Cusick et al. 1985). In galagos, the number of such neurons in the hand representation remains high (Cusick and Kaas 1986), while they are much reduced in New World monkeys and even more sparsely distributed in Old World monkeys (Killackey et al. 1983).

One interpretation of such species differences is that in mammals with few cortical areas, each area has multiple functions (see Kaas 1989a). Area 17 in rats, for example, must distribute information broadly, since rats have few visual areas (Kaas 1989b; Kaas and Krubitzer 1991). Thus, area 17 is not only broadly connected with the other hemisphere, but with motor and somatosensory and auditory areas of the same hemisphere (see Paperna and Malach 1991). Mammals with more areas can have more specialized fields, and delay some functions until later stages in sensory processing. For example, in the three successive stages of somatosensory processing in monkeys, the hand representation of area 3b has sparse callosal connections, area 1 (which is activated by area 3b inputs) has more, and area 2 (activated by area 1 inputs) has the most (Killackey et al. 1983). Thus, ipsilateral processing is emphasized in the early stages of the hierarchy, while the interhemispheric transfer of information becomes successively more important. Such specializations depend on having multiple sensory areas. Because humans are likely to have the greatest number of cortical areas, callosal and other connection patterns may be most specialized in humans.

MODULAR SUBSYSTEMS WITHIN FIELDS MAY DIFFER IN CALLOSAL CONNECTIONS.

A few areas of cortex in primates are clearly subdivided into interdigitating arrays of different classes of processing modules or cortical columns (see Kaas 1991). Such columns are distinguished by neurons with distinct response properties and patterns of connections. Thus, a single area can act as several by having several partially isolated sets of columns. In area 17 of all primates, for example, cortex can be divided into dot-like "blobs" of tissue expressing high amounts of the metabolic enzyme, cytochrome oxidase (CO). Neurons in these blobs have specialized properties, typically being selective for color and non-selective for stimulus orientation (e.g., Livingstone and Hubel 1988). Blobs are further distinguished from interblob regions by unique inputs from the interlaminar neurons of the lateral geniculate nucleus (Weber et al. 1983), and a distinct pattern of connections with other visual areas (DeYoe and Van Essen 1988; Krubitzer and Kaas 1990b). Notably, the lateral connections between blobs are more extensive than the lateral connections between interblob regions (Cusick and Kaas 1988). This would allow blob neurons to be influenced by stimuli over larger regions of visual space than interblob regions. In a similar manner, interhemispheric connections between blobs are more extensive than for the interblob regions (Cusick et al. 1984). Thus, blob neurons are also influenced by a larger region of the ipsilateral as well as the contralateral hemifield.

As another example of subregions of sensory representations differing in callosal connections, somatosensory cortex of rats and mice have arrays of barrel-like clusters of larger neurons separated by a matrix of smaller, less densely packed neurons. Neurons in the barrels receive activating thalamic inputs from the ventroposterior nucleus, while matrix neurons receive callosal inputs (Olavarria et al. 1984.). Somatosensory cortex in primates is organized somewhat differently, but there is evidence for broader regions of densely packed, larger neurons separated by narrow regions of smaller neurons and less myelination in area 3b. In the small-brained marmoset monkeys where this parcellation is more obvious, less myelinated strips of cortex are found immediately lateral and immediately medial to the hand representation in area 3b. Neurons in these regions are less responsive to cutaneous stimuli than in other parts of the representation, and these regions receive dense callosal inputs (Krubitzer and Kaas 1990a.)

Differences in callosal connections for types of modules within area 18 also exist (see Cusick and Kaas 1988), but we currently have only a limited understanding of the nature of modular organization within most cortical fields. Thus, it is uncertain if most areas have two or more sets of modules, and if different sets have contrasting patterns of callosal connections. The major point is that different types of modules, because they have different functional roles, can be expected to differ in callosal connections.

CONCLUSIONS

1. Callosal connections are widespread, and they appear to have several functional roles. Some connections are homotopic and they contribute to midline receptive fields, and they may mediate "midline fusion," as traditionally proposed. However, the majority of connections are heterotopic, and they probably contribute to large, antagonistic surrounds of receptive fields. These include mirror-image connections

between pairs of fields and feedback callosal connections from higher to lower levels. Other callosal connections, most probably the feedforward type, may contribute to large, bilateral excitatory receptive fields.

2. Callosal connections are unevenly distributed. In primary visual, somatosensory, and motor areas of primates, large regions of cortex are specialized for processing ipsilateral thalamic and cortical inputs, and callosal connections are sparse or absent. This specialization is possible because higher primates have more processing areas than generalized mammals, and early stations in processing hierarchies need not be multipurpose fields. In generalized mammals, the interhemispheric transfer of information is one of the many functions of primary areas. Callosal connections are also unevenly distributed because areas of cortex are modularly organized, with some sets of modules more involved in the interhemispheric transfer of information than others. To the extent that cortical areas are more sharply differentiated into functionally distinct sets of modules in higher primates, callosal connections may be more unevenly distributed.

REFERENCES

Allman, J.M., Meizin, F., and McGuinnes, E., 1986, Stimulus specific responses from beyond the classical receptive field: Neuro-physiological mechanisms for local-global comparisons in visual neurons, *Ann. Rev. Neurosci.* 8:407-430.

Choudhury, P.B., Whitteridge, D. and Wilson, M.E., 1965, The function of the callosal connections of the visual cortex, *Q.J. Exp. Physiol.* 50:214-219.

Clarke, S. and Miklossy, J., 1990, Occipital cortex in man: organization of callosal connections, related myelo-and cytoarchitecture, and putative boundaries of functional visual areas, *J. Comp. Neurol.* 298:188-214.

Conti, F., Fabri, M., and Manzoni, T., 1986, Bilateral receptive fields and callosal connectivity of the body midline representation in the first somatosensory area of primates, *Somatosens. Res.* 3:273-289.

Cusick, C.G. and Kaas, J.H., 1986, Interhemispheric connections of cortical sensory and motor representations in primates, *in*: "Two Hemispheres One Brain: Functions of the Corpus Callosum," F. Lepore, M. Ptito and H.H. Jasper, eds., Alan Liss, New York.

Cusick, C.G. and Kaas, J.H., 1988, Cortical connections of area 18 and dorsolateral visual cortex in squirrel monkeys, *Visual Neurosci.* 1:211-237.

Cusick, C.G. and Kaas, J.H., 1988, Surface view patterns of intrinsic and extrinsic cortical connections of area 17 in a prosimian primate, *Brain Research* 458:386-288.

Cusick, C.G., Gould, H.J. III, and Kaas, J.H., 1984, Interhemispheric connections of visual cortex of owl monkeys (*Aotus trivirgatus*), marmosets (*Callithrix jacchus*) and galagos (*Galago crassicaudatus*), *J. Comp. Neurol.* 230:311-336.

Cusick, C.G., MacAvoy, M.G., and Kaas, J.H. 1985, Interhemispheric connections of cortical sensory areas in tree shrews, *J. Comp. Neurol.* 235:111-128.

DeYoe, E.A., and Van Essen, D.C., 1988, Concurrent processing streams in monkey visual cortex, *Trends Neurosci.* 5:219-226.

Ebner, F.F., 1967, Afferent connections to neocortex in the opossum (*Didelphis vivginiana*), *J. Comp. Neurol.* 129:241-268.

Ebner, F.F. and Myers, R.E., 1965, Distribution of corpus callosum and anterior commissure in cat and raccoon, *J. Comp. Neurol.* 124:353-366.

Felleman, D.J. and Kaas, J.H., 1984, Receptive field properties of neurons in the middle temporal visual area (MT) of owl monkeys, *J. Neurophysiol.* 52:488-513.

Gould, H.J., and Kaas, J.H., 1981, The distribution of commissural terminations in somatosensory areas I and II of the grey squirrel, *J. Comp. Neurol.* 196:489-504.

Gould, H.J., Cusick, C.G., Pons, T.P., and Kaas, J.H., 1986, The relationship of corpus callosum connections to electrical stimulation maps of motor, supplementary motor, and frontal eye fields in owl monkeys, *J. Comp. Neurol.* 247:297-325.

Gould, H.J., Weber, J.T., and Rieck, R.W., 1987, Interhemispheric connections in the visual cortex of the squirrel monkey (*Saimiri sciureus*), *J. Comp. Neurol.* 256:14-28.
Granger, E.M., Masterton, R.B. and Glendenning, K.K., 1985, Origin of interhemispheric fibers in a callosal opossum (with comparison to callosal origins in rat), *J. Comp. Neurol.* 241:82-98.
Guillemot, J.P., Richer, L., Prevost, L., Ptito, M., and Lepore, F., 1987, Receptive field properties of somatosensory callosal fibers in the monkey, *Brain Res.* 402:293-302.
Heath, C.J. and Jones, E.G., 1971, Interhemispheric pathways in the absence of a corpus callosum, *J. Anat.* 109:253-270.
Hofman, M.A., 1985, Size and shape of the cerebral cortex in mammals. I. The cortical surface, *Brain Behav.* Vol 3. 27:28-40.
Jones, E.G. and Powell, T.P.S., 1969, Connections of the somatic sensory cortex of the rhesus monkey. II Contralateral cortical connections, *Brain*, 92:717-730.
Jouandet, M.L. and Gazzaniga, M.S., 1979, Cortical field of origin of the anterior commissure of the rhesus monkey, *Exp. Neurol.* 66:381-397.
Kaas, J.H., 1982, The segregation of function in the nervous system: Why do sensory systems have so many subdivisions? *In*: "Contributions to Sensory Physiology," W.P. Neff, ed., Academic Press, New York, pp. 210-240.
Kaas, J.H., 1983, What, if anything, is S-I? The organization of the "first somatosensory area" of cortex, *Physiol. Rev.*, 63:206-231.
Kaas, J. H., 1987, The organization of neocortex in mammals: Implications for theories of brain function, *Annual Review of Psychology*, 38:124-151.
Kaas, J.H., 1989a, The evolution of complex sensory systems in mammals, *J. of Exper. Biol.* 146:165-176.
Kaas, J.H., 1989b, Why does the brain have so many visual areas? *J. Cognitive Neurosci.* 1:121-135.
Kaas, J.H., 1991, The parcellation of somatosensory cortex: modules, columns, and somatotopic segregations, *in*: "Information Processing in the Somatosensory System," O. Franzein and J. Westmon, eds., Stockton Press, New York, pp. 211-219.
Kaas, J.H., and Krubitzer, L.A., 1991, The organization of extrastriate visual cortex, *in*: "Neuroanatomy of Visual Pathways and their Retinotopic Organization," Vol. III of "Vision and Visual Dysfunction," J. Cronly-Dillon, gen. ed., The MacMillan Press, London, pp. 302-323.
Karol, E.A. and Pandyda, D.N., 1971, The distribution of the corpus callosum in the rhesus monkey, *Brain*, 94:471-486.
Kennedy, M. and Dehay, C., 1988, Functional implications of the anatomical organization of the callosal projections of visual areas V1 and V2 in the macaque, *Behav. Brain Res.* 19:225-236.
Kennedy, H., Dehay, C. and Bullier, J., 1986, Organization of the callosal connections of visual areas V1 and V2 in the macaque monkey, *J. Comp. Neurol.* 247:398-415.
Kennedy H., Meissirel, C. and Dehay, C., 1991, Callosal pathways and their compliancy to general rules governing the organization of corticocortical connectivity, *in*: "Vision and Visual Dysfunction," Vol. 3, "Neuroanatomy of the Visual Pathways and Their Development," B. Dreher and S.R. Robinson, eds., MacMillian Press, London, pp. 324-359.
Killackey, H.P., Gould, H.J.,III, Cusick, CG., Pons, T.P. and Kaas, J.H., 1983, The relation of corpus callosum connections to architectonic fields and body surface maps in sensorimotor cortex of new and old world monkeys, *J. Comp. Neurol.* 219:384-419.
Krubitzer, L.A. and Kaas, J.H., 1990a, Cortical connections of MT in four species of primates: Areal, modular, and retinotopic patterns, *Visual Neurosci.* 5:165-204.
Krubitzer, L.A. and Kaas, J.H., 1990b, Convergence of processing streams in the extrastriate cortex of monkeys, *Vis. Neurosci.* 5:609-613.
Krubitzer, L.A. and Kaas, J.H., 1990c, The organization and connections of somatosensory cortex in marmosets, *J. Neurosci.* 10:952-974.
Land, E.H., Hubel, D.H., Livingstone, M.S., Perry, S.H. and Burns, M.M., 1983, Color-generating interactions across the corpus callosum, *Nature*, 303:616-618.
Lepore, F., Ptito, M. and Guillemot, J.P., 1986, The role of the corpus callosum in midline fusion, *in*: "Two Hemispheres One Brain: Functions of the Corpus Callosum," F. Lepore, M. Ptito and H.H. Jasper, eds., Alan Liss, New York, pp. 211-229.
Leuthke, L.E., Krubitzer, L.A., and Kaas, J.H., 1989, Connections of primary auditory cortex in the New World monkey, *Saguinus*, *J. Comp. Neurol.* 285:487-513.

Livingstone, M.S. and Hubel, D.H., 1988, Segregation of form, color, movement, and depth: Anatomy, physiology, and perception, *Science*, 240:740-749.

Manzoni, T., Barbaresi, P. and Conti, F., 1984, Callosal mechanisms for the interhemispheric transfer of hand somatosensory information in the monkey, *Behav. Brain Res.* 11:155-170.

Manzoni, T., Barbaresi, P., Conti, F., and Fabri, M., 1989, The callosal connections of the primary somatosensory cortex and the neurol bases of midline fusion, *Exp. Brain Res.* 76: 251-266.

Maunsell, J.H.R. and Van Essen, D.C., 1983, The connections of the middle temporal visual area (MT) and this relationship to a cortical hierarchy in the macaque monkey,*J. Neurosci.* 3:2563-2586.

Morel, A. and Kaas, J.H., 1992, Subdivisions and connections of auditory cortex in owl monkeys, *J. Comp. Neurology* 318:27-63.

Myers, R.E., 1962, Commissural connections between occipital lobes of the monkey, *J. Comp. Neurol.* 118:1-16.

Myers, R.E., 1965, The neocortical commissures and interhemispheric transmission of information, *in:* "Functions of the Corpus Callosum," E.G. Ettinger (ed.), Boston, Little Brown, pp. 1-17, 133-193.

Newsome, W.T. and Allman, J.M., 1980, Interhemispheric connections of visual cortex in the owl monkey, *Aotus trivirgatus* and the bushbaby, *Galago senegalensis. J. Comp. Neurol.* 194:209-233.

Olavarria, J., Van Slugters, R.C., and Killackey, H., 1984, Evidence for the complementary organization of callosal and thalamic connections within rat somatosensory cortex, *Brain Res.*, 291:364-368.

Pandya, D.N., Karol, E.A. and Lele, P.P., 1975, The distribution of the anterior commissure in the squirrel monkey, *Brain Res.* 49:177-180.

Pandya, D.N., Karol, E.A., and Heibronn, D., 1971, The topographic distribution of interhemispheric projections in the corpus callosum of the rhesus monkey, *Brain Res.* 32, 31-43.

Paperna, T. and Malach, R., 1991, Patterns of sensory intermodality relationships in the cerebral cortex of the rat, *J. Comp. Neurol.* 308:432-456.

Payne, B.R., Pearson, H.E., and Berman, N., 1984, Role of corpus callosum in functional organization of cat striate cortex, *J. Neurophysiol.*, 52:570-594.

Pritzel, M., Kretz, R., and Rager, G., 1988, Callosal projections between areas 17 in the adult tree shrew (*Tupaia belangeri*), *Exp. Brain Res.* 72:481-493.

Robinson, C.J. and Burton, H., 1980, Somatotopic organization in the second somatosensory area of M. *Fascicularis, J. Comp. Neurol.* 192:43-67.

Rocha-Miranda, C.E., Bender, D.B., Gross, L.G., and Mishkin, M., 1975, Visual activation of neurons in inferotemporal cortex depends on striate cortex and forebrain commissures, *J. Neurophysiol.* 38:474-491.

Rockland, K.S. and Pandya, D.N., 1979, Laminar origins and terminations of cortical connections of the occipital lobe in the rhesus monkey, *Brain Res.* 179:3-20.

Sesma, M.A., Casagrande, V.A., and Kaas, J.H., 1984, Cortical connections of area 17 in tree shrews, *J. Comp. Neurol.*, 230:337-351.

Shanks, M.F., Rockel, A.J. and Powell, T.P.S., 1975, The commissural fibre connections of the primary somatic sensory cortex, *Brain Res.* 98:166-171.

Shoumura, K., Ando, T., and Kato, K., 1975, Structural organization of "callosal" OBg in human corpus callosum agenesis, *Brain Res.* 93:241-252.

Sur, M., Nelson, R.J., and Kaas, J.H., 1982, Representation of the body surface in cortical areas 3b and 1 of squirrel monkeys: Comparisons with other primates, *J. Comp. Neurol.* 211-177-192.

Van Essen, D.C. and Zeki, S.M., 1978, The topographic organization of rhesus monkey prestriate cortex, *J. Physiol.* 277:193-226.

Van Essen, D.C., Newsome, W.T., and Bixby, J.L., 1982, The pattern of interhemispheric connections and its relationship to extrastriate visual areas in the macaque monkey, *J. Neurosci.* 2:265-283.

Weber, J.T., Huerta, M.F., Kaas, J.H. and Harting, J.K., 1983, The projections of the lateral geniculate nucleus of the squirrel monkey: Studies of the interlaminar zones and layers, *J. Comp. Neurol.* 213:135-145.

Weller, R.E. and Kaas, J.H., 1981, Cortical and subcortical connections of visual cortex in primates, *in*: "Cortical Sensory Organization, Vol. 2: Multiple Visual Areas," C.N. Woolsey, ed., Humana Press, Clifton, NJ, pp. 121-155.

Weller, R.E., Sur, M., and Kaas, J.H, 1987, Callosal and ipsilateral cortical connections of the body surface representations in S-I and S-II of tree shrews, *Somatosensory Research* 5:107-133.

Zeki, S.M., 1973, Comparison of the cortical degeneration in the visual regions of the temporal lobe of the monkey following sections of the anterior commissure and the splenium, *J. Comp. Neurol.* 148:167-176.

Zeki, S.M., and Sandeman, D.R., 1976, Combined anatomical and electrophysiological studies on the boundary between the second and third visual areas of rhesus monkey cortex, *Proc. R. Soc. Lond. (Biol.)*, 194:555-562.

ORGANIZATION AND DEVELOPMENT OF INTERHEMISPHERIC CONNECTIONS OF THE PREFRONTAL CORTEX IN RHESUS MONKEY

Michael L. Schwartz, Ph.D.

Yale University School of Medicine
Section of Neurobiology
New Haven, CT 06437

The corpus callosum is the principal pathway for the communication of sensory and cognitive information between cortical areas of the two cerebral hemispheres (Sperry 1982; Gazzaniga 1985; Sperry 1974) and has been the subject of considerable clinical interest due to its role as the primary substrate for the interhemispheric spread of discharges in generalized epileptic seizures (Reeves and O'Leary 1985; Wilson et al. 1975). Studies of callosal connections within sensory and motor regions of the cerebral cortex over the past 30 years have provided important insight into the functional, organizational and developmental features of these pathways. In contrast, the study of callosal connections for association regions of the cerebral mantle has not kept pace. In this chapter, I will present data from ongoing studies of corticocortical connectivity in which we have focused on the organization and development of interhemispheric connections between areas of the monkey prefrontal association cortex. I will first describe general features of the topographic, laminar, and columnar organization of these connections in the mature monkey and will conclude with results of our studies examining the prenatal development and maturation of these connections.

The prefrontal cortex (PFC) is of particular interest due to its prominence both anatomically and functionally in primates. In humans, this region of the frontal lobes comprises 24-29% of the entire surface of the cerebral cortex (Brodmann 1912; Rakic 1982) and is involved in many aspects of cognitive activity that are considered to be distinctively human (Goldman-Rakic et al. 1983; Goldman-Rakic 1987). We have used the rhesus monkey as a model for understanding the functional organization and development of this region since the cytoarchitectonic differentiation and proportion of cortical surface occupied by the prefrontal cortex in the monkey is greater than that found in any non-primate species. Moreover, damage to this area in mature monkeys produces a pattern of cognitive and behavioral deficits that are in many ways analogous to those seen in human infants and in adults with frontal lobe damage (Goldman-Rakic 1987; Diamond and Goldman-Rakic 1989).

The prefrontal cortex in humans and in monkeys can be divided into a number of distinctive cytoarchitectonic areas that are characterized by unique connectional and functional features (Brodmann 1909; Walker 1940; Goldman-Rakic 1987; Barbas 1992). The present chapter will focus on Walker's area 46 (Walker 1940); a subdivision of the prefrontal cortex in monkeys that occupies the dorsolateral surface of the frontal lobes including the cortex lining the banks of the principal sulcus. This area is characterized by a well-demarcated granular layer IV; a distinctive feature of the dorsolateral prefrontal cortex in primates that is not present in any region of the prefrontal cortex in non-primate species (Akert 1964). Additionally, clinical studies of humans with damage encompassing this portion of the prefrontal cortex (Milner 1982; Guitton et al. 1985; Freedman and Oscar-Berman 1986) and behavioral studies of monkeys with experimental lesions of this region (for review see Goldman-Rakic 1987) reveal striking parallels in the resulting constellation of cognitive deficits.

CALLOSAL CONNECTIONS IN THE PREFRONTAL CORTEX OF MATURE MONKEYS

Areal topography. Fibers connecting areas of the prefrontal cortex of the two hemispheres are found within the genu and rostral portions of the body of the corpus callosum (Pandya et al. 1971; Barbas and Pandya 1984). The axons within these regions are topographically distributed. However, their topography is not purely a reflection of the relative tangential positions of their cells of origin, but is also determined by the degree of architectonic differentiation of the cytoarchitectonic area from which they originate. Thus, axons from areas with similar patterns of architectonic differentiation tend to occupy adjacent positions in the callosum although they may occupy non-adjacent regions on the cortical surface (Barbas and Pandya 1984).

In the last decade, a number of studies have applied sensitive tract tracing techniques to map both the targets and sources of callosal axons connecting prefrontal area 46 with regions of the contralateral hemisphere. Two principles of callosal organization have emerged from these studies: 1) connections between area 46 and cytoarchitectonic areas of the contralateral hemisphere are reciprocal, and 2) all cortical areas that have a callosal connection with area 46 also maintain an ipsilateral associational projection with area 46 of the same hemisphere. The majority of callosal connections with area 46 are restricted to cytoarchitectonic areas of the contralateral prefrontal cortex. These include the homotopic area 46, as well as heterotopic areas 8a, 8b, 9, 10, and 45 on the dorsolateral surface, orbital areas 11 and 12, and cingulate area 24 on the medial surface of the hemisphere (Jacobson and Trojanowski 1977b; Schwartz and Goldman-Rakic 1984; McGuire et al. 1991). The density of these connections is greatest between area 46 of the two hemispheres (Schwartz and Goldman-Rakic 1984; McGuire et al. 1991), is somewhat diminished for areas 8A, 8B, 24 and 45 and is relatively modest for areas 9, 10, 11, and 12. Only a small subset of prefrontal areas lacks callosal connections with area 46 of the opposite hemisphere. These include areas 13 and 14 on the orbital surface, and medial area 25. Callosal connections between area 46 and regions of the contralateral hemisphere outside the prefrontal cortex are sparse, and are limited to premotor area 6 and to the supplementary motor area (SMA) (Jacobson and Trojanowski 1977a; Schwartz and Goldman-Rakic 1984; McGuire et al. 1991).

Columnar and Laminar Organization. Studies of the microarchitecture of the dorsolateral prefrontal cortex in monkeys indicate that the afferent, efferent and local circuit elements of this area are organized in a modular pattern in both the radial and tangential domains (Goldman and Nauta 1977; Goldman-Rakic and Schwartz 1982; Schwartz and Goldman-Rakic 1984; Schwartz et al. 1988). This modularity was first appreciated in an autoradiographic study of the callosal and associational projections of area 46 by Goldman and Nauta (1977). Following injection of ^{3}H-leucine and ^{3}H-proline into the dorsolateral PFC of one hemisphere, they observed that the pattern of terminal and axonal labeling in the homotopic area of the opposite hemisphere appeared as a series of vertically oriented stripes or columns. These stripes were roughly 0.5 mm in width, extended through all six cortical layers and alternated with unlabeled regions of approximately the same dimensions. Subsequent studies focusing on the topography of projections to area 46 from areas of the ipsilateral prefrontal cortex, the posterior parietal cortex and the SMA have reported a similar modularity in the pattern of afferent terminations (Leichnetz 1980; Goldman-Rakic and Schwartz 1982; Schwartz and Goldman-Rakic 1984; Selemon and Goldman-Rakic 1988; McGuire et al. 1991). Thus, the columnar distribution and regular periodicity of callosal terminations within the dorsolateral PFC appears to be a general feature that characterizes all cortico-cortical afferent terminations within prefrontal areas of the association cortex.

Consistent with the disjunctive organization of afferent terminations within the PFC, cortico-cortical projection neurons of area 46 are also unevenly distributed. For example, neurons of area 46 that project to the homotopic region of the opposite hemisphere wax and wane in density across the cortical surface. However, few regions of the PFC are truly devoid of labeled neurons. For this reason, the true modularity and periodicity of efferent neuronal populations is often not fully appreciated in the absence of careful quantitative examination of the density of labeled cells across the cortical surface. For example, generating a histogram of variations in the density of identified efferent populations across the cortical surface readily reveals that projection neurons have a repeating modular distribution (Fig. 1). This distribution shares the cyclical dimensions of the more sharply defined afferent territories. This is best demonstrated by the results of our HRP tracing studies in which areas containing the highest density of retrogradely labeled callosal projection neurons were also found to correspond with anterogradedly labeled columns of callosal terminations (Fig. 2) (Schwartz and Goldman-Rakic 1984; Goldman-Rakic and Schwartz 1982). A similar registration in the periodicity's of projection neuron density and afferent termination zones is evident for cortico-cortical connections of the PFC with other areas of the same and opposite hemisphere.

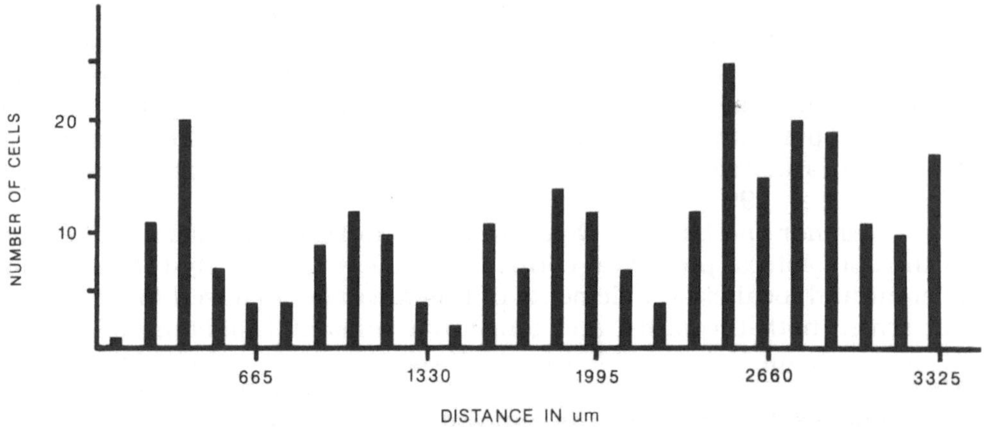

Figure 1. Histogram representing the tangential distribution of retrogradely labeled callosal neurons in the dorsal bank of principal sulcus (PS) from a representative 40 μm coronal section. The injection of the fluorescent tracer fast blue was made into the homotopic area 46 of the opposite hemisphere. The dorsal bank of the PS was divided into a series of adjacent 133 μm wide bins extending from the pial surface to the layer VI /white matter border. The number of retrogradely labeled callosal projection neurons in each bin was determined and is indicated by the height of each black bar in the histogram. Adjacent bars represent adjacent cortical bins. Note the regularly repeating waxing and waning in callosal neuron number across the surface of the cortex.

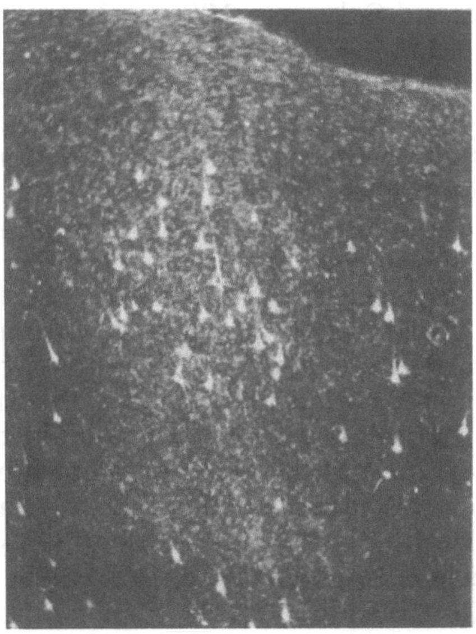

Figure 2. Correspondence of anterograde and retrograde HRP labeling in the dorsal bank of the PFC following an injection in area 46 of the opposite hemisphere. Note that the fine dust-like anterograde labeling of terminals and axons is coincident with the region of densest retrograde labeling of callosal projection neurons.

Similarities in the periodicity of afferent and efferent modules interconnecting area 46 with cortical areas of the ipsilateral and contralateral hemisphere raise the issue of how these different modules might be related. For example, what source(s) of input and which projection neuron populations occupy cortical regions between callosal projection columns? Do different cortico-cortical connections occupy unique or overlapping modular territories within the PFC? We initially addressed these questions using an anterograde double-labeling paradigm to simultaneously identify the terminal territories of two sources of cortico-cortical input that converge in the dorsolateral PFC. For this purpose, rhesus monkeys were unilaterally injected with ^{3}H-leucine and ^{3}H-proline into the dorsal bank of the principal sulcus (area 46) in order to label columns of callosal axons and terminals in the cortex of the principal sulcus of the contralateral hemisphere. Horseradish peroxidase (HRP) was then injected into the posterior bank of the intra-parietal sulcus of the opposite hemisphere to identify the termination zones of this associational projection within the PFC (Fig. 3). Alternate sections through the principal sulcus of the PFC were processed autoradiographically or for HRP visualization and the distributions of labeled afferents were reconstructed. The principal finding of this study was that callosal and associational terminals from these two sources occupy unique and interdigitated columnar territories within the PFC (Fig. 3).

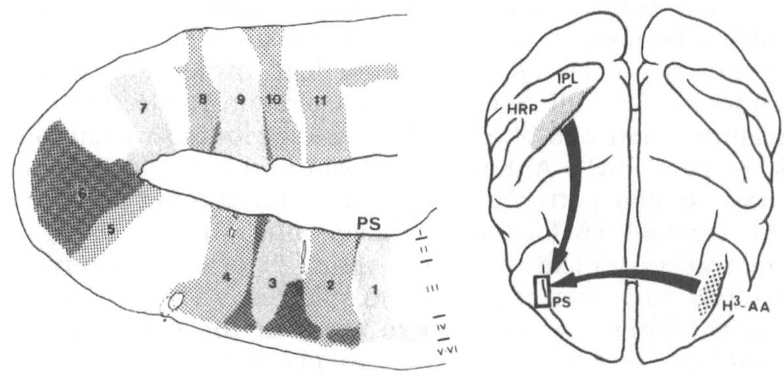

Figure 3. Composite diagram of two adjacent 50μm coronal sections cut through the convergence zone in the principal sulcus (PS; see inset). Autoradiographically labeled callosal fiber columns (2,4,5,8,10,11) are indicated by course stipple; associational fiber columns arising from neurons in the posterior bank of the intra-parietal sulcus (IPL) labeled by anterograde transport of HRP (1,3,6,7,9) are indicated by fine stipple. Blood vessels in HRP reacted sections are indicated by dashed lines; those from autoradiograms are indicated by continuous lines. Several blood vessels penetrating layers I and II orthogonally were used to align adjacent sections. (Reprinted from Goldman-Rakic and Schwartz, 1982, Interdigitation of contralateral . . ., *Science* 216:755-757 with permission from the American Association for the Advancement of Science.)

Using a similar double labeling strategy with retrogradely transported fluorescent dyes, we have also examined the relationship between efferent neuron populations within area 46. In these studies, the tracer fast blue was injected into the cortex lining the principal sulcus of one hemisphere and diamidino-yellow was injected into the posterior bank of the intra-parietal sulcus of the opposite hemisphere. The distribution of labeled neurons containing either fast blue, diamidino-yellow, or both tracers were then charted in area 46 of the un-injected prefrontal cortex. An example of the pattern of labeling found in these studies is illustrated in Figure 4. As was expected from the results of previous single labeling studies, the tangential distribution of callosal and associational projection neuron populations showed considerable overlap and were not separated by sharp boundaries. Despite the intermixing of these efferent populations, both had periodic variations in neuron density across the cortical surface that were inversely related. Thus, areas of peak callosal neuron density corresponded to areas containing relatively few associational neurons, while areas of low callosal neuron density were coincident with regions containing the highest densities of associational projection neurons. Taken together with our results concerning the modularity of afferent terminations, these data indicate that information carried over callosal and associational pathways are topographically segregated at the stages of initial input and final output in prefrontal cortical areas. Therefore, the ultimate integration of information between these modules must rely heavily upon the local circuit connections of the PFC.

In addition to the spatially disjunctive distribution of cortico-cortical efferents and afferents in the tangential domain, these connections also exhibit a very distinctive pattern of laminar or radial termination and origin. For example, retrograde tracing studies using either HRP or fluorescent dyes indicate that callosal and associational neurons within area 46 have a bilaminar distribution. Approximately 80% of all callosal and 70% of associational cortico-cortical projection neurons are distributed within layer III (Schwartz and Goldman-Rakic 1984). The bulk of the remaining neurons of each class is distributed within layer V (19 and 14%, respectively). Although, callosal and associational projection neurons have similar patterns of laminar origin and are often intermixed across the cortical surface (see above), our studies using multiple fluorescent tracers indicate that less than 1% of these neurons project simultaneously to areas of both hemispheres. The low incidence of neurons with bilateral projections to areas of the two hemispheres in mature monkeys, underscores the independence and segregation of information communicated by callosal and associational pathways to and from the dorsolateral PFC.

Callosal neuron morphology. Consistent with the results of previous studies in the prefrontal cortex (Jacobson and Trojanowski 1977b), and other areas of the monkey cerebral cortex (Jones and Wise 1977; Rockland and Pandya 1979), our studies using retrograde tracers have generally confirmed that the vast majority of callosal projections of area 46 arise from neurons with a pyramidal morphology (Schwartz and Goldman-Rakic 1984). These data are in agreement with the long-standing view that callosal projections exert an excitatory influence on target neurons of the contralateral hemisphere (Toyama et al. 1974; Toyama and Matsunami 1976). However, careful analyses of the morphological features of callosal projection neurons in our studies have suggested that a small population of these projections may originate from non-pyramidal neurons found predominately within layers V and VI. In these studies we examined the morphological features of HRP labeled neurons with reaction product extending into proximal portions

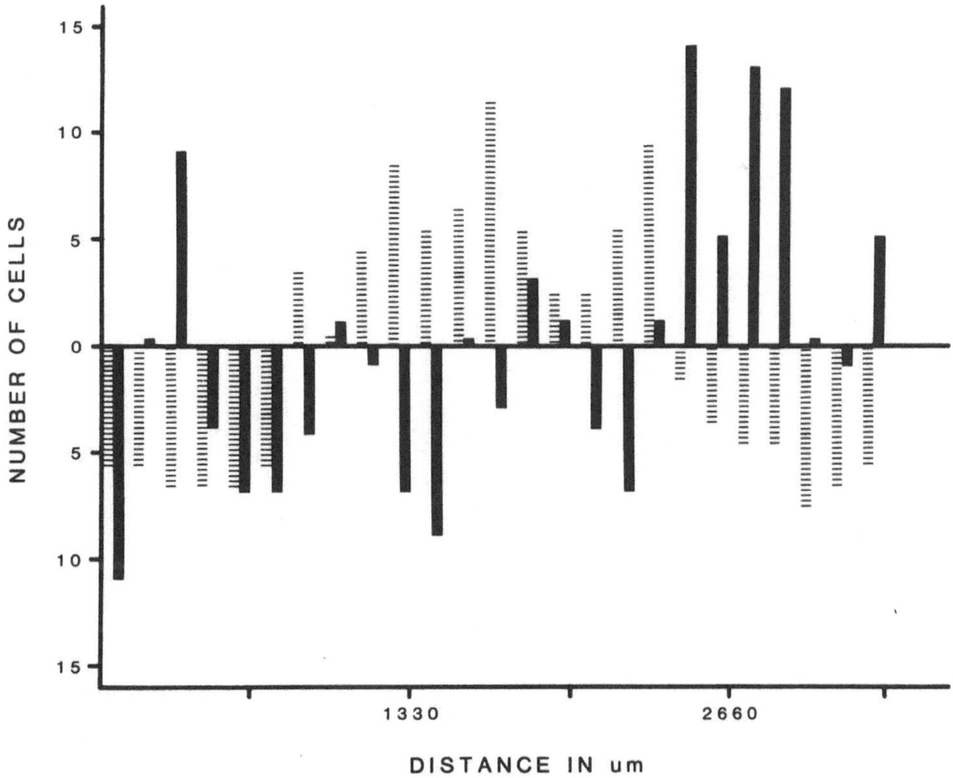

Figure 4. Relative distributions of callosal and associational projection neurons in the principal sulcus of the PFC. The number of neurons labeled by each fluorescent tracer was determined for adjacent 133 μm wide cortical slabs (bins) extending from the pial surface to the border of layer VI with the underlying white matter. Adjacent hatched and solid bars in the histogram represent the values for each cell class in a single bin or unit area of cortex. The value for each bin represents the number of callosal neurons (solid bars) and ipsilateral neurons (hatched bars) represented as deviations from their respective means. The line at zero denotes the mean; downward deviations signify values below the mean, and upward deviations signify density values greater than the mean. (Reprinted from Schwartz and Goldman-Rakic, 1984, Callosal and intrahemispheric connectivity . . ., *J. Comp. Neurol.* 226:403-420 with permission from the Journal of Comparative Anatomy.)

of the dendritic tree following an injection of tracer into the homotopic area of the opposite hemisphere. Neurons of uncertain morphology due to insufficient dendritic labeling were considered as pyramids. Thus, our analysis was conservatively biased toward the pyramidal phenotype. Despite this bias, we classified approximately 15% of all labeled layer V neurons and 60% of labeled layer VI neurons as having a non-pyramidal morphology (Schwartz and Goldman-Rakic 1984). We have pursued this issue by using a double-labeling strategy in the ferret prefrontal cortex. Cholera toxin conjugated to colloidal gold was used to retrogradely label callosal projection neurons and was combined with immunocytochemistry to simultaneously label GABA-containing neurons. The major finding of these studies was that a small population of infragranular callosal projection neurons was also immuno-cytochemically identified as GABA-containing neurons (Fig. 5). These data suggest that a limited population of callosal projections have a direct inhibitory influence on neurons of the opposite hemisphere and may explain the results of physiological studies demonstrating inhibitory interactions related to callosal connections (Asanuma and Okuda 1962).

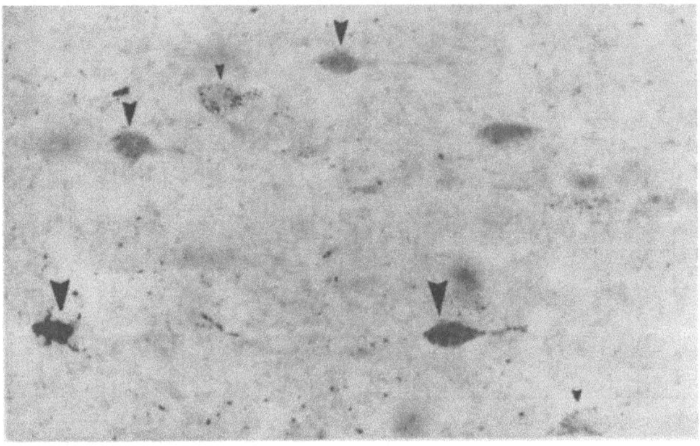

Figure 5. Photomicrograph of callosal and GABA-containing neurons in infragranular cortical layers of the ferret prefrontal cortex. Callosal neurons are retrogradely labeled by a conjugate of cholera toxin -- colloidal-gold (CT-gold) which was injected into the prefrontal cortex of the opposite hemisphere (small arrowheads). GABA-immunoreactive neurons are indicated by medium sized arrowheads. Double-labeled neurons containing both CT-gold and GABA are indicated by large arrowheads.

DEVELOPMENT OF CALLOSAL CONNECTIONS IN THE PREFRONTAL CORTEX OF FETAL MONKEYS

Many of the important milestones in the development of cortical organization and connectivity take place prior to birth in both humans and monkeys. The timing of these events has made them inaccessible to experimental examination until the recent development of prenatal surgical techniques (Rakic and Goldman-Rakic 1984). With the perfection of these surgical methods it has now become feasible to design studies that incorporate surgical interventions to perturb normal development and to apply modern tract-tracing methodologies to the study of connectional organization at different stages of fetal development. In this section I will describe the results of studies that have combined sensitive retrograde tract tracing techniques and prenatal neurosurgery to explore the timing and specificity of developing callosal projections of the dorsolateral prefrontal cortex of the monkey.

An important issue that must be addressed in developmental studies of cortico-cortical connections is how one defines the target and source of a projection. In the retrograde tracing studies I will describe, our goal has been to characterize the distribution of callosal projection neurons whose axons have entered the dorsolateral cortex of the opposite hemisphere and could therefore be said to project to this cortical area (cortex-specific callosal neurons). To address this point we confined our tracer injections to the layers of the emerging cortical plate and were careful not to allow their encroachment into the prospective white matter and transient development zones that are present below the cortex during early stages of cortical development. Previous studies have demonstrated that these subcortical areas contain growing axons, fibers of passage passing to and from other cortical areas, and form a waiting compartment for axonal populations prior to their growth into appropriate cortical targets (Wise and Jones 1978; Goldman-Rakic 1981; Kostovic and Rakic 1990; Shatz et al. 1988). Thus, the restriction of our injections to the cortical plate was essential, since the objective of our studies was to characterize the distribution and development of callosal projections that actually innervate, rather than those approaching or waiting under prefrontal area 46.

We began these studies by examining the distribution of callosal projections to the dorsolateral PFC in fetal monkeys at mid-gestation, embryonic day 82 (E82) of the 165 day gestational period. By this age, most neurons of layers III-VI that will give rise to the callosal connections of area 46 have been generated and have migrated, or are in the process of migrating to their final resting positions within the cortical layers (Schwartz et al. 1991). Animals in our earliest fetal group received injections of retrogradely transported fluorescent dyes or HRP into the cortical layers of the dorsolateral prefrontal cortex between E82 and E89. We confirmed that the lateral spread of tracer within the cortex did not extend into areas outside the dorsolateral PFC by examining the pattern of retrograde thalamic labeling in each case. Thalamo-cortical axons are already within their cortical targets at these early fetal ages and the restriction of retrogradely labeled neurons to lateral portions of the mediodorsal thalamic nucleus in these cases was consistent with the pattern of thalamic projections to area 46 in mature monkeys (Goldman-Rakic and Porrino 1985; Giguere and Goldman-Rakic 1988). Fetuses with injections of tracer that were restricted to the cortical layers had only a small number of retrogradely labeled neurons in the opposite hemisphere (usually only 1-5 per 40 μm section). The few labeled neurons were generally confined to the cortical layers of the dorsolateral cortex homotopic to the injection sites and to regions of the prospective cingulate cortex on the medial surface of the hemisphere. Both are regions that maintain projections to the dorsolateral prefrontal cortex of the opposite hemisphere in mature monkeys. Although we consistently identified labeled neurons in these cases, the extremely small number of identified neurons indicated that these fetal ages are near the very beginning of callosal axon ingrowth into the PFC. In contrast to the paucity of neurons labeled by cortically restricted injections, even small injections placed below the cortical plate at these ages labeled a substantial population of neurons. While most of these were restricted to the cortical layers, many were also present within the developmentally transient subplate and intermediate zones lying below the cortical plate. The greater density of labeled neurons in cases with injections extending below the cortical plate, is consistent with the pattern of labeling which would be expected based upon the results of anterograde tracing studies of the monkey PFC (Goldman-Rakic 1981). These studies indicate that the majority of callosal axons at these ages wait below the cortical plate for a period of weeks prior to entering their cortical targets. Thus, while callosal axons are present within the contralateral hemisphere prior to E90, few have actually innervated their cortical targets.

We had expected that injections of retrograde tracers into the subcortical waiting compartment should label a greater number of neurons in the cortical layers of the opposite hemisphere than the cortically restricted injections. However, we were surprised that many retrogradely labeled cells were also present within the subcortical subplate and intermediate zones of the opposite hemisphere. These data raised the question of whether these cells represent a sub-population of developmentally transient subplate cells extending axons into the contralateral hemisphere or whether they may be presumptive layer III callosal neurons extending axons into the opposite hemisphere during their migration into the cortical plate. Although previous studies have suggested that axogenesis is a relatively late event in neuronal development that occurs only after neurons have completed their migration into the cortical plate (Morest 1970; Berry and Rogers 1965), several pieces of data favor the possibility that some of these retrogradely labeled cells may indeed be migrating neurons. First, many retrogradely labeled callosal neurons in subcortical areas of fetuses injected at E89 or before have a morphology resembling that of migrating cortical neurons (Rakic 1972; Schwartz et al. 1991). Second, the density of labeled callosal neurons within these subcortical regions is sharply reduced between E84 and E104 coincident with an increasing density of neuronal labeling in layer III. Finally, the reduction in labeled subcortical neurons by E104 coincides with the time at which the majority of neurons in the lower half of layer III has completed their migration into the cortical plate, but is still at a time when subplate neurons are at their peak density (Schwartz et al. 1991; Schwartz and Meinecke 1992). We directly tested the possibility that these neurons may represent a population of cells extending axons toward the contralateral PFC during migration using a double labeling paradigm combining ^{3}H-thymidine birthdating and retrograde tracing with fluorescent dyes. Briefly, we injected fetuses between E74 and E76 with ^{3}H-thymidine to label prospective layer III neurons that were being generated at this time. We subsequently injected the PFC of one hemisphere with fluorescent tracers at E84, a time when these neurons would be in the last half of their migration toward the cortical plate. Examination of the cerebral wall contralateral to the tracer injections revealed many neurons singly labeled with either ^{3}H-thymidine or retrograde tracer, which were distributed among all cortical layers, the subplate and intermediate zones. However, a smaller number of neurons were labeled with both ^{3}H-thymidine and retrograde tracer (Fig. 6). These neurons were most prominent in layers III-IV of the cortical plate, but small numbers of double-labeled neurons were also found in layers V and VI, as well as in the subplate and intermediate zones below the cortex. The presence of double-labeled neurons within the intermediate and subplate zones provided direct evidence that some migrating neurons extend an axon into the opposite hemisphere. These data further suggest that some features of axonal projections may be specified for cortical neurons well in advance of their attainment of their final positions within the laminae of the cerebral cortex.

In contrast to the limited number of neurons innervating the cortical layers of the contralateral PFC in fetuses with cortically restricted injections prior to E90, the results of our analysis of fetuses injected after E90 suggest that the major wave of callosal axon growth into the six layers of the dorsolateral prefrontal cortex takes place between E100 and E124. For example, injections of HRP restricted to the cortical layers of

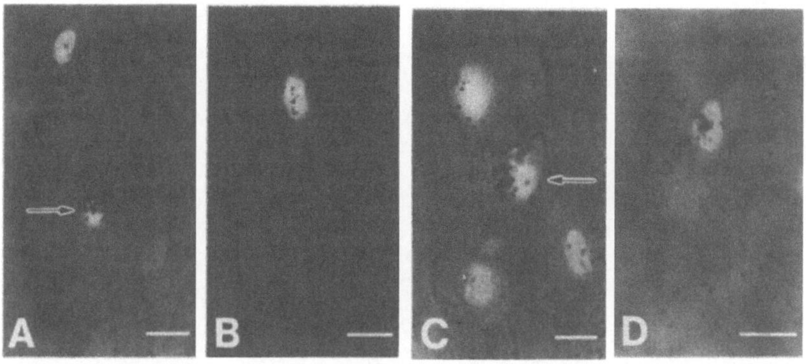

Figure 6. Double-labeled neurons: ^{3}H-thymidine injected at E74 and E76 and retrograde fluorescent tracers injected at E84. A) DY labeled subplate/intermediate zone neuron heavily labeled with ^{3}H-thymidine; B) DY and ^{3}H-thymidine labeled neuron in the upper subplate zone; C) DY and ^{3}H-thymidine labeled neuron at the border of layer VI and the subplate; D) DY and ^{3}H-thymidine labeled neuron in layer III. Scale bars equal 10μm. (Reprinted from Schwartz, Rakic, and Goldman-Rakic, 1991, Early phenotype expression . . ., *Proc. Natl. Acad. Sci.* 88:1354-1358 with permission.)

prospective area 46 at E111 labeled many neurons in both homotopic and heterotopic areas of the opposite hemisphere (Fig. 7). Although the immaturity of cortical sulcal and lamination patterns at these early ages limit the precision with which one can identify specific cortical areas, comparisons of the topographic distribution of labeled neurons with animals at older fetal ages and with the pattern found in mature monkeys reveal very similar distributions (Fig. 7). At the earliest ages for which we could conclusively identify cytoarchitectonic areas of the PFC (fetuses injected between E124- E137), the topography of projections was similar to that seen at earlier fetal ages and resulted in retrograde labeling of callosal neurons in all cytoarchitectonic areas of the prefrontal cortex that have callosal projections to area 46 in mature monkeys. Conversely, we found no ectopic labeled cells in cytoarchitectonic areas of the prefrontal cortex of the opposite hemisphere that do not project to area 46 in adults. Thus, the data from these retrograde tracing studies lead us to conclude that: 1) the constellation of cytoarchitectonic areas that have callosal projections to the PFC of the opposite hemisphere appears to be stable from E111 through adulthood; 2) the pattern of projections between different areas of the PFC in opposite hemispheres is highly specified at the time of axonal growth into the cortical layers; and 3) callosal axons from divergent areas of the opposite hemisphere innervate the contralateral PFC in unison. Since most of these fibers wait for some period of time below the cortex of the PFC prior to growing into the cortical plate (Goldman-Rakic 1981), the synchrony of ingrowth for

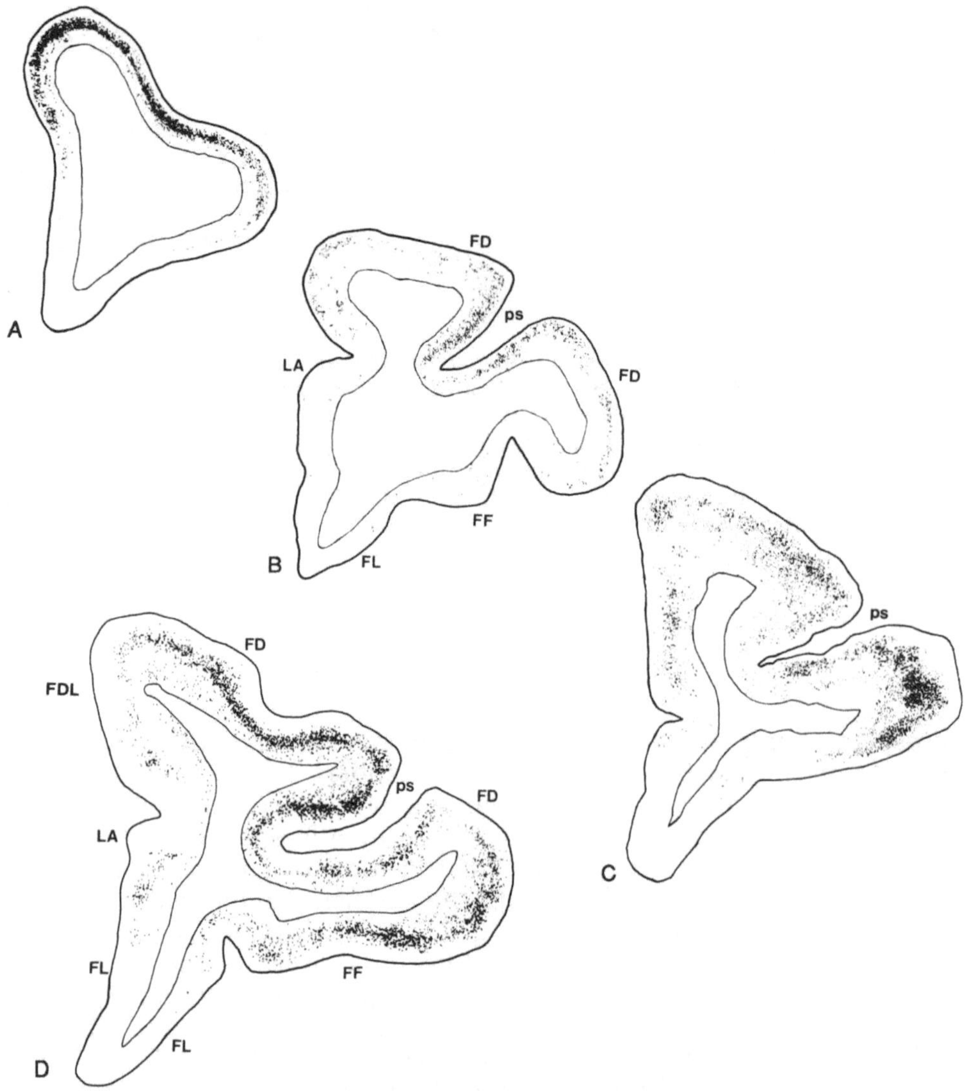

Figure 7. Frontal sections of the PFC from monkeys at E111 (A), E135 (B), postnatal day 10 (P10;C)) and adulthood (>3 years of age;D). Anterior-posterior levels represented at the different ages are from areas homotopic to the injection site of the opposite hemisphere and represent roughly equivalent regions of the PFC based upon landmarks such as sulcal location and comparable positions between the frontal pole and the head of the caudate nucleus. All labeled neurons are represented in each plot. Note the similarities in the distribution and areas of concentration for callosal neurons at the different fetal ages and the similarity with the distribution found in the adult monkey. Letter designations for von Bonin and Bailey's {#379} cytoarchitectonic areas LA, FD, FF and FL are shown for the E135 and adult cases. Walker's area 46 and von Bonin and Bailey's area FD are approximately coextensive. ps, principal sulcus.

afferents of diverse sources suggests that the waiting compartment may be important for coordinating their entrance into the cortical layers, possibly to insure competitive interactions between the different afferent populations for the establishment of synaptic and columnar territories.

Another aspect of callosal topography that we have examined is the emergence of the intra-areal columnar organization of callosal projection neurons in area 46 (Schwartz and Goldman-Rakic 1990; Schwartz and Goldman-Rakic 1991). This analysis of the emergence of modular features for the callosal neuron distribution focused on areas of the greatest and most complete callosal neuron labeling. Thus, we determined the distribution of labeled neurons across the cortex lining the banks and rims of the principal sulcus homotopic to the injection of tracer in the opposite hemisphere. The pattern of labeling at each age examined between E100 and maturity revealed a band of labeled callosal neurons extending across the entire breadth of the cortex lining the principal sulcus (Fig. 7). As in mature monkeys, the density of neuronal labeling within this band waxed and waned at each fetal age. Quantitative analyses of these variations were made for each age and representative histograms for fetuses injected at E111, E135 and P12 are illustrated in Figure 8 . Although the distribution of labeled neurons displayed periodic variations in density across the cortical surface at all ages, the tangential dimensions and regularity that characterize this modular organization in mature animals is not apparent until somewhere between E136 and the early postnatal period. These data raise the question of how the inverse relationship between callosal and associational projection fields emerges during this period. We addressed this issue by using the same double labeling paradigm we previously used in mature monkeys. The retrogradely transported fluorescent dye fast blue was injected into the cortex of the principal sulcus of one hemisphere, and a second fluorescent tracer, diamidino yellow was injected into the posterior bank of the intra-parietal sulcus in the opposite hemisphere (Schwartz and Goldman-Rakic 1982; Schwartz and Goldman-Rakic 1990). In order to exam the spatial relationship of associational and callosal projection neurons we generated histograms of their densities across the cortical surface of area 46 (Fig. 9). These data revealed that despite the irregularity of the individual distributions of these two efferent populations, there is a tendency for areas of high callosal neuron density to coincide with regions of low associational neuron density, and vice versa. Thus, although the basic inverse relationship between these efferent projection domains is apparent during the fetal period, the refinement of their periodic distributions continues into the early postnatal period. These data are consistent with the data of LaMantia and Rakic (LaMantia and Rakic 1990) demonstrating that the number of axons within the corpus callosum continues to undergo modification and attrition throughout the early postnatal period in monkeys.

A final feature of callosal neuron organization that we have examined is the laminar distribution of these cells at different stages of fetal development. Following injections restricted to the cortex between E82 and E89, the few labeled neurons identified were concentrated within prospective layers V and VI of the cortical plate. These data indicate that some of the earlier generated neurons of the infragranular layers innervate the contralateral PFC at a time when neurons destined for layer III are still being generated and migrating to their cortical positions. At the time of greatest ingrowth of callosal axons into the cortex of the opposite hemisphere (E100-E124), the majority of callosal projection neurons are present within layer III of area 46. Quantitative analysis of the number of labeled neurons within supragranular and infragranular cortical laminae at different fetal and postnatal ages revealed that the proportion of supragranular neurons contributing to the callosal projections of area 46 peaks at approximately E124 (94%) and gradually diminishes to the level found in adults (80-84%) during the late fetal and early postnatal period (Fig. 10). It is unlikely that this reduction in the proportion of callosal neurons is due to an increase in the number of infragranular callosal neurons since these cells are generated prior to the cells of layer III, and thus their axons are likely to innervate the opposite cortex prior to, or in synchrony with axons of later generated layer III neurons. Rather, it seems reasonable that the limited over-representation of layer III neurons to the callosal projection of area 46 is reduced by a selective, and possibly exclusive, susceptibility to cell death or callosal axon loss by the supragranular layers. Similar observations on the selective susceptibility of later generated neurons to attrition by cell death or

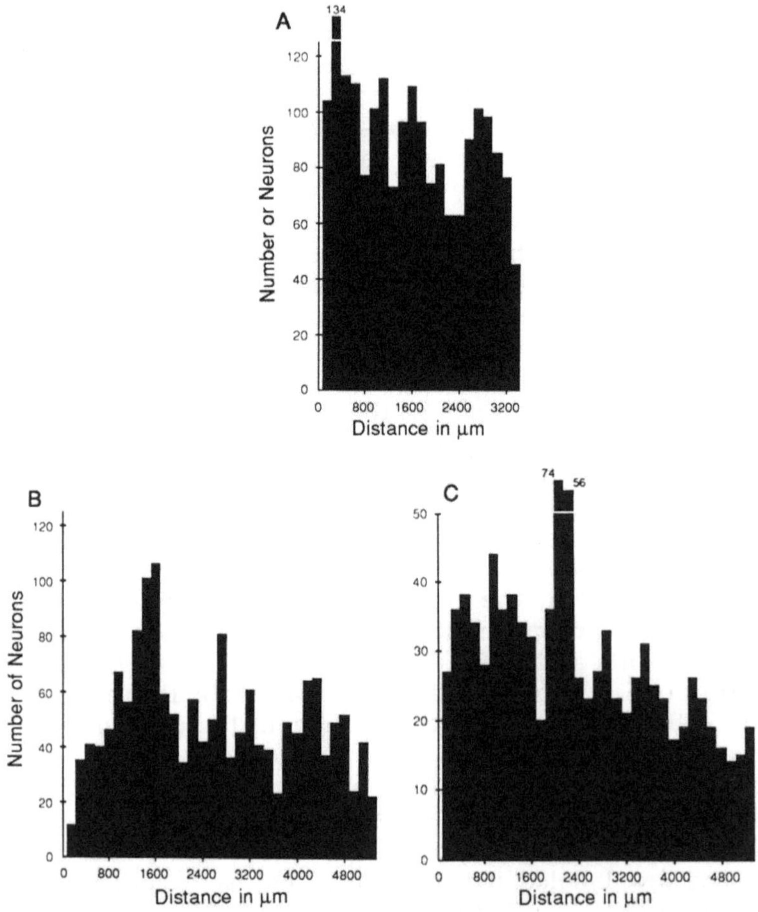

Figure 8. Histograms representing the tangential distribution of callosal neurons at E111 (A), E135 (B) and P12 (C). The bins in each histogram represent the number of labeled callosal neurons in a region of cortex 160 µm wide and spanning the entire cortical depth from the white matter to the pial surface. While all ages exhibited a somewhat columnar distribution of retrogradely labeled neurons, the regular periodicity and gradual waxing and waning which characterize the adult distribution was only well established in the P12 monkey.

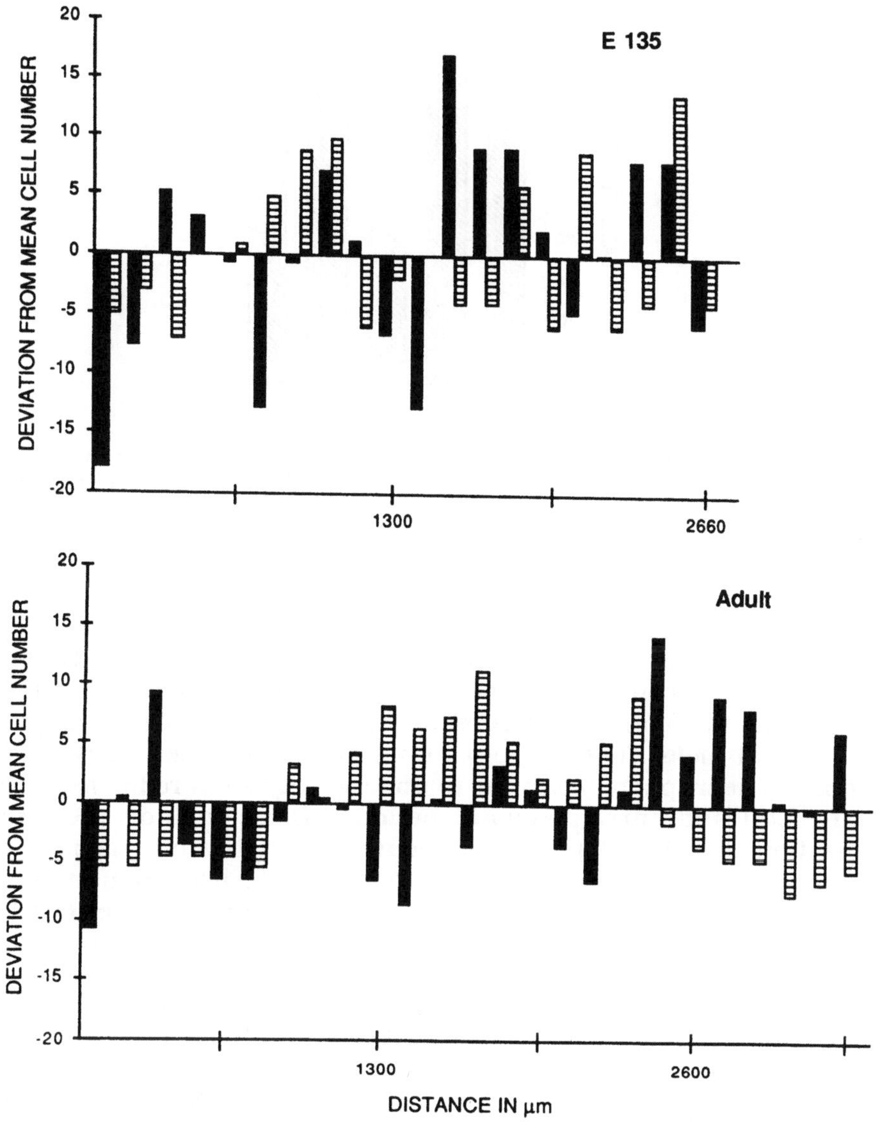

Figure 9. Relative distributions of callosal and associational projection neurons in the principal sulcus of the PFC at E135 (top panel). The relationship between the densities of cells projecting to the posterior bank of the ipsilateral intra-parietal sulcus, hatched bars, and those projecting to the cortex of the contralateral principal sulcus are represented in the same graphic conventions as used in Figure 4. As before, both populations are plotted as deviations from their mean density across the dorsal bank of the principal sulcus.

axonal losses have been made in studies of callosal connections in developing rat sensory cortex (Ivy and Killackey 1981; Finlay and Slattery 1983; Olavarria and Van Sluyters 1985).

In summary, the conclusion we draw from these studies of the development of cortex-specific callosal projections for prefrontal area 46 is that many features of the areal and intra-areal organization of callosal projections are specified prior to, or just after, birth. Of these features the areal topography of connections between different cytoarchitectonic regions of the two hemispheres is specified as early as the axons of

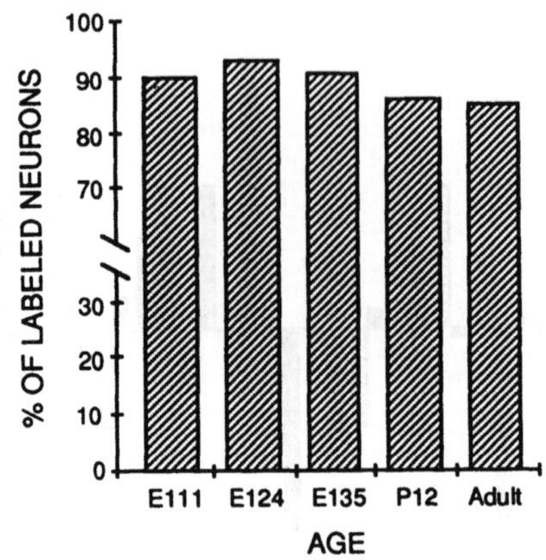

Figure 10. Histogram of the percentage of all labeled callosal neurons that were located in supragranular layer III from animals injected at E111, E124, E135, P12 and in an adult monkey. (Reprinted from Schwartz and Goldman-Rakic, 1991, Prenatal specification of callosal . . ., *J. Comp. Neurol.* 307:144-162 with permission from the Journal of Comparative Anatomy.)

callosal neurons have entered the cortex of the contralateral hemisphere and are essentially unchanged between E111 and maturity. Still other aspects of the modular organization of callosal connections, while present at the time of major ingrowth of callosal axons into their targets within the opposite hemisphere, continue to undergo a gradual refinement during the late fetal and early postnatal period. These include the laminar organization of callosal projection neurons and the emergence of the regularly repeating periodicity of callosal neuron densities that characterizes the mature monkey PFC. The early prenatal specification of callosal neuron organization in the primate PFC is consistent with the prenatal specification of callosal neuron distributions in other areas of the developing primate cortex, e.g., the visual (Dehay et al. 1988; Chalupa et al. 1989) and somatosensory (Killackey and Chalupa 1986) cortices. These data also, suggest a greater level of specificity in the initial patterning of these cortico-cortical connections than is present in many non-primate species (for review, see Schwartz and Goldman-Rakic 1990 and Kennedy and Dehay 1993).

This work was supported by grants from NSF (BNS-8617585) and NIH (NS22807-06).

REFERENCES

Akert, K., 1964, Comparative anatomy of the frontal cortex and thalamo-cortical connections, *in:* "The Frontal Granular Cortex and Behavior," J.M. Warren and K. Akert, eds., McGraw-Hill, New York, pp. 372-396.

Asanuma, A. and O. Okuda, 1962, Effects of transcallosal volleys on pyramidal tract activity of cat, *J. Neurophysiol.* 25:198-208.

Barbas, H., 1992, Architecture and cortical connections of the prefrontal cortex in the rhesus monkey, *Adv. Neurol.* 57:91-115.

Barbas, H. and Pandya, D.N., 1984, Topography of commisural fibers of the prefrontal cortex in the rhesus monkey, *Exp. Br. Res.* 55:187-191.

Berry, M. and Rogers, A.W., 1965, The migration of neuroblasts in the developing cerebral cortex, *J. Anat.* 99:691-709.

Brodmann, K., 1909, "Vergleichende Lokalisationslehre der Grosshirnrinde," Verlag J.A. Barth, Leipzig.

Brodmann, K., 1912, Neue ergebnisse über die vergleichende histologische lokalisation der grosshirnrinde mit besonderer berüksichtigung des stirnhirns, *Anat. Anz.* 41:157-216.

Chalupa, L.M., Killackey H.P., Snider, C.J., and Lia, B., 1989, Callosal projection neurons in area 17 of the fetal rhesus monkey, *Dev. Brain Res.* 46:303-308.

Dehay, C., Kennedy, H., Bullier, J., and Berland, M., 1988, Absence of interhemispheric connections of area 17 during development in the monkey, *Nature* 331:348-350.

Diamond, A. and Goldman-Rakic, P.S., 1989, Comparison of human infants and rhesus monkeys on Piaget's AB task: Evidence for dependence on dorsolateral prefrontal cortex, *Exp. Br. Res.* 74:24-40.

Finlay, B.L. and Slattery, M., 1983, Local differences in the amount of early cell death in neocortex predict adult local specializations, *Science* 219:1349-1351.

Freedman, M. and Oscar-Berman, M., 1986, Bilateral frontal lobe disease and selective delayed-response deficits in humans, *Behav. Neurosci.* 100:337-342.

Gazzaniga, M.S., 1985, Some contributions of split-brain studies to the study of human cognition, *in:* A.G. Reeves, ed., "Epilepsy and the Corpus Callosum," Plenum Press, New York, pp. 341-348.

Giguere, M. and Goldman-Rakic, P.S., 1988, Mediodorsal nucleus: areal, laminar, and tangential distribution of afferents and efferents in the frontal lobe of rhesus monkeys, *J. Comp. Neurol.* 277:195-213.

Goldman, P.S. and Nauta, W.J.H., 1977, Columnar distribution of cortico-cortical fibers in frontal association, limbic and motor cortex of the developing rhesus monkey, *Brain Res.* 122:393-414.

Goldman-Rakic, P.S., 1981, Development and plasticity of primate frontal association cortex, *in:* F.O. Schmitt, F.G. Worden, G. Adelman, and S.G. Dennis, eds., "The Organization of the Cerebral Cortex," M.I.T. Press, Cambridge, MA, pp. 69-97.

Goldman-Rakic, P.S., 1987, Circuitry of the prefrontal cortex and the regulation of behavior by representational knowledge, *in:* F. Plum and V. Mountcastle, eds., "Handbook of Physiology. Vol. 5," American Physiological Society, Bethesda, MD, pp. 373-417.

Goldman-Rakic, P.S., Isseroff, A., Schwartz. M.L., and Bugbee, N.M., 1983, The neurobiology of cognitive development, *in:* P. Mussen ed., "Handbook of Child Psychology: Biology and Infancy Development," Wiley, New York, pp. 281-344.

Goldman-Rakic, P.S. and Porrino, L.J., 1985, The primate mediodorsal (MD) nucleus and its projection to the frontal lobe, *J. Comp. Neurol.* 242:535-560.

Goldman-Rakic, P.S. and Schwartz, M.L., 1982, Interdigitation of contralateral and ipsilateral columnar projections to frontal association cortex in primates, *Science* 216:755-757.

Guitton, D., Buchtel, H.A., and Douglas, R.M., 1985, Frontal lobe lesions in man cause difficulties in suppressing reflexive glances and in generating goal-directed saccades, *Exp. Br. Res.* 58:455-472.

Hubel, D.H. and Wiesel, T.N., 1977, Functional architecture of macaque monkey and visual cortex, *Proc. R. Soc. Lond. B.* 198:1-59.

Ivy, G.O. and Killackey, H.P., 1981, The ontogeny of the distribution of callosal projection neurons in the rat parietal cortex, *J. Comp. Neurol.* 195:367-389.

Jacobson, S. and Trojanowski, J.Q., 1977a, Prefrontal granular cortex of the rhesus monkey. II Interhemispheric cortical afferents, *Brain Res.* 132:235-246.

Jacobson, S. and Trojanowski, J.Q., 1977b, Prefrontal granular cortex of the rhesus monkey. I Intrahemispheric cortical afferents, *Brain Res.* 132:209-233.

Jones, E.G. and Wise, S.P., 1977, Size, laminar and columnar distribution of efferent cells in the sensory-motor cortex of monkeys, *J. Comp. Neurol.* 175:391-438.

Kennedy, H. and Dehay, C., 1993, Cortical specification of mice and men,*Cereb. Cortex* 3:171-186.

Killackey, H.P. and Chalupa, L.M., 1986, Ontogenetic change in the distribution of callosal projection neurons in the postcentral gyrus of the fetal rhesus monkey, *J. Comp. Neurol.* 244:331-348.

Kostovic, I. and Rakic, P., 1990, Developmental history of the transient subplate zone in the visual and somatosensory cortex of the macaque monkey and human brain, *J. Comp. Neurol.* 297:441-470.

LaMantia, A.-S. and Rakic, P., 1990, Axon overproduction and elimination in the corpus callosum of the developing rhesus monkey, *J. Neurosci.* 10:2156-2175.

Leichnetz, G.R., 1980, An intrahemispheric columnar projection between two cortical multisensory convergence areas (inferior parietal lobule and prefrontal cortex): an anterograde study in *macaque* using HRP gel, *Neurosci. Lett.* 18:119-124.

McGuire, P.K., Bates, J.F., and Goldman-Rakic, P.S., 1991, Interhemispheric integration: I. Symmetry and convergence of the corticocortical connections of the left and right principal sulcus (PS) and the left and the right supplementary motor area (SMA) in the rhesus monkey, *Cerebral Cortex* 1:390-407.

Milner, B., 1982, Some cognitive effects of frontal-lobe lesions in man, *Phil. Trans. R. Soc. Lond. B.* 298:211-226.

Morest, D.K., 1970, A study of neurogenesis in the forebrain of opposum pouch young, *Z. Anat. Entwickl. Gesch.* 130:265-305.

Olavarria, J. and Van Sluyters, R.C., 1985, Organization and postnatal development of callosal connections in the visual cortex of the rat, *J. Comp. Neurol.* 239:1-26.

Pandya, D.N., Karol, E.A., and Heilbronn, D., 1971, The topographic distribution of interhemispheric projections in the corpus callosum of the rhesus monkey, *Brain Res.* 32:31-43.

Rakic, P., 1972, Mode of cell migration to the superficial layers of fetal monkey neocortex, *J. Comp. Neurol.* 145:61-84.

Rakic, P., 1982, Early developmental events: Cell lineages, acquisition of neuronal positions, and areal and laminar development, *Neurosciences Res. Prog. Bull.* 20:439-451.

Rakic, P., and Goldman-Rakic, P.S., 1984, Use of fetal neurosurgery for experimental studies of structural and functional brain development in nonhuman primates, *in:* "Prenatal Neurology and Neurosurgery," R.A. Thompson, J.R. Green, and D. Johnson, eds., Spectrum Press, New York, pp. 1-15.

Reeves, A.G. and O'Leary, P.M., 1985, Total corpus callosotomy for control of medically intractable epilepsy, *in:* "Epilepsy and the Corpus Callosum," A.G. Reeves, ed., Plenum Press, New York, pp. 269-280.

Rockland, K.S. and Pandya, D.N., 1979, Laminar origins and terminations of cortical connections of the occipital lobe in the rhesus monkey, *Brain Res.* 179:3-20.

Schwartz, M.L. and Goldman-Rakic, P.S., 1982, Single cortical neurons have axon collaterals to ipsilateral and contralateral cortex in fetal and adult primates, *Nature* 299:154-156.

Schwartz, M.L. and Goldman-Rakic, P.S., 1984, Callosal and intrahemispheric connectivity of the prefrontal association cortex in rhesus monkey: relation between intraparietal and principal sulcal cortex, *J. Comp. Neurol.* 226:403-420.

Schwartz, M.L. and Goldman-Rakic, P.S., 1990, Development and plasticity of the primate neocortex, *Clinics in Perinatology* 17(1):83-102.

Schwartz, M.L. and Goldman-Rakic, P.S., 1991, Prenatal specification of callosal connections in rhesus monkey, *J. Comp. Neurol.* 307:144-162.

Schwartz, M.L. and Meinecke, D.L., 1992, Early expression of GABA-containing neurons in the prefrontal and visual cortices of rhesus monkeys, *Cerebral Cortex* 2:16-37.

Schwartz, M.L., Rakic, P., and Goldman-Rakic, P.S., 1991, Early phenotypic expression of cortical neurons: evidence that a subclass of migrating neurons have callosal axons, *Proc. Natl. Acad. Sci. USA* 88:1354-1358.

Schwartz, M.L., Zheng, D.S., and Goldman-Rakic, P.S., 1988), Periodicity of GABA-containing cells in primate prefrontal cortex, *J. Neurosci.* 8(6):1962-1970.

Selemon, L.D. and Goldman-Rakic, P.S., 1988, Common cortical and subcortical targets of the dorsolateral prefrontal and posterior parietal cortices in the rhesus monkey: evidence for a distributed network subserving spatially guided behavior, *J. Neurosci.* 8:4049-4068.

Shatz, C.J., Chun, J.M., and Luskin, M.B., 1988, The role of the subplate in the development of the mammalian telencephalon, *in:* "Cerebral Cortex," A. Peters and E.G. Jones, eds., Plenum Press, New York, pp. 35-58.

Sperry, R., 1982, Some effects of disconnecting the cerebral hemispheres, *Science* 217:1223-1226.

Sperry, R.W., 1974, Lateral specialization in the surgically separated hemispheres, *in:* "The Neurosciences Third Study Program," F.O. Schmitt and F.G. Worden, eds., MIT Press, Cambridge, pp. 5-19.

Toyama, K. and Matsunami, K., 1976, Convergence of specific visual and commissural impulses upon inhibitory interneurons in cat's visual cortex, *Neuroscience* 1:107-112.

Toyama, K., Matsunami, K., Ohno, T., and Tokashiki, S., 1974, An intracellular study of neuronal organization in the visual cortex, *Exp. Br. Res.* 21:45-66.

Walker, A.E., 1940, A cytoarchitectural study of the prefrontal area of the macaque monkey, *J. Comp. Neurol.* 73:59-86.

Wilson, D.H., Culver, C., Waddington, M., and Gazzaniga, M., 1975, Disconnection of the cerebral hemispheres: an alternative to hemispherectomy for the control of intractable seziures, *Neurology* 25:1149-1153.

Wise, S.P. and Jones, E.G., 1978, Developmental studies of thalamocortical and commissural connections in the rat somatic sensory cortex,*J. Comp. Neurol.* 178:187-208.

ONTOGENY OF VISUAL CALLOSAL PROJECTIONS IN PRIMATES

Henry Kennedy, Ph.D. and Colette Dehay, Ph.D.

INSERM Unité 371 "Cerveau et Vision"
18 avenue du Doyen Lépine
69500 Bron
France

Understanding the development of the cerebral cortex constitutes a major challenge in neurobiology. The mature neocortex is characterized by a large number of areas each of which can be distinguished by its physiology, cytoarchitecture and the pattern of connections formed with other cortical and sub-cortical structures. A focal point of interest in the field of corticogenesis are the mechanisms which ensure the emergence of the anatomical features which distinguish individual areas. One possibility is that the morphogenesis of the cortex proceeds by an early phase in which the cortex appears essentially uniform. Where in the adult the connections formed by neurons are characteristic of each cortical region, an epigenetic developmental theory would conceive that neurons in all parts of the immature cortex would form similar connections and the adult pattern of connectivity would be achieved by a process of selective elimination of inappropriate connections. Establishing precise connections would require a mechanism that first of all recognizes then removes the unwanted connection. Theories which postulate an early stage of diffuse connectivity can be appealing because they raise the possibility that the elimination or sorting out of the inappropriate connections might come under the control of activity dependent cell-cell interactions (Stryker 1991). An alternative view is that connections are orderly from the onset, and complexity achieved by increasing the number of connections. This would mean that the adult pattern of connectivity is the consequence of mechanisms which allow growing axons to follow precisely determined tracts and to identify and connect with their appropriate targets.

There have been numerous studies of the corpus callosum in part because of its easy access. Callosal projecting neurons have a clearly defined target so that axonal tracers can be injected in one hemisphere and the distribution of callosal axon terminals and cell bodies in the

Epilepsy and the Corpus Callosum II, Edited by A.G. Reeves
and D.W. Roberts, Plenum Press, New York, 1995

other hemisphere can be examined. However what does the development of callosal connections tell us about principles governing the development of cortical connections in general? In other words, how representative are interhemispheric connections of cortical pathways?

Cortical pathways can be placed in one of three categories: association pathways (linking neurons in neocortical areas within a hemisphere), intrinsic connections (linking neurons within a cortical area) and callosal connections (linking neurons in the two hemispheres). Could it be that as a category callosal connections fulfill a unique psychophysical function and that they have evolved particular anatomical characteristics to be able to carry out that function? There are a number of reasons for thinking that this is not the case. For example, it has been argued that callosal connections have a very special relationship with the cortical representation of the body midline. Hence in the visual system, callosal connections would be expected to be centered in that part of the visual system which is represented in both hemispheres (the vertical meridian) and it has been proposed that this pathway provides the morphological substrate for perceptual continuation across the vertical meridian (Whitteridge 1965). Similar arguments have been proposed for the somatosensory system (Manzoni et al. 1989). The concept of a unique function with regards to the body midline is at odds with a number of anatomical observations which have been made by a large number of authors in many species: increasingly higher sensory and motor areas have callosal connections that show no concentration around the representation of the body midline. Hence it has been argued elsewhere that callosal connections comply to general rules governing the organization of cortical connectivity and only a brief summary will be given here (Kennedy et al. 1991). Callosal connections can be broken down into three categories: (1) anterior directed heterotopic projections from a lower to higher area; (2) posterior directed heterotopic projections from a higher to a lower area; and (3) homotopic projections. In terms of information processing, these pathways can be assimilated to the following pathways: (1) feed-forward association connections; (2) feedback association connections; and (3) intrinsic connections. This is supported by anatomical features shared by these pathways. Measurements of the divergence and convergence shown by these pathways reveal similar values between the appropriate association and callosal pathways. Further, the cell bodies and the terminals are of the same type in both sets of pathways. There are also psychophysical findings which support the notion that association and callosal pathways fulfill similar functions. It has been clearly established that the disconnection syndrome found in patients who have undergone surgical section of the corpus callosum is not found when the callosum is congenitally absent (see Lassonde 1986 for references). This is almost certainly due to developmental plasticity since early commissurectomy led to much smaller deficits in kittens, which was curiously related to an overall reduced rate of learning. This led Lassonde (1986) to compare inter- and intrahemispheric processing in patients which had agenesis of the corpus callosum and patients which had undergone section of the callosum. The results of these experiments were straightforward: they showed that the callosal pathway exerts a facilitatory influence on transmission in interhemispheric pathways. Such callosal modulation of association pathways requires that these two pathways have a common target which has actually been shown to be the case in the monkey (Kennedy and Dehay 1986) .

The Old World simian is of particular interest for the study of the development of visual callosal connections because of the remarkably precise configuration of the pathway in the adult. In the Old World monkey, terminals of callosal projecting fibers in area 17 or V1 are confined to a narrow strip of cortex lying adjacent to area V2 (Zeki 1970; Van Essen and Zeki 1978). Callosal projections are reciprocal in area V1 and neurons in area V1 that project to the contralateral hemisphere are confined to the strip of callosal recipient cortex lying adjacent to area V2 (Kennedy et al. 1986). The callosally connected part of V1 extends 1-2.5 mm from the V2 border (Kennedy et al. 1986) . In Old World monkeys, the laminar location of callosal connections is stricter than that found in non-primates. Terminals of callosal afferents are located in layers 4B and 5, and callosal projecting neurons are scarce and limited to supragranular layers.

It has recently been shown in cats, rodents and lagomorphs that the restricted adult pattern of callosal connectivity emerges during postnatal life from a precocial organization in which callosal projecting neurons have an extended distribution (Wise and Jones 1976; Innocenti et al. 1977; Ivy et al. 1979; Chow et al. 1981; Cusick and Lund 1982; Ivy and Killackey 1982; Rhoades and Fish 1983; Miller and Vogt 1984). Given that there is late decrease in the numbers of fibers in the corpus callosum of the monkey it is possible that there is a developmental restriction of callosal connections at the V1/V2 border (La Mantia and Rakic 1990).

Further, unlike non-primates, the V1/V2 border is very precise so that one can measure in microns the location of callosal connections with respect to the V1/V2 border thus making it possible to obtain a highly reliable quantification of developmental changes of the tangential distribution of callosal connections.

It has been claimed that the post-natal elimination of callosal connections from area V1 in rodents and carnivores is influenced by neural activity in the ascending visual system (Lund et al. 1978; Lund and Mitchell 1979 a,b; Innocenti and Frost 1980; Rhoades and Dellacroce 1980; Cusick and Lund 1982; Berman and Payne 1983; Rhoades and Fish 1983; Miller and Voght 1984; Olavarria and Van Sluyters 1984). However, there are confusing contradictions in the literature. Whereas enucleation affects the distribution of callosal projecting neurons in both cats and rats, visual input has only been reported to influence the distribution in cats (Lund et al. 1978; Lund and Mitchell 1979a; Innocenti and Frost 1980; Cusick and Lund 1982; Berman and Payne 1983). These results can be taken to mean that in the cat, visual mediated neural activity is involved in the formation of the adult pattern of callosal connections. However, this would mean that there is a somewhat unusual species difference between carnivores and rodents. The alternative is that the reported effect of visual experience in cats is in some way flawed perhaps due to the use of non-optimum axonal tracing techniques.

The primate is of particular interest with respect to the importance of visual experience on the development of the visual system. Unlike carnivores and rodents, monkeys are born with open eyes and display some visually guided behavior. Comparison of the developmental stages of monkeys, carnivores and rodents shows that monkeys are born at a relatively late stage of development (Dehay and Kennedy 1988; Kennedy and Dehay 1992). Nevertheless neural connections are not fully formed (O'Kusky and Colonnier 1982) and abnormal experience can severely perturb the visual system (LeVay et al. 1980). An indication of the impact of experience on development is provided by comparing the organization of callosal projecting neurons of the young inexperienced animal (i.e., at birth) with the adult. Monkeys are ideal for this because unlike cats and rodents they are born more mature at birth and genetic processes would have a greater opportunity to be expressed before birth.

Using axonal tracers we were able to show that in a number of respects the callosal connectivity of visual areas V1 and V2 showed an adult-like organization at birth. Area V1 callosal projecting neurons were as rare in the newborn as in the adult where in both cases they were restricted to a 1-2 mm strip along the V1/V2 border with similar laminar locations. In area V2 both the laminar location and the patchy distribution were similar in neonates and adults. In both cases callosal connections extended a maximum of 8 mm in to V2 and were concentrated in regions of high cytochrome-oxidase activity.

The fact that the tangential distribution of callosal projecting neurons appears very adult-like in the newborn monkey makes it very different from that found in other species. However, this could merely reflect the fact that monkeys are more mature at birth. If one considers the moment of eye opening and not the moment of birth as the significant developmental event then there is very much less variability. Cats, rats and rabbits open their eyes 11-17 days after birth and in each species the tangential distribution of callosal projecting neurons is largely mature at the moment of eye opening. The sheep is the only animal beside the monkey that has been reported to have a restricted adult-like distribution at the moment of birth (Rao 1979). The sheep like the monkey has good visuomotor coordination on the day of birth, the optic media are clear and single neurons in the visual cortex respond briskly to visual stimulation (Kennedy et al. 1983) .

One possibility is that inappropriate callosal connections in area 17 are symptomatic of a less sophisticated visual system. If this were the case one would expect sheep to have such neurons, since this animal has a visual system far more similar to that of the cat than the monkey. It would seem therefore that the most plausible explanation for the similarities of callosal connections of sheep and monkey is that they are both born relatively mature with open eyes.

The finding that the callosal projections in the newborn Macaque have a restricted distribution identical to that found in the adult raises the possibility that the development of this pathway differs radically from that found in all other species studied including rodents, lagomorphs and carnivores. If the development of this cortical pathway is to resemble that reported in non-primates then one would predict that callosal projecting neurons would have a widespread distribution prenatally. To examine this, we have used axonal tracers with maximum sensibility to trace the pattern of connectivity from very early stages prior to axogenesis to birth in Old World monkeys (Dehay et al. 1988) .

What these studies show is that right from the moment the cortical plate is formed and before axogenesis begins (at about 100 days of conception -E100) the limits of the primary visual area V1 can be clearly detected using Nissl staining. We have looked at the overall distribution of callosal connections at regular 3-8 days intervals between E100 and birth at E165. By plotting the position of labeled cells in area V1 and V2 it has been possible to quantify the changing developmental pattern of connections, both in terms of density and distance of the labeled neurons from the V1/V2 border. These results show that callosal neurons in the visual system have an invarient laminar distribution throughout development. Callosal projecting neurons were not found to be more numerous nor do they extend significantly further into area V1 at any prenatal stage.

In area V2, however, there were developmental changes in the tangential distribution of callosal connections. Whereas in the adult callosal connections only extend part of the way through area V2, in the prenatal animal they continued as a more or less continuous band of labeled neurons from area 17 out to area 19. Further, the clustered pattern of neurons was not detected in area V2. These results could mean that there has been an elimination of inappropriate connections. However, in the adult callosal connections run in bands orthogonal to the V1/V2 border (Kennedy, Barone and Dehay, unpublished). As yet we have not been able to measure the change in the density profile in this direction. It could turn out that callosal connections show a waxing and waning and that a simple decrease in the numbers of neurons coupled with cortical expansion could be responsible for the emergence of the discontinuities found in the adult.

These results show that the ontogeny of the primary visual cortex area V1 in the primate is unique amongst mammals and is characterized by an absence of callosal connections throughout corticogenesis. This absence of transient callosal connections could be related to the particular status of the primate area 17. Whereas in primates the relay neurons in the lateral geniculate nucleus project uniquely to area 17 (Bullier and Kennedy, 1983), in non-primates the primary thalamic nuclei have multiple cortical targets. Zeki (1978) has suggested that the functional specialization of the primate area 17 is to integrate and dispatch information to the multiple extrastriate visual areas. One consequence therefore of the organization of the geniculo-striate pathway in primates is that the burden on area 17 as a structure for dispatching information to extrastriate areas is considerably greater than in non-primates.

The absence of transient callosal connections in the primate has a number of theoretical consequences. For instance, it demonstrates very clearly that exuberant connectivity is not an integral part of growing up: one can build a cortical area, and a remarkably sophisticated one at that, without transient connections either to the contralateral hemisphere or to cortical areas belonging to other sensory modalities (Dehay and Kennedy, unpublished). However, it is a little more difficult to relate the uniqueness of the development of the primate primary visual area to a precise developmental strategy. One possibility is that the organization of the geniculo-striate pathway in the primate is such that it cannot afford to maintain transient connections of the type seen in lower mammals. This would seem to imply that transient connections do not play a meaningful role during development and have to be abandoned in more sophisticated structures. Another possibility is that the primate cortex enjoys a very different type of transient connectivity. Recent results support the latter hypothesis. In the adult monkey, projections from extrastriate cortex to area 17 are particularly extensive (Kennedy and Bullier 1985). These caudal-directed connections can be thought of as feed-back projections and have been shown to terminate outside of layer 4. Although feed-back projections originate from both supra- and infragranular layers, we have been able to show that as one moves further rostrally away from area 17, there is an increase in the relative proportion of infragranular layer neurons which project to area 17. In fact, the precise laminar distribution of area 17 afferents is highly characteristic for each individual cortical area (Kennedy and Bullier 1985; Kennedy et al. 1989). Results in the newborn animal show that large numbers of supragranular layer neurons project to area 17. These projections from supragranular layers also originate from very rostral cortical areas where, in the adult, projections are almost uniquely from infragranular layers, indicating that these connections constitute a veritable transient pathway.

The disappearance of transient projections from the supragranular layers takes place over a protracted period of postnatal development and constitutes an intensive period of remodeling of corticocortical connections. What is clear is that this remodeling of cortical connections is a feature unique to the developing primate. Similar axonal tracing studies in the cat show that there is no major reorganization of the laminar distribution of area 17 afferents during development (Barone, Kennedy and Dehay, unpublished). Interspecies comparison therefore gives some clues as to the developmental significance of transient cortical connections. To summarize: the primate has evolved a very unique set of transient connections. Unlike non-primates there are no interconnections between cortical areas which are not interconnected in the adult. This appears to allow the visual cortex to form a large number of transient connections between structures which show stable connections in the adult.

To conclude, what the development of callosal pathways tells us is that many aspects of the adult cortical connectivity are present from very early stages. The adult pattern of connectivity develops independently from visual experience. However this does not mean that deprivation might not have a harmful effect, preliminary results from experiments looking at the callosal connections of long-term visually deprived monkeys suggest that under certain conditions there can be a failure to maintain callosal connections (Dehay, Vital-Durand and Kennedy, unpublished). There are also indications that neural activity in the ascending pathways may play a role in the shaping of callosal pathways. Bilateral enucleation one month before birth has been shown to lead to a much denser callosal connectivity in area 18 at birth compared to the normal neonate (Dehay et al. 1988). However, here we need to be cautious before assuming that these results reflect a lack of the normal elimination of connections. It is known that enucleation leads to a drastic decrease in the areal dimensions of area 17 (Rakic et al. 1991; Dehay et al. 1991) coupled with an increase in cell density (Dehay et al. 1991). It is thought that massive depletion of LGN neurons following enucleation is in part responsible for the areal reduction and increase in cell density of area 17. Area 17 provides a major input to area 18 so that it is highly conceivable that area

18 will also show reduced dimensions and increased packing densities. If this is the case, then one would predict that enucleation could lead to increased densities of callosal-projecting neurons without the necessity of evoking increased levels of axon elimination.

What is clear is that the acallosal nature of area 17 is not the consequence of the elimination of large numbers of inappropriate connections. However, this does not mean that selective elimination is entirely absent. It has been shown to occur to some extent in the somatosensory system (Chalupa and Killackey 1986) and in the visual system the patchy distribution of callosal projecting neurons which is present at birth emerges from an earlier prenatal condition when callosal projecting neurons appear to have a more uniform distribution. However, the relative importance of this selective elimination in terms of percentages of cells could be quite low as has been shown for the developmental segregation of callosal and association pathways (Meissirel et al. 1990). It would seem therefore that a highly sophisticated system such as is found in the primate where there is strict segregation of pathways does not depend on an increased contribution form epigenetic mechanisms but rather shows a large degree of early specification. The question remains whether early specification is a characteristic feature of primate development. The alternative is that the prolonged development and more clearly defined patterns of connections facilitate a quantitative approach to developmental changes in neural organization and enables the early patterning of connections to be experimentally measured more easy to detect. However, the acallosal nature of primate area 17 suggests that there is a degree of early specification which is characteristic if not unique, to primates as has been suggested in the frontal cortex (Schwartz and Goldman-Rakic 1982; Schwartz and Goldman-Rakic 1991). Similar conclusions have been reached from studies of the development of the primate retina (Chalupa and Lia 1991), so that it would seem that high degrees of segregation of pathways in the adult animal requires an early specification of connections during development and a high degree of order in early stages of formation.

REFERENCES

Berman, N.E.,and Payne, B.R., 1983, Alterations in connections of the corpus callosum following convergent and divergent strabismus, *Brain Res.* 274:201-212.

Bourdet, C., Dehay, C., and Kennedy, H., 1988, Characteristics of inter-and intra-hemispheric transient pathways in the young kitten, *J. Physiol.* (Lond.) 396:67.

Bullier, J. and Kennedy, H., 1983, Projection of the lateral geniculate nucleus on to cortical area V2 in the macaque monkey, *Exp. Brain Res.* 53:168-172.

Chow, K.L., Baumbach, H.D., and Lawson R.,1981, Callosal projections of the striate cortex in the neonatal rabbit, *Exp. Brain Res.* 42:122-126.

Cusick, C.G. and Lund, R.D., 1982, Modification of visual callosal projections in rats, *J. Comp. Neurol.* 212:385-398.

Chalupa, L.M., Lia, B., 1991, The nasotemporal division of retinal ganglion cells with crossed and uncrossed projections in the fetal rhesus monkey, *J. Neurosci.* 11:191-202.

Chalupa, L.M., and Killackey, H.P., 1986, Process elimination underlies ontogenetic change in the distribution of callosal projection neurons in the postcentral gyrus of the fetal rhesus monkey, *Proc. Natl. Acad. Sci.* (USA) 86:1076-7079.

Dehay, C.,and Kennedy, H., 1988, The maturational status of thalamocortical and callosal connections of visual areas V1 and V2 in the newborn monkey, *Behav. Brain Res.* 29:237-224.

Dehay, C., Kennedy, H., Bullier, J. and Berland, M.,1988, Absence of interhemispheric connections of area 17 during development in the monkey, *Nature*, 331:348-350.

Dehay, C., Horsburgh, G., Berland, M., Killackey, H., and Kennedy, H., 1991, The effects of bilateral enucleation in the primate fetus on the parcellation of visual cortex, *Dev. Brain Res.* 62:137-141.

Innocenti, G.M., Fiore, L., and Caminiti, R., 1977, Exuberant projection into the corpus callosum from the visual cortex of newborn cats, *Neurosci. Lett.* 4:237-242.

Innocenti, G.M. and Frost, D.O., 1980, The postnatal development of visual callosal connections in the absence of visual experience of the eyes, *Exp. Brain Res.* 39:365-375.

Ivy, G.O., Akers, R.M., and Killackey, H.P., 1979, Differential distribution of callosal projection neurons in the neonatal and adult rat, *Brain Res.* 173:532-537.

Ivy, G.O. and Killackey, H.P.,1982, Ontogenetic changes in the projections of neocortical neurons, *J. Neurosci.* 2:735-743.

Kennedy, H., Martin, K.A.C., and Whitteridge, D., 1983, Receptive field characteristics of neurons in newborn lambs and adult sheep, *J. Neurosci.* 10:295-300.

Kennedy, H. and Bullier, J., 1985, A double-labelling investigation of the afferent connectivity of the afferent connectivity to cortical areas V1 and V2, *J. Neurosci.* 5:2815-2830.

Kennedy, H., Bullier, J., and Dehay, C., 1985, Cytochrome oxidase activity in the striate cortex and lateral geniculate nucleus of the newborn and adult macaque monkey, *Exp. Brain Res.*. 57:208-212.

Kennedy, H., Dehay, C., and Bullier, J., 1986, The organization of the callosal connections of visual areas V1 and V2 in the macaque monkey, *J. Comp. Neurol.* 247: 398-415.

Kennedy, H., Bullier, J., and Dehay, C., 1989, Transient projection from the superior temporal sulcus to area 17 in the newborn macaque monkey, *Proc. Natl. Acad. Sci. USA.* 86:8093-8097.

Kennedy, H., Meissirel, C., and Dehay, C., 1991, Callosal pathways in primates and their compliancy to general rules of cortical connectivity, *in:* "Neuroanatomy of the Visual Pathways and their Development," B. Dreher and S. Robinson, eds., Vol. 3, "Vision and Visual Dysfunction, " J. Cronly-Dillon, ed., McMillan Press, pp. 324-359.

Kennedy, H. and Dehay, C., 1993, Cortical specification of mice and men, *Cerebral Cortex* 3(3):171-186.

La Mantia, A.S., Rakic, P., 1990, Axon overproduction and elimination in the corpus callosum of the developing rhesus monkey, *J. Neurosci.* 10:2156-2175.

Lassonde, M., 1986, The facilitatory action of the corpus callosum, *in:* "Two Hemispheres, One brain," F. Lepore, M. Ptito, and H. Jasper, eds., Alan R. Liss, New York, pp. 385-402.

LeVay, S., Viesel, T.N. and Hubel, D.H.,1980, The development of ocular dominance columns in normal and visually deprived monkeys, *J. Comp. Neurol.* 19:1-51.

Lund, R.D., Mitchell, D.E. and Henry, G.H.,1978, Squint-induced modification of callosal connections in cats, *Brain Res.*144:169-172.

Lund, R.D. and Mitchell, D.E.,1979a, The effects of dark rearing on visual callosal connections of cats, *Brain Res.*167:172-175.

Lund, R.D. and Mitchell, D.E.,1979b, Asymmetry in the visual callosal connections in strabismic cats, *Brain Res.* 167:176-179.

Miller, M.W. and Vogt, B.A., 1984, Heterotopic and homotopic callosal connections in rat visual cortex, *Brain Res.* 297:75-89.

Manzoni ,T., Barbaresi, P., Conti, F., and Fabri, M., 1989, The callosal connection of the primary somatosensory cortex and the neural bases of midline fusion, *Exp. Brain Res.* 76:251-266.

Meissirel, C., Dehay, C., Berland, M., Kennedy, H., 1990, Incidence of visual cortical neurons which have axon collaterals projecting to both cerebral hemispheres during prenatal primate development, *Dev. Brain. Res.* 56:123-126.

O'Kusky, J. and Colonnier, 1982, Postnatal changes in the number of neurons and synapses in the visual cortex (area 17) of the macaque monkey: A sterological analysis in normal and monocularly deprived animals, *J. Comp. Neurol.* 210:291-306.

Olavarria, J. and Van Sluyters, R.C., 1984, Callosal connections of the posterior neocortex in normal-eyed, congenitally anophthalmic, and neonatally enucleated mice, *J. Comp. Neurol.* 230:249-268.

Rakic, P., Suner, I., and Williams, R.W., 1991, A novel cytoarchitectonic area induced experimentally within the primate visual cortex, *Proc. Natl. Sci. USA*, 88:2083-2087.

Rao, V.M., 1979, Interhemispheric connections between primary visual areas in adult sheep and newborn lamb, *J. Physiol.* 296:65.

Rhoades, R.W. and Dellacroce, D.D., 1980, Neonatal enucleation induces an asymmetric pattern of visual callosal connections in hamsters, *Brain Res.* 202:189-19 .

Rhoades, R.W. and Fish, 1983, Bilateral enucleation alters visual callosal but not corticotectal or corticogeniculate projections in hamsters, *Exp. Brain Res.*. 51:451-462.

Schwartz, M.L. and Goldman-Rakic, P.S., 1982a, Single cortical neurons have axon collateral to ipsilateral and contralateral cortex in fetal and adult primates, *Nature* 299:154-156.

Schwartz, M.L. and Goldman-Rakik, P.S., 1991, Prenatal specification of callosal connections in Rhesus Monkey, *J. Comp. Neurol.* 307:144-162.

Stryker, M.P., 1991, Activity-dependent reorganization of afferents in the developing mammalian visual system, *in:* "Development of the Visual System," D.M-K. Lam and C.J. Shatz, eds., Proceedings of the Retina Research Foundation Symposia, MIT Press, pp. 267-289.

Van Essen and Zeki, S.M., 1978, The topographic organization of rhesus monkey prestriate cortex, *J. Physiol.* 277:193-226.

Wise, S.P. and Jones, E.G., 1976, The organization and postnatal development of the commissural projection of the somatic sensory cortex of the rat, *J. Comp. Neurol.* 168:313-343.

Whitteridge, D., 1965, *in:* "Functions of the Corpus Callosum," E.G. Ettlinger, ed., pp. 115-120, Churchill, London.

Zeki, S.M., 1970, Interhemispheric connections of prestriate cortex in monkey, *Brain Res.* 19:63-75.

Zeki, S.M., 1978, Functional specialisation in the visual cortex of the rhesus monkey, *Nature* 274:423-428.

EXPERIMENTAL MANIPULATIONS OF THE ORGANIZATION OF INTERHEMISPHERIC PROJECTIONS

Robert Rhoades, Ph.D.

Department of Anatomy
Medical College of Ohio
P.O. Box 10008
Toledo, Ohio 43699

Visual callosal projections have been used extensively to study how central nervous system pathways develop normally and the ways in which development can be altered by perinatal manipulations. Given the experimental attention that has been paid to this system, it is surprising that we know relatively little about the morphological characteristics of individual visual callosal axons at any stage of either normal or altered development.

We have used Phaseolus vulgaris-leukoagglutinin (PHAL) to trace individual callosal axons in normal adult hamsters and adults that sustained removal of both eyes at birth. We have also employed the carbocyanine dye, Di-I, to assay the development of single occipital callosal axons in developing hamsters. In normal adults, injections of PHAL into the region of the 17-18a border produced dense labelling in all layers in the region of the contralateral 17-18a border. Larger injections resulted in callosal labelling that extended across the lateral one-half of area 17, primarily in layers I and V. Individual callosal axons from normal adult hamsters had very simple terminal arbors. They gave off short collaterals in the infragranular layers and branched more extensively in the uppermost part of layer II-III and in lamina I.

The development of the visual callosal projection in hamsters is very rapid and straight forward. Anterograde labelling with Di-I demonstrated labelled axons in the anterior commissure and rostral part of the corpus callosum on P-O. Labelled fibers extended into the white matter underlying the occipital cortex on P-1 and entered the cortical plate on P-2 with some axons reaching as far as the marginal layer. It thus appears that callosal axons in hamster do not "wait" in the white matter for more than one day before entering the cortex. Many developing callosal axons had short branches in the white matter, but generally extended only a single collateral into the cortical gray matter. Callosal axons in perinatal animals branched very little within the cortex and generally resembled fibers labelled with PHAL in adult hamsters.

The "visual" callosal projection expands in animals that sustain enucleations on the day of birth. Small injections of PHAL into the occipital cortex of neonatally enucleated adult hamsters resulted in labelled axons throughout most of areas 17 and 18a in the contralateral hemisphere. The terminal arbors of most individual callosal axons in eyeless hamsters were not appreciably different from those in sighted animals. However, over one-fourth of the fibers reconstructed through all cortical laminae in these animals had much more widespread branches than any of the axons recovered from normal hamsters.

The results from these blinded animals provided new information about the way in which the expanded tangential distribution of occipital callosal neurons in eyeless hamsters gives rise to the abnormally extensive distribution of callosal axons in these animals. There are at least two ways in which such a projection might develop. The first is that the expanded distribution of callosal neurons projected to the contralateral hemisphere in a "point-to-point" manner. The fact that single focal injections of PHAL produced widespread labelling in the contralateral occipital cortices of enucleated animals indicates that this proposition is not correct. The second possibility is that individual callosal axons in enucleated hamsters might develop terminal arbors that were more extensive than those of such axons in sighted animals. Reconstruction of individual axons indicated that this did, in fact, occur for a substantial number of the fibers reconstructed from the blinded animals.

REFERENCES

1. S. Clarke and G.M. Innocenti, Organization of immature intrahemispheric connections, *J. Comp. Neurol.* 25:1-22 (1986).
2. C.G. Cusick and R.D. Lund, Modification of visual callosal projections in rats, *J. Comp. Neurol.* 212:385-398 (1982).
3. S.E. Fish, R.W. Rhoades, C.A. Bennett-Clarke, B.A. Figley, and R.D. Mooney, Organization of visual callosal projections in normal and eyeless hamsters: tracing the projections of single axons with Phaseolus vulgaris leukoagglutinin and Di-I, *Eur. J. Neurosci.* 3:1255-1270 (1991).
4. C. Hedin-Pereira, R. Lent, and S. Jhaveri, The development of callosal and corticocortical innervation in the neocortex of the hamster, *Brazilian J. Med. Biol. Res.* 21:499-502 (1988).
5. G.M. Innocenti, Growth and reshaping of axons in the establishment of visual callosal connections,*Science* 212:824-827 (1981).
6. G.M. Innocenti, L. Fiore, and R. Caminiti, Exuberant projection into the corpus callosum from the visual cortex of newborn cats,*Neurosci. Lett.* 4:237-242 (1977).
7. G.M. Innocenti and D.O. Frost, Effects of visual experience on the maturation of the efferent system to the corpus callosum, *Nature* 280:231-234 (1979).
8. R.D. Mooney, R.W. Rhoades, and S.E. Fish, Neonatal superior collicular lesions alter visual callosal development in hamster, *Exp. Brain Res.* 55:9-25 (1984).
9. J. Olavarria, R. Malach, and R.C. Van Sluyters, Development of visual callosal connections in neonatally enucleated rats, *J. Comp. Neurol.* 260:231-348 (1987).
10. R.W. Rhoades and D.D. DellaCroce, Neonatal enucleation induces an asymmetric pattern of visual callosal connections in hamsters, *Brain Res.* 202:189-195 (1980).

II

EXPERIMENTAL EPILEPSY

MIDLINE SUBCORTICAL STRUCTURES FOR TRANSHEMISPHERIC ICTAL AND INTERICTAL TRANSMISSION

Juhn A. Wada, O.C., M.D., Med. Sc.D., F.R.C.P.(C)

Professor and Director
Divisions of Neuroscience and Neurology
University of British Columbia
Vancouver, B.C. Canada V6T 2A1

INTRODUCTION

Current understanding of the fundamental mechanisms of epilepsy is largely based on experimental animal studies. In these studies, the intrinsic difference between chronic epileptic brain and intact non-epileptic brain must be considered. The use of normal brain is important for understanding the capacity of cerebral structures to produce electroclinical manifestations of epilepsy. However, chronic or acquired epileptic brain in animal models such as kindling has now been shown to be significantly different from normal brain in terms of synaptic function with certain alteration of morphology.[1] The presence of comparable morphological changes has also been confirmed in temporal lobe specimens removed for therapeutic purposes in medically refractory epilepsy.[2-7] To study the pathophysiology of epilepsy, it would therefore be desirable to use a chronic animal model having a potential of developing a recurrent spontaneous seizure with persistent epileptogenic susceptibility either acquired or genetically dictated. In this regard, kindling[8] is an ideal model of partial epilepsy with secondarily generalized seizures. Kindling causes permanent reorganization of brain function leading to spontaneous emission of partial or generalized seizures.[9-11] In this model, interaction of acquired and genetically dictated factors may be studied by comparing the pattern and outcome of kindling between animal species with or without predisposition to epileptic seizures such as Senegalese baboons, Papio papio and Rhesus monkeys, respectively.[12,13]

In addition, the anatomical difference which exists between human and sub-human mammalian brains also requires some consideration. Although a marked similarity in brain organization across species is assuring, the significance of species differences such as the evolutionary

Epilepsy and the Corpus Callosum II, Edited by A.G. Reeves and D.W. Roberts, Plenum Press, New York, 1995

progression and expansion of the corpus callosum (CC) and the reverse tendency to regression of both the anterior and the hippocampal commissures, must be appreciated if one is to make a valid interpolation of mechanisms from subhuman studies. This is particularly relevant when one assesses the significance of potential anatomical routes of transhemispheric transmission of epileptic information across the midline. Among these routes, the forebrain commissures, particularly the CC have been the subject of intensive experimental and clinical studies, while others have not received much attention.

Results of studies on the potential role played by the forebrain commissures in both acute and chronic models of epilepsy in rats, cats, monkeys and baboons suggested a major, if not exclusive, role of the CC in seizure bilateralization and bisynchronization.[14-22] In this regard, the effect of surgical bisection of the forebrain commissures on developing seizure was consistent with the observation in animals with congenital agenesis of the forebrain commissures[20] and the potential relevance of the animal findings to those of human epilepsy was discussed, with particular emphasis on potential CC dysfunction as an aspect of pathophysiology underlying intractable generalized convulsive seizure in man.[21]

However, kindling-induced generalized convulsions can occur under certain circumstances even when the forebrain commissures are completely bisected.[22] This finding suggests the availability of a subcortical pathway across the midline. Identification of such a pathway and characterization of the conditions under which such a pathway is used are important for better understanding of the mechanisms of transhemispheric interictal and ictal transmission and potential subcortical integration of forebrain processes. Furthermore, such knowledge would contribute significantly to the clinical process of identifying appropriate surgical candidates for callosal section. In this paper, findings obtained in our laboratory during recent years on the subcortical midline structures, i.e., the massa intermedia and the brain stem will be selectively reviewed.

A. MASSA INTERMEDIA (MI)

The MI in man lacks commissural fibres, is generally small and variable in size and is frequently absent. Therefore, one can conclude that this structure is not of significance in human epilepsy. However, the MI is rather large in all subhuman mammalian brains, including those of monkeys and baboons used in our experimental study of epilepsy. Since an obvious morphological difference implies functional difference, elucidation of the significance of the MI in the experimental models of epilepsy is of importance for the appropriate interpretation and interpolation of animal data to human problem. In the past, the MI has been reported to have an active role in transthalamic seizure bilateralization in both acute[15,23] and chronic models of experimental epilepsy.[24] In the rodent amygdaloid (AM) kindling model of epilepsy, complete and selective midsagittal section of the forebrain commissures, including the MI was reported to have no effect on either AM kindling or transhemispheric positive transfer effect (PTE), which is conceived to be evidence of increased excitability at distant brain sites. The finding suggested that AM kindling induces a lasting seizure facilitatory transhemispheric trace through the brain stem.[25] In order to gain further insight into the role of the MI in the transhemispheric mechanisms of AM kindling, the following studies were undertaken in rats, cats and primates.

A-I. Rodent Studies

Two studies, sectioning/lesioning and stimulation, were done.

1. Sectioning and lesioning studies (Ehara and Wada, unpublished data)

Three groups of animals: intact MI (control n=9), midsagittally sectioned MI (section n=9), and ibotenic acid lesioned MI (IBO n=9), were subjected to AM kindling at the primary and contralateral secondary sites to see the degree of transhemispheric PTE. The animals were then subjected to primary site retest to see the extent of interference effect.

The mean number of AM stimulations required for kindled stage 5 seizure development at the primary and secondary site AM was: control: 13 and 6, section: 11 and 4, and IBO: 14 and 15, respectively. Thus, at the contralateral AM, PTE was present in both the control and midsagittal sectioned groups but was completely absent in the IBO group. At the primary site retest, the mean number of stimulations required to recall kindled stage 5 seizure was: control: 2, section: 2 and IBO: 5. Thus, an interference effect was present in all three groups, though its degree was somewhat greater in the IBO group. It is clear that midsagittal section of the MI has no effect, while destruction of the MI eliminates PTE. A greater degree of interference in the IBO group is most likely due to the complete absence of PTE. These findings confirmed the pioneering observation of MacIntyre that MI section does not eliminate PTE.[25] In addition, our studies showed that cellular aggregates in the MI are critically involved in the transhemispheric mechanism of PTE in rodent AM kindling. Although midline nuclei are known to project to the limbic system and widespread cortical areas, these projections are almost entirely unilateral.[26-31] Also, the intralaminar system is known to have a bilateral connection across the midline[32] and transhemispheric corticosubcortical connections are present via the MI.[33-35] Since midline section of the MI does not interfere with PTE, it is clear that the role of the MI in the organization of the mechanism underlying transhemispheric PTE is in a vertical rather than horizontally oriented network of the neuronal system.

2. Stimulation Studies

In order to complement the results of the sectioning and lesioning studies summarized above, the following stimulation studies were done.

i. Kindling of the MI.[36] Seventeen animals underwent kindling of the MI. Later histological examination showed the kindling electrode tip to be located in the various nuclei of the MI. It was found that the MI has a relatively high after discharge threshold which sometimes increased during the course of kindling. In addition, there was significant seizure stage instability with frequent regression to earlier stages as in the case of neocortical kindling.

With the commencement of kindling, regardless of the presence or absence of after discharge, 8/17 animals abruptly dropped to the floor with the fore limbs pronated and hind limbs caudally extended, followed by a violent and synchronous beating movement of the forelimbs. In four animals whose electrode tips were located in the nucleus of rhomboid or reuniens, violent jumping and/or running was observed. Fourteen (of 17) animals which developed after discharge underwent MI kindling. Behavioral manifestations were similar to those of limbic kindling but with frequent regression to earlier seizure stages. This seizure stage instability remained unchanged even after stage 5 seizure was reached with a mean of 29.5 stimulations. During the early phase of kindling, two animals showed a generalized seizure which was quite different from the AM kindled stage 5 seizure with falling on their side or back with tonic caudal extension of both fore and hind limbs and tonic neck flexion. This was followed by a secondary bisymmetrical clonic convulsion.

Stage 5 seizure in 14 animals was reached with a mean of 29.5 stimulations (range 8-54). The criterion for the establishment of kindling was reached in 13/14 animals with a mean of 40.2 stimulations (range 18-70). One animal failed to reach this criterion. A generalized seizure triggering threshold with an "all-or-none" property could not be established in any of these animals.

Upon completion of MI kindling, eight animals were subjected to AM kindling. All the animals developed a generalized stage 5 seizure with a mean of 3.4 stimulations (range 1-12 with standard deviation 3.8). When the same strain of animals were subjected to AM kindling without prior MI kindling, the mean number of stimulations required for the first final stage seizure was 8.0 (range 5-12, n=6, standard deviation 2.5). Therefore, MI kindling is a time consuming process but it facilitated subsequent AM seizure development (P<0.02, students T-test).

ii. Intra-MI injection of the excitatory amino acid, N-Methyl-D-Aspartate (NMDA).[37] Thirty-two animals with a cannula implanted into the MI were used. Ten of these animals were used as control and were injected with injection of 0.01 M phosphate-buffered saline in a concentration of 25 nmol/1 ul. The experimental group (n=16) received 25 nmol/1 ul. of NMDA. Electroclinical observation began two minutes prior to injection and the drug was delivered through a 1 nmol Hamilton syringe over two minutes through the implanted cannula. The injection cannula was kept in position for an additional two minutes to prevent back flow. Both behavior and EEG were continuously monitored. Neither behavioral nor EEG change was observed following intra-MI injection of phosphate-buffered saline in the control group. In contrast, 20-30 seconds after the commencement of the NMDA injection, all the animals developed wild struggling, tachycardia, tachypnea, and biting of the holding cage, emitting shrill cries, without evidence of EEG discharge in the hippocampus or AM. This was followed by immobility lasting for 30-60 seconds with sustained EEG discharge in the hippocampus and/or AM. Beginning at about three minutes following the injection, oral automatism, followed by extension of all limbs was observed in 11 animals. The five animals whose cannula tips were located in the nucleus reuniens displayed wild running and jumping, followed by tonic generalized convulsion lasting for 5-10 seconds. Similar running and jumping without tonic convulsion were observed in the five animals whose cannula tips were located in the central medial and rhomboid nuclei. None of the wild running/jumping and tonic seizures were associated with an EEG seizure. Intra-MI injection of NMDA induced sustained limbic seizure activity in the hippocampus and/or AM bilaterally or unilaterally and recurred fairly frequently but eventually dissipated within 30-60 minutes. These findings suggest that the NMDA receptor in the MI participates in the generation and expression of convulsive seizure of limbic system origin.

iii. Limbic kindling following intra-MI injection of N-Methyl-D-Aspartate.[38] Since the above study suggested that an NMDA receptor in the MI participates in the mechanism of limbic seizure initiation, examination of the effect of intra-MI injection of NMDA on subsequent AM and hippocampal kindling was made in 38 animals. The animals were divided into control (phosphate-buffered saline injection) hippocampal kindling, n=9, and AM kindling, n=8; and experimental (NMDA injection) hippocampal kindling, n=12, and AM kindling, n=9. Both the experimental and control groups twice received an intra-MI injection of 0.5 umol of either NMDA or phosphate-buffered saline, respectively, with a 14 day inter-injection interval.

No electroclinical change occurred following the phosphate-buffered saline injection. In contrast, the NMDA group developed a remarkable electroclinical change about 30 seconds following the injection with restless, hissing or howling behaviour for 30-60 seconds. This was followed by a period of staring immobility, associated with the development of a sustained EEG seizure discharge in the hippocampus or the AM. Ten out of 21 animals developed wild running and jumping followed by tonic seizures lasting for 5-10 seconds. In four of the 10 animals, these evolved to a fatal recurrent generalized tonic/clonic convulsion. Wild locomotion without a convulsion developed in 3/21 animals. Ten to 20 minutes after the injection, head nodding progressing to clonic generalized convulsion was observed in 9/21 animals. None of the latter showed an initial tonic convulsion after the injection. After a 14 day interval, the remaining 34 animals receiving the second NMDA (n=18) or phosphate-buffered saline (n=16) injection developed affects identical to those described above after the first injection.

When the animals subsequently underwent kindling at either the hippocampus or the AM, only the experimental group showed a significant facilitation of the kindling rate. Thus, the mean latency to the first stage 5 seizure (number of stimulations) in the experimental group was significantly shorter in both hippocampal and AM kindling (22.3 and 6.9 stimulations, compared with 41.9 and 10.6 in the control group, respectively). Among the animals in the hippocampal kindling group, four with the NMDA injection site localized in the nucleus reuniens developed stage 5 seizure with an average of 15.5 stimulations, which was significantly faster than the other four animals with cannula tips outside the reunien nucleus that required an average of 29.0 stimulations. In the AM kindling group, five animals receiving the NMDA injection into the reuniens nucleus reached stage 5 seizures with a mean of six stimulations, whereas the three animals receiving NMDA outside of the reuniens required 8.33 stimulations, but this was not considered significant.

At the secondary site AM kindling, the latency to stage 5 seizure development in both AM and hippocampal kindling was faster in the experimental group (a mean of 2.3 and 1.3 stimulations in the hippocampal and AM groups, respectively) than in the control group (a mean of 3.9 and 3.5 stimulations in hippocampal and AM, respectively). This was of statistical significance in AM kindling. The chronology of both AM and hippocampal seizure development showed facilitation throughout stages 2-5, suggesting that intra-MI injection of NMDA selectively reduced the latency to head nodding. The results suggested that an NMDA receptor in the MI, particularly the reuniens nucleus, participates in modulation of temporal limbic excitability and seizure development.

A-II. Feline Studies

Studies on the effect of selective mid-sagittal MI section, electrolytic MI lesioning, or ibotenic acid (IBO) MI lesioning upon AM kindling and a stimulation study with intra-MI injection of N-methyl-D-aspartate were carried out.

1. Sectioning and Lesioning Studies

i. Sectioning study.[39] One group of animals underwent selective and complete stereotaxic mid-sagittal MI section prior to the commencement of AM kindling (prior group) and another group underwent the same surgery upon completion of the primary site AM kindling (post group). In both groups of animals, the anatomical integrity of all the midline structures, particularly the forebrain commissures, was preserved, with the exception of the MI.

Prior group (n=9). The mean number of AM stimulations required for kindled stage 6 seizure development was: primary site 26, secondary contralateral site 6, primary site retest 1.4. Some change in the developing seizure pattern was noted with the missing of stages 3 and 5 and an increasing number of axial rotations during stage 4. However, both PTE and an interference effect were present as in intact animals.[40]

Post-group (n=5). The mean of the AM stimulations required for kindled stage 6 seizure development at the primary site prior to the MI section was 30. AM re-stimulation following MI section immediately recalled the kindled seizure with an increased number of axial rotations during stage 4. In addition, the after discharge (AD) duration was increased, suggesting that the MI sectioning caused some delay of secondary generalization. The mean number of AM stimulations required for stage 6 development at the contralateral (secondary) site AM and primary site retest was 7 and 1.8, respectively. Stages 3 and 5 were missing with an increased number of axial rotations during stage 4 and some increase of AD duration was evident. Therefore, transhemispheric PTE was present at the secondary AM site but interference was also present at both the secondary site and the primary site re-test. In addition, at the primary site re-test, one animal developed electroclinically well separated but sequentially-arranged two hemispheric "split seizures", i.e., beginning with an ipsilateral hemispheric seizure and then developing a contralateral one upon completion of the former. Thus, the results of selective midline interruption of the MI suggested that the intra-MI bilateral connection has no role in transhemispheric PTE and interference. However, regardless of its timing, MI section caused modification of the established seizure pattern with some delay of secondary generalization. Furthermore, development of sequential "split seizures" was observed in one animal; therefore, transhemispheric commissural system via the MI has some role in the bilateral patterning involved in secondary generalization.

ii. Electrolytic lesioning study.[41,42] Two groups of animals were used; electrolytic lesioning of the MI prior to commencement of AM kindling (prior group), and the same upon completion of the primary site AM kindling (post group).

Prior group (n=9). During AM kindling, the animals showed a pattern of seizure development strikingly similar to that of forebrain bisected cats. The development of contralateral hemiconvulsive seizures was coincident with the onset of EEG seizure discharge in the ipsilateral motor cortex. Unlike the forebrain bisected cats, however, the MI lesioned animals eventually developed bisymmetrical clonic convulsions with initial contralateral and then ipsilateral clonic involvement. The mean AM stimulations required for kindled stage 6 seizure development was 33 which is identical to the rate of kindling at the primary site of intact animals (Hiyoshi and Wada, n=9, unpublished data). At the secondary AM site (contralateral), the mean of AM stimulations required for kindling was 17 which was significantly slower than four in intact animals.[40] When compared with primary site kindling, kindling at the secondary site showed a longer AD duration, slower seizure generalization and a larger number of axial rotations. When the primary site was retested, a kindled seizure was recalled with a mean of three stimulations which was significantly different from the one stimulation required in the control group. Thus, lesioning of the MI prior to AM kindling prevented the development of PTE at the contralateral secondary site AM, while the interference effect remained unchanged.

Post-group (n=8). When primary site AM kindled animals were re-stimulated at the primary site following MI lesioning, all the animals responded with the established pattern of bisymmetrical generalized tonic clonic convulsion. Subsequent secondary site AM kindling showed a distinctly modified pattern of seizure development identical to that observed during primary site kindling of the prior MI lesioned group described above, requiring a mean of 16 stimulations, in contrast to four stimulations in the control animals for stage 6 seizure development. Therefore, transhemispheric PTE was absent. In addition, two animals showed development of sequential electroclinical "split seizures" with initial ictal involvement of the ipsilateral and then the contralateral hemispheres, independently. On primary site retest, a mean of two stimulations recalled the kindled convulsion with a longer AD duration. Furthermore, two additional animals showed the pattern of "split seizures," i.e., two entirely independent electroclinical seizure involving initially the ipsilateral and then the contralateral hemisphere. These findings suggested that electrocoagulation of the MI, partially destroying both the neurons and bilateral pathways, can eliminate transhemispheric PTE, modify established seizure patterns, and delay secondary generalization causing "split seizures," leaving the interference effect intact. The findings further suggested that the transhemispheric PTE in AM kindling is a primary site hemisphere dependent process, and that the MI plays a critical role in enabling the secondary site AM to access the kindled primary site neuro-circuitry.

iii. Ibotenic acid lesioning study.[43] Two groups of animals were prepared. The IBO group (n=6) received intra-MI injection of IBO 5 ug ul (dissolved in phosphate buffer 50 nM, pH 7.4), while the control group (n=6) received an intra-MI injection of phosphate buffer prior to commencement of AM kindling. There was no difference in the AD threshold, generalized seizure triggering threshold and the AD duration between the control and IBO groups. The mean number of AM stimulations required for final stage 6 seizure development was also similar between the control and IBO groups (26 and 24, respectively). However, there was a significant modification of the seizure pattern in the IBO group with the absence of stages 3 and 5, and an increased number of axial rotations during stage 4. At the secondary site AM, a similar change in the seizure pattern was noted. In addition, three animals developed "split seizures" with two entirely independent but sequential ipsilateral-contralateral hemispheric seizures. The mean of the AM stimulations required for kindled stage 6 development at the secondary site was 23 in the IBO group and 6 in the control group. Therefore, PTE was completely absent in the IBO group. At the primary site retest, both the IBO and control groups required a mean of two stimulations to recall kindled seizure. Therefore, the interference effect remained intact in both the IBO and control groups.

2. Stimulation study [44]

The potential role of excitatory amino acids in the initiation and generalization of epileptic seizures has been demonstrated in experimental models.[45] As described earlier, intra-MI injection of NMDA in rats caused generalized seizures associated with EEG discharge in the limbic system, in addition to wild running and jumping. Results of subsequent limbic kindling strongly suggested its participation in limbic seizure development. In this study, we investigated the effect of intra-MI injection of NMDA on limbic seizure development in 14 cats weighing 3.8-4.6 kg. NMDA (250 nM/1 ul dissolved in phosphate buffer 50 nM, pH=7.4) was given in a volume of 2 ul at a rate of 0.5 ul per minute through implanted cannula by means of Hamilton syringe.

Intra-MI NMDA injection in the naive animals caused profound behavioral changes with salivation, urination, defecation, and loud howling almost immediately after the commencement of injection. This was followed by wildly agitated running behaviour lasting for about 30 seconds with subsequent staring immobility associated with repeated recurrence of sustained EEG seizure discharge in the hippocampus and AM. In two animals, repetitive facial twitching and head nodding developed coincident with the EEG seizure in the MI, AM and hippocampus. Comparison of AM kindling in animals with or without prior intra MI NMDA injection showed no significant difference in the electroclinical parameters between these groups.

Intra-MI NMDA injection in the AM kindled cats caused them to develop bilateral sustained EEG discharges in the hippocampus with partial seizures displaying immobility, facial twitching, salivation and head nodding in four animals while the remaining four animals developed generalized convulsive seizures.

In the same group of AM kindled animals, hippocampal stimulation at 1 mA produced an identical response to that of intra-MI injection of NMDA, i.e., those four animals with partial seizures responded with stage 2 partial seizures, while the remaining four animals developed generalized convulsive seizures. The four animals with a partial (stage 2) seizure response underwent hippocampal kindling and subsequently received intra-MI injection of NMDA. All the animals then responded with kindled generalized convulsions. These findings suggest that NMDA in the MI modulates limbic excitability and participates in the precipitation of hippocampal onset generalized convulsions.

Results of the above series of feline studies are consistent with the findings of the rodent studies and suggest that a) the midline transthalamic pathway has some role in bilateral patterning involved in secondary generalization; b) the cellular aggregates in the MI are critical for transhemispheric mechanisms of PTE; c) the role of the MI in PTE is oriented towards a vertical, rather than a horizontal, intrahemispheric neuronal network; and d) a NMDA receptor in the MI is involved in modulation of temporal limbic onset generalized seizures.

A-III. Primate Studies

Two studies had been done in Senegalese baboons, Papio papio: AM kindling in animals with selective midsagittal MI section (n=3) and electrolytic MI lesioning (n=6). None of the animals had spontaneous seizures and they showed only weak and unstable photosensitivity prior to the kindling.

1. Sectioning study (Tsuchimochi and Wada, unpublished data)

Three animals were stimulated once a day for 130 days at the primary site AM (left). All the animals reached stage 4 asymmetrical generalization with a mean of 13 stimulations (range 8-16). Only one animal (W250) reached a stage 5 seizure and on the 65th day, while the other two animals (W248, W249) remained in a stable stage 4 seizure state. The MI was then midsagittally sectioned through a stereotaxically-introduced wire knife. Two weeks following the surgery, the animals underwent secondary site AM (right) kindling. One animal (W250) consistently responded with stage 5 seizures from day one. The remaining two animals failed to progress beyond stage 2 despite a larger number of AM stimulations than those required for stage 3 seizure development at the primary site. When the primary site was re-tested, a kindled seizure was recalled with 1-3 stimulations. Histological examination showed that the MI was completely bisected in all the animals. The finding suggested that midline separation of the MI does not modify either the PTE or the seizure suppressive effect expected at the secondary site in AM kindling of intact Papio papio.[11]

2. Electrolytic lesioning study (Sakai and Wada, unpublished data)

MI lesion was produced either prior to (prior group) or following primary site AM kindling (post-group). All the animals were stimulated once a day for 130 days at the primary site. The secondary site AM was stimulated after a two week rest interval until stages 4/5 seizures were elicited.

Prior group (n=3). The mean number of AM stimulations required for the development of stage 4 asymmetrical generalization was 11.7 (range 6-22). Only one animal (W227) reached stage 5 on the 122nd day. At the secondary site, all three animals responded with asymmetrical generalized convulsions upon the first stimulation. Therefore, PTE was present. When the primary site was re-tested, a kindled seizure was recalled with 1-3 stimulations.

Post group (n=3). The mean number of AM stimulations required for the primary site stage 4 asymmetrical generalization was 17.5 (range 12-32). Only one animal (W229) developed stage 5 bisymmetrical convulsions (84th day). At the secondary site AM, one animal (W226) developed a stage 4 seizure on the third day while another animal (W230) developed the same on the 38th day. One animal (W229) failed to progress beyond stage 2 after 130 stimulations. Therefore, both the PTE and seizure suppressive effects were present as in intact animals. At the primary site re-test, a mean of 4 stimulations (range 1-6) were required to re-establish primary site kindled seizures.

The above findings suggested that neither midline section nor electrolytic lesioning of the MI affected the PTE at the secondary site AM. However, histological examination showed that while the midline section was complete in the sectioning group the extent of the destruction in the electrolytic lesioning group was small and restricted only to the dorsal third of the MI. It is clear that the intra-MI transhemispheric pathway does not play a role for PTE in primates but the significance of perikarya in the MI for primate AM kindling remains unclarified.

Comments

Findings in all three species of animals, i.e., rodents, cats and primates, are congruent in that midsagittal section of the MI does not eliminate PTE. However, the results of electrolytic or ibotenic acid lesioning both cats and rats suggest that the perikarya rather than the commissural pathways in the MI are crucial for the mechanism of transhemispheric PTE in AM kindling. In addition, we found that the PTE was lost when MI was lesioned upon completion of the primary site AM kindling. This finding suggests that PTE is dependent on the kindled primary site intra-hemispheric neurocircuitry for which perikarya in the MI play an important role. Therefore, the presence of PTE at the contralateral homotopic (mirror) site does not necessarily represent enhanced excitability (intrinsic functional change) at that site. Rather, it reflects ready accessibility from the mirror site to the kindled primary site hemispheric mechanism, and suggests that perikarya in the MI play an important role in this process. Furthermore, evidence has been obtained that a NMDA receptor in the MI modulates temporal limbic excitability and convulsive seizure development.

In the past, loss of transhemispheric PTE has been found following either midsagittal section of the corpus callosum[46,47] or lesioning of the nucleus dorsalis medialis, thalamic reticular nucleus, or midbrain reticular formation.[48] Since the forebrain projection of the MI neurons is always unilateral including bilaterally situated cell groups, and since a morphological basis exists for thalamic and callosal interaction,[49] the CC system may be conceived as an extension of the thalamo-cortical pathways to the contralateral hemisphere. This would explain why the PTE may be disrupted following CC bisection.[46] On the other hand, an effective site to eliminate PTE is not limited to forebrain structures. Indeed, there has been evidence to suggest that the brain stem is important for bilateral transmission of ictal and inter-ictal information in AM kindling. Thus, a unique pattern of sustained discharge develops in the midbrain reticular formation prior to seizure bilateralization and generalization.[50] Similarly, cooling or lesioning of the midbrain reticular formation ipsilateral to the kindled AM markedly elevates the generalized seizure triggering threshold[51] and eliminates PTE at the contralateral homotopic AM site.[48] Therefore, PTE requires accession from the secondary site AM to the kindled intrahemispheric neurocircuitry extending from the cerebral cortex to the nonspecific mesial thalamus to the brain stem.

B. BRAIN STEM

In order to gain further insight into the potential role of the brain stem in kindling and PTE mechanisms, the effect of either midline lesion or midsagittal section of the brain stem in AM kindling is currently being examined in rats, cats and primates. Our study is incomplete but preliminary findings in rodents and cats are briefly described below.

B-I. Rodent studies

1. Initial study.[52] Two groups of animals with (n=9) and without (n=5) brain stem midsagittal section performed stereotaxically were subjected to AM kindling. The mean number of AM stimulations required for primary and secondary site AM kindling was 10 and 9 in the sectioned group and 15 and 7 in the control group, respectively. Thus, the control group showed an expected PTE but the bisected group did not. There was no significant difference in AD duration, seizure pattern or generalized seizure triggering threshold between the two groups. However, when the histological extent of the midsagittal section was examined, five animals with an absence (less than 20% saving) of PTE showed the midline prepontine section to include the interpeduncular nucleus. The remaining four animals which showed more than 20% saving had their section more dorsally located, leaving the interpeduncular nucleus intact. This finding suggested the possibility that midline decussating fibres or cellular aggregates including the interpeduncular nucleus, or both are involved in the mechanism of PTE, i.e., accession to the kindled neurocircuitry from the secondary site AM.

2. Subsequent studies (Chiba & Wada, unpublished data). In order to test the hypothesis that PTE is dependent on the anatomical integrity of the midline brain stem, a much larger bisection of the brain stem was performed in three groups of animals: ventral group (n=11), dorsal group (n=9), and control group (n=5). The ventral group had a complete bisection of the interpeduncular nucleus, while the dorsal group had extensive bisection beginning in the quadrigeminal plate downward excluding the interpeduncular nucleus. The control group had identical surgery with a burr hole opening without section.

Subsequent AM kindling resulted in stage 5 seizures in both the control and ventral groups with a mean of 12 and 10.5 stimulations. At the secondary site AM, the control and ventral groups readily reached stage 5 with a mean of 4.2 and 3.8 stimulations, respectively. In the dorsal group 4/9 animals developed stage 5 seizures at the primary site with a mean of 10.1 stimulations. These animals reached stage 5 with a mean of 3.8 stimulations at the secondary site. On the other hand, the remaining 5/9 animals failed to progress beyond stage 4 at the primary site and again failed to progress beyond stage 4 at the secondary site despite a mean of 17.3 stimulations. Thus, PTE remained intact in animals with ventral bisection of the interpeduncular nucleus, while animals with bisection dorsal to that nucleus were less likely to reach stage 5.

Among the midline brain stem structures bisected, the interpeduncular nucleus is one of the most prominent nuclear masses. Since lesioning but not midsagittal bisection of the MI caused loss of PTE in our earlier study, we tested the hypothesis that the perikarya of the interpeduncular nucleus and its projections are involved in PTE by the placement of electrolytic lesion in six animals. Five control animals underwent identical surgical procedures but DC current was not delivered. At the primary site AM, the control group developed stage 5 with a mean of 7.6 stimulations (range 6-10) while the lesion group required a mean of 18.4 stimulations (range 11-23). At the secondary site AM, the control group reached stage 5 with a mean of 2.6 stimulations (range 1-4), while the lesion group either failed to progress beyond stage 4 (4/6 animals) or stage 2 (2/6 animals) despite a mean of 21.4 stimulations. When the primary site was re-tested, stage 5 seizure was recalled with a mean of 1.6 stimulations (range 1-3) in the control group and 4.6 stimulations (range 2-9) in the lesion group. This finding confirms an earlier report that lesioning of the interpeduncular nucleus retards AM seizure development.[53] In addition, our study showed failure of AM kindling at the secondary site. The fact that primary site AM kindling in the lesioned group was delayed but resulted in stage 5 suggests that the lesioned area is not critical for AM seizure development. However, failure of AM kindling at the secondary site AM suggests an absence of PTE. The much larger number of stimulations required at the primary site re-test in the lesioned group is probably due to enhanced interference as a consequence of the absent PTE. Preliminary results of the histological examination indicated a rather large area of midline midbrain lesioning extending from the tegmentum to the ventral area involving the interpeduncular nucleus.

The findings obtained so far suggest that brain stem mechanisms participate not only in the development of kindled stage 5 seizure but also in PTE.

B-II. Feline Studies

Our preliminary observations (Ishibashi and Wada, unpublished data) were based on five AM kindled cats with selective midline MI section mentioned earlier (see AII-1-i). As a terminal observation, the effect of stereotaxic mid-sagittal brain stem section on kindled seizures was examined. These animals were well kindled with a stable stage 6 seizure response to AM stimulation at a generalized seizure triggering threshold, both at the primary and the secondary site AM. As already mentioned, there was no significant difference between these MI sectioned and control animals as to AM seizure development at either the primary or the secondary site. When these animals underwent primary site AM re-stimulation following the midline brain stem section, all the animals required an additional mean of 4.6 (range 2-6) AM stimulations to re-activate a stage 6 seizure response at the primary site. One animal (#970) required 18 stimulations to re-establish a stable stage 6 seizure state. Another animal (#971) developed independent, sequential "split seizures" with ipsilateral and then contralateral hemispheric stage 4 seizure manifestations. At the secondary site AM, a mean of 5.6 (range 2-13) stimulations were required for re-activation of stage 6 seizures. One animal (#973), however, failed to re-establish a stable stage 6 seizure state with frequent regression to an earlier stage alternating with sequential "split seizures" even after 60 daily stimulations. Histological examination showed a varying degree of midline section in the mid-brain with #973 having the largest area of bi-section but leaving the interpeduncular nucleus intact.

With the above preliminary information, we extended our study of midline brain stem bisection upon AM kindling (Hamada and Wada, unpublished data). Twelve animals underwent extensive midsagittal brain stem bisection between A1 - A6 covering the sagittal area 5-6 mm. from the ventral midline brain stem surface. This bisection covered the entire anteroposterior extent of the interpeduncular nucleus. Seven animals survived the surgery and were subjected to AM kindling. At the time of this writing both primary and secondary site AM kindling are completed in 5 animals. In these animals, a mean of 44.6 stimulations (range 14-93) were required for stage 6 seizure development at the primary site AM. At the secondary site AM, one animal (#998) failed to progress beyond stage 2 despite 70 stimulations in contrast to 31 stimulations required for stage 6 development at the primary site. The remaining four animals reached stage 6 with a mean of 39.25 stimulations. This is in contrast to a mean of 4.3 stimulations required in control animals.[40] At the primary site retest, kindled seizures were recalled with 1-2 stimulations in all 5 animals. At the time of this writing, histological examination has not been completed.

Comments

Preliminary results of mid-sagittal brain stem bi-section in rodent AM kindling is inconclusive. However, lesioning of the midline midbrain eliminated PTE for AM kindling at the secondary site. On the other hand, in feline species, erosion and regression of kindled generalized seizures to partial/"split seizures" following mid-sagittal midbrain section in bilaterally AM kindled cats at both the primary and the secondary site implies significant disruption of established ictal transhemispheric mechanisms. In one cat (#970) kindled generalized seizure stability was lost after the mid-sagittal brain stem section and required considerable re-kindling while another cat (#973) never regained a stable generalized seizure state, often having "split seizures" despite over 60 re-kindling stimulations. "Split seizures" have been documented to occur in our laboratory on AM kindling of forebrain bisected animals.[19,45] Results of our ongoing study suggest that midsagittal brain stem section can eliminate PTE induced by AM kindling although the precise spatial extent of such bisection is not known yet at the time of this writing. Therefore, the findings obtained so far suggest that yet to be identified midline structures in the brain stem play a significant role in:

a) secondary generalization of partial onset seizures and maintenance of kindled generalized convulsive seizure state, and

b) in enabling ready access from the secondary site AM to the primary site the kindled neurocircuitry for generalized convulsive seizure generation.

GENERAL SUMMARY AND COMMENT

The potential importance of midline brain structures for seizure generalization in patients with epilepsy is well recognized. Indeed, a great deal of experimental and clinical observations have identified the corpus callosum as the major, if not exclusive, anatomical substratum for seizure bilateralization and bisynchronization. However, it is also recognized that seizure generalization can occur in the absence of the forebrain commissures. Obviously, the brain stem must play a pivotal role in this regard but no specific information about the identity and functional characteristics of such pathways is available to this date.

On the other hand, our understanding of fundamental mechanisms of epilepsy is largely based on the findings obtained in animal models. However, one must pay attention to a) the presence or absence of intrinsic excitability, either acquired or genetically dictated, in various animal models; and b) existence of varying degrees of morphological similarity and dissimilarity between human and subhuman mammalian brains

We have chosen the kindling model of epilepsy as a valid chronic model of human partial onset epilepsy. Furthermore, our investigation was in part directed towards obtaining clearer understanding of significance of the morphological dissimilarity of midline brain structures, particularly the massa intermedia of human and subhuman mammalian brains, in order for us to be able to make appropriate interpolation of animal data to the human problem of epilepsy. In this paper, results of our recent studies on the massa intermedia and the brain stem are selectively reviewed.

Despite a morphological massiveness of the midline thalamus in rats, cats and subhuman primates, results of our study consistently indicated that vertical intrahemispheric (rather than horizontal transthalamic) connections with the cerebral cortex, temporal limbic system and the brain stem are importantly involved in the kindling effect. This finding suggests that the presence of significant morphological differences in the massa intermedia of the human and subhuman mammalian brain does not pose serious difficulty for interpolation of animal data regarding interhemispheric transfer of ictal and interictal information. In this regard, the relevance of the distant effect of kindling, i.e., repeated partial seizures needs to be reassessed. Once kindling is attained at one brain site such as AM, contralateral AM can be kindled swiftly. Thus, a mean rate of AM kindling at primary/secondary sites in our laboratory has been 14/7 in rats, 25/4 in cats, 192/96 in Rhesus monkeys, 87/37 in Papio hamadryas and 72/1 in Papio papio. This effect was called positive transfer effect and is presumed to indicate enhanced excitability induced by primary site kindling at a distant secondary site. However, results of our past studies consistently indicated that the positive transfer effect can be lost if either the corpus callosum is bisected, or the non-specific thalamus or mid-brain reticular formation is lesioned. In this paper, we described loss of the positive transfer effect upon completion of AM kindling following either destruction of perikarya in the midline thalamus or the brain stem and following midline bisection of the brain stem. Thus, it is clear that the positive transfer effect per se does not represent increased excitability at a distant secondary site. Rather, it suggests ready accession from a distant site to the vertically oriented intrahemispheric kindled neurocircuitry extending from the cerebral cortex to the midline thalamus and the brain stem. However, we are also aware of the fact that increased excitability at a distant brain site capable of initiating independent seizures can develop following prolonged kindling.[52] Therefore, the question of under what specific circumstances such an independent distant effect takes place must be pursued. If one believes that kindling is a consequence of repeated seizures leaving a permanent trace behind each seizure, a precise neurobiological definition of kindling induced "trace" is of immense clinical importance since kindling occurs in all the subhuman mammalian species including four different species of primates. It would be unreasonable to assume that the human brain is exempt from such consequences and therefore future research in this direction is critical for prevention, treatment and management of epilepsy.

Finally, preliminary results of our brain stem bisection study in AM kindled cats indicated the importance of midline brain stem structures for secondary generalization and maintenance of generalized seizures. Precise definition of the transhemispheric subcortical pathway and characterization of the condition under which such pathway is utilized for seizure generalization are critical for our understanding of not only the mechanisms of transhemispheric interictal and ictal transmission but also for gaining insight into mechanisms of potential subcortical integration of forebrain processes.

REFERENCES

1. T. Sutula, H. Xiao-Xian, I. Cavazos, and G. Scott, Synaptic reorganization in the hippocampus induced by abnormal functional activity, *Science* 239:1147-1150 (1988).

2. N.C. Delanerolle, J. H. Kim, R.J. Robbins, and D. D. Spencer, Hippocampal interneuron loss and plasticity in human temporal lobe epilepsy, *Brain Res.* 495:387-395 (1987).
3. T.L. Babb, W.R. Kupfer, and J.K. Pretorius, Recurrent excitatory circuits of "sprouted" mossy fibers into the fascia dentata of human hippocampal epilepsy, *Epilepsia* 29:674 (1988).
4. T.K. Babb, W.R. Kupfer, and J.K. Pretorius, Synaptic reorganization of mossy fibers into inner molecular layer in human epileptic fascia dentata, *Soc. Neurosci. Abstr.* 14:881 (1988).
5. A. Represssa, O. Robain, E. Tremblay, and Y. Ben-Ari, Hippocampal plasticity in childhood epilepsy, *Neurosci. Lett.* 99:351-355 (1989).
6. C.R. Houser, J.E. Miyashiro, R.E. Swartz, G.O. Walsh, J.R. Rich, and A.V. Delgado-Escueta, Altered patterns of dynorphic immunoreactivity suggest mossy fiber reorganization in human hippocampal epilepsy, *J. Neurosci.* 10(1):267-82 (1990).
7. T. Sutula, G. Cascino, J. Cavazos, I. Parada, and L. RamiRez, Mossy fiber synaptic reorganization in the epileptic human temporal lobe, *Ann. Neurol.* 26:321-330 (1989).
8. G.V. Goddard, D.C. McIntyre, and C.K. Leech, A permanent change in brain function resulting from electrical stimulation, *Exp. Neurol.* 25:295-330 (1969).
9. J.A. Wada, M. Sato, and M.E. Corcoran, Persistent seizure susceptibility and recurrent spontaneous seizures in kindled cats, *Epilepsia* 15:464-478 (1974).
10. J. A. Wada, T. Osawa, and T. Mizoguchi, Recurrent spontaneous seizure state induced by prefrontal kindling in Senegalese baboons, Papio papio, *Can. J. Neurol. Sci.* 2:447-492 (1975).
11. J.A. Wada, and T. Osawa, Spontaneous recurrent generalized seizure state induced by daily amygdaloid stimulation in Senegalese baboons, Papio papio, *Neurology* 26(3):273-286 (1976).
12. J.A. Wada, T. Mizoguchi, and T. Osawa, Secondarily generalized convulsive seizure induced in rhesus monkeys, *Neurology* 28:1026-1036 (1978).
13. J.A. Wada, Clinical relevance of kindling: species, brain sites and seizure susceptibility, *in:* "Limbic Mechanisms," Livingston and Hornykiewicz, eds., Plenum Press, New York (1978).
14. J. Musgrave and P. Gloor, The role of the corpus callosum in bilateral interhemispheric synchrony of spike and wave discharge in feline generalized penicillin epilepsy, *Epilepsia* 21:369-378 (1980).
15. J.A. Kusske and J.L. Rush, Corpus callosum and propagation of afterdischarge to contralateral cortex and thalamus, *Neurology* 28:905-912 (1978).
16. E.M. Marcus, Generalized seizure models and the corpus callosum, *in:* "Epilepsy and the Corpus Callosum," A.G. Reeves, ed., Plenum Press, New York (1985).
17. J.A. McCaughran, Jr., The role of forebrain commissures in kindled seizure development, *in:* "Epilepsy and the Corpus Callosum," A.G. Reeves, ed., Plenum Press, New York (1985).
18. J.A. Wada and S. Komai, Effect of anterior 2/3 callosal bisection upon bisymmetrical and bisynchronous generalized convulsions kindled from amygdala in epileptic baboon, Papio papio, *in:* "Epilepsy and the Corpus Callosum," A.G. Reeves, ed., Plenum Press, New York (1985).
19. J.A. Wada and M. Sato, Generalized convulsive seizure state induced by daily electrical stimulation of the amygdala in split brain cats, *Epilepsia* 16:417-430 (1975).
20. T. Hiyoshi and J.A. Wada, Feline agenesis of the corpus callosum, *Epilepsia* 28 (4):395-398 (1987).
21. J.A. Wada, Anterior 2/3 Callosal bisection: comparative observations in animals and man, *in:* "Fundamental Mechanisms of Human Brain Function," J. Engel Jr. et al., eds., Raven Press, New York (1987).
22. J.A. Wada and T. Mizoguchi, Effect of forebrain bisection upon amygdaloid kindling in epileptic baboon, Papio papio, *Epilepsia* 25(3):278-287.
23. R.C. Collins, C. Kennedy, L. Sokoloff, and F. Plum, Metabolic anatomy of focal motor seizure, *Arch. Neurol.* 33:536-542 (1976).
24. K. Ono, K. Mori, H. Baba, K. Seki, and J.A. Wada, A new chronic model of partial onset generalized seizure induced by low frequency cortical stimulation: its

relationship to the kindling phenomenon, *in:* "Kindling 3," J.A. Wada, ed., Raven Press, New York (1986).
25. D.C. McIntyre, Split-brain rat: transfer and interference of kindled amygdala convulsions, *in:* "Kindling," J.A. Wada, ed., Raven Press, New York (1975).
26. E.G. Jones and R.Y. Leavitt, Retrograde axonal transport and the demonstration of non-specific projections to the cerebral cortex and striatum from thalamic intralaminar nuclei in the rat, cat and monkey, *J. Comp. Neurol.* 154:349-378 (1974).
27. G. Macchi, A. Quattrini, P. Chinzari, G. Marchesi, and G. Capocchi, Quantitative data on cell loss and cellular atrophy of intralaminar nuclei following cortical and subcortical lesions, *Brain Res.* 89:43-59 (1975).
28. G. Macchi, M. Bentivoglio, C. D'Atena, P. Rossini, and E. Tempesta, The cortical projections of the thalamic intralaminar nuclei restudied by means of the HRP retrograde axonal transport, *Neurosci. Lett.* 4:121-126 (1977).
29. K. Niimi, N. Niimi, and Y. Okada, Thalamic afferents to the limbic cortex in the cat studied with the method of retrograde axional transport of horseradish peroxidase, *Brain Res.* 145:225-238 (1978).
30. K. Niimi, H. Matsuoka, T. Aisaki, and Y. Okada, Thalamic afferents to the prefrontal cortex in the cat traced with horseradish peroxidase, *J. Hirnforsch.* 22:221-241 (1981).
31. R.T. Robertson and S.S. Kaitz, Thalamic connections with limbic cortex. 1. Thalamocortical projections, *J. Comp. Neurol.* 195:501-525 (1981).
32. M.E. Scheibel and A.B. Scheibel, Structural organization of nonspecific thalamic nuclei and their projection towards cortex, *Brain Res.* 6:60-94 (1967).
33. H. Kuenzle, Thalamic projections from the precentral motor cortex in Macaca fascicularis, *Brain Res.* 105:253-312 (1985).
34. M. Molinari, D. Minciacchi, M. Bentivoglio, and G. Macchi, Efferent fibers from the motor cortex terminate bilaterally in the thalamus of rats and cats, *Exp. Brain Res.* 57:305-312 (1985).
35. G.T. Sakai and D. Tanaka, Jr., Contralateral corticothalamic projections from area 6 in the racoon, *Brain Res.* 299:371-375 (1984).
36. N. Mori and J.A. Wada, Kindling of the massa intermedia of the thalamus in rats, *Brain Res.* 575:148-150 (1992).
37. Y. Hirayasu and J.A. Wada, Convulsive seizures in rats induced by N-methyl-D-aspartate injection into the massa intermedia, *Brain Res.* 577:36-40 (1992).
38. Y. Hirayasu and J.A. Wada, N-methyl-D-aspartate injection into the massa intermedia facilitates development of limbic kindling in rats, *Epilepsia* 33(6):965-970 (1992).
39. M. Ishibashi and J.A. Wada, Division of the massa intermedia has no effect on amygdaloid kindled seizure in cats, *Epilepsia* 31(5):632 (1990).
40. A. Wake and J.A. Wada, Transfer and interference in amygdaloid kindling in cats, *J. Neurol. Sci.* 4:5-11 (1977).
41. T. Hiyoshi and J.A. Wada, Midline thalamic lesion and feline amygdaloid kindling. I. Effect of lesion placement prior to kindling, *Electroenceph. Clin. Neurophysiol.* 70:325-338 (1988).
42. T. Hiyoshi and J.A. Wada, Midline thalamic lesion and feline amygdaloid kindling, II. effect of lesion placement upon completion of primary site kindling, *Electroenceph. Clin. Neurophysiol.* 70: 339-349 (1988).
43. Y. Ehara and J.A. Wada, Midline thalamus and amygdaloid kindling, *in:* "Kindling 4," J.A. Wada, ed., Plenum Press, New York (1990).
44. Y. Ehara and J.A. Wada, NMDA injection into the massa intermedia precipitates convulsive seizure, *Epilepsia* 30:707 (1989).
45. S. Piredda and K. Gale, Role of excitatory amino acid transmission in the genesis of seizures elicited from the deep prepiriform cortex, *Brain Res.* 377:205-210 (1986).
46. J.A. Wada, T. Nakashima, and Y. Kaneko, Forebrain bisection and feline amygdaloid kindling, *Epilepsia* 23:521-530 (1982).
47. H. Fukuda, J.A. Wada, D. Riche, and R. Naquet, Role of the corpus callosum and hippocampal commissure on transfer phenomenon in amygdaloid-kindled cats, *Exp. Neurol.* 98:189-197 (1987).

48. J.A. Wada, Secondary cerebral functional alterations examined in the kindling model of epilepsy, *in:* "Secondary Epileptogenesis," A. Mayersdorf and R.P. Schmidt, eds., Raven Press, New York (1982).
49. E.G. Jones, Anatomy, development and physiology of the corpus callosum, *in:* "Epilepsy and the Corpus Callosum," A.G. Reeves, ed., Plenum Press, New York (1985).
50. J.A. Wada and M. Sato, Generalized convulsive seizures induced by daily electrical stimulation of the amygdala in cats, *Neurology* 24:565-574 (1974).
51. J.A. Wada and M. Sato, Effects of unilateral lesion in the midbrain reticular formation upon kindled amygdaloid convulsion in cats, *Epilepsia* 16:693-697 (1975).
52. Y. Hirayasu and J.A. Wada, The effect of brain stem bisection prior to the amygdaloid kindling in rats, *Brain Res.* 610:354-357 (1993).
53. R.F. Ackermann and J. Engel, Jr., Lesions of the interpeduncular nucleus retard development of amygdaloid-kindled seizures in rats, *Neurosci. Abstr.* 4:139 (1978).
54. T. Hiyoshi, M. Seino, N. Kakegawa, T. Higashi, K. Yagi, and J.A. Wada, Evidence of secondary epileptogenesis in amygdaloid overkindled cats: electroclinical documentation of spontaneous seizures, *Epilepsia* 34(3):408-415 (1993).

FOREBRAIN COMMISSURES AND LIMBIC KINDLING

Dan C. McIntyre, Ph.D.

Department of Psychology
Life Science Research Building
Carleton University, Ottawa
Ontario, Canada, K1S 5B6

INTRODUCTION

Because of its anatomical prominence, but without the benefit of corroborating data, the corpus callosum (CC) was assumed previously to play a critical role in interhemispheric communication. Clear confirmation as to its true role in information transmission, however, was not provided until the 1950's (e.g., Myers and Sperry 1953). In earlier experiments, it was proposed that the CC contributed to epileptogenesis, as well as seizure transmission (Erickson 1940; Curtis 1940), but its role was determined not to be essential, since other forebrain commissures could carry seizure interhemispherically (Fox, Fisher and Desalva 1948; Frost, Baldwin, and Wood 1958; Poblete, Ruben, and Walker 1959).

It is now quite clear that the CC provides the major communication link between the cerebral hemispheres. As effective as it is in the interhemispheric integration and dissemination of normal enabling activity, it appears equally capable of integrating and transmitting abnormal disabling activity. This is particularly true of propagating seizures. For this reason, it has become a surgical target to arrest the interhemispheric spread of propagating seizure.

In the surgical attack against intractable human epilepsy, the therapeutic outcome of early CC bisections was not very satisfying (Van Wagenen and Herren 1940). Consequently, commissurotomy, as a tool for seizure management, lay fallow for the next 20 years (Bogen and Vogel 1963). However, more recently, in humans, commissurotomy involving either complete or partial bisection of the CC has indicated considerable success in reducing the severity, and to a lesser extent, the frequency of otherwise intractable seizures (see Spencer and Spencer, in this volume).

Epilepsy and the Corpus Callosum II, Edited by A.G. Reeves
and D.W. Roberts, Plenum Press, New York, 1995

To further understand the pathways over which seizures propagate, as well as the mechanisms governing focal seizure genesis, various experimental models of epilepsy have been developed. Evaluation of these models, particularly with surgical treatment of human patients in mind, has included assessment of seizure propagation in the presence of various kinds and degrees of forebrain commissurotomy. There are several experimental models available which approximate human complex partial seizures with secondary generalization. Of those models, I believe kindling is the best.

Kindling, of course, is the progressive increase in both electrographic and behavioral seizures resulting from intermittent, low-intensity, tetanizing stimulation of any one of several forebrain sites (Goddard, McIntyre, and Leech 1969). After several stimulations, depending upon the structure being kindled, the successive intensification of focal seizure activity progressively recruits more widespread neural circuits and, eventually, triggers behavioral convulsions. With many additional repetitions, spontaneous seizures ultimately are observed (Wada, Sato, and Corcoran 1974; Pinel, Mucha, and Phillips 1975; Wada and Osawa 1976).

The changes underlying the progression in kindled seizure development are largely permanent, and can be seen in all vertebrate nervous systems (Racine 1978), including man (Sato, Racine, and McIntyre 1990). The statement that 'man can not be kindled' and, therefore, 'kindling may not be a good model of human epilepsy' (e.g., Niedermeyer 1990), is based on the scarcity of reports of kindling in man, and the observation that the rhesus monkey kindles slowly. However, it also selectively ignores the fact that the baboon, Papio papio, kindles quite easily (Wada and Osawa 1976). With the above in mind, however, I must tell the reader that over the past 20 years, six neurologists have indicated to me or my associates that during the repeated stimulation of human brain they have inadvertently "kindled" their patients. The legal ramifications of such a treatment outcome must surely dampen enthusiasm for publication. I suspect it is for this reason that a paucity of published reports of kindling in humans exists.

What follows is a review of various experiments conducted in my laboratory over the past 20 years assessing the involvement of several forebrain commissures in the development and transmission of seizures and/or status epilepticus initiated from a kindled site in either the amygdala or hippocampus of the rat.

AMYGDALA KINDLING IN THE BISECTED BRAIN

The technique I developed for producing commissurotomies reflected the need to section in the midline, while avoiding damage to the superior sagittal sinus and the cingulum. To achieve this, I fashioned a #22 scalpel blade into the shape of a scythe, leaving its cutting edge intact (Figure 1). This modification allowed the blade tip to be inserted beneath the skull and sinus, through a deep longitudinal groove drilled in the midline of the skull, immediately anterior to the frontal cortex and above the olfactory bulbs. The tip and cutting edge of the blade were advanced ventralward between the bulbs, and moved posteriorward between the cinguli. After achieving the appropriate posterior penetration, the commissures were sectioned with a downward arcing motion. Different commissures, or combinations of commissures, were sectioned with variations of this procedure (see McIntyre 1975, for a more complete description).

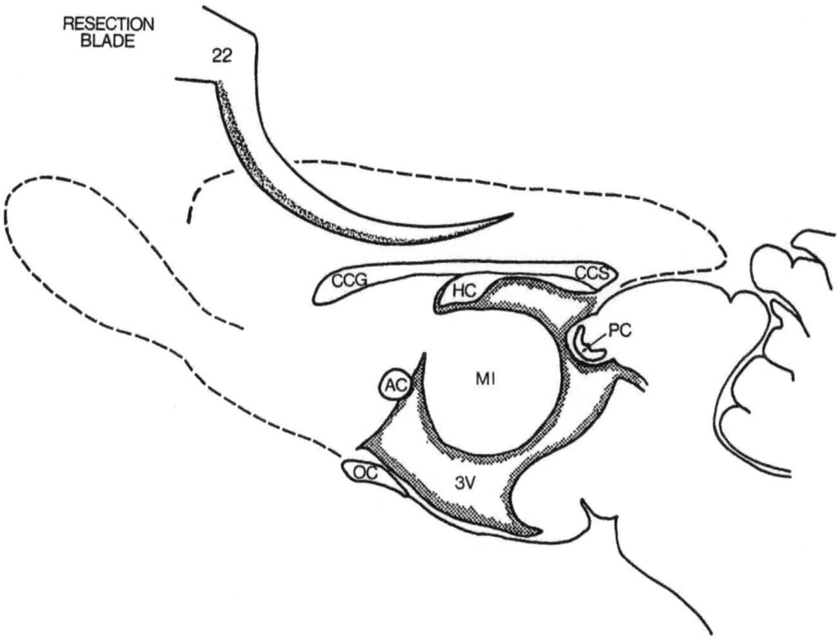

Figure 1. Schematic diagram of a mid-sagittal view of the forebrain commissures and the surgical approach to an anterior forebrain commissurotomy. See text for abbreviations.

Table 1. Primary and secondary site amygdala kindling rates (mean number of stimulations with AD and their range) to the first stage-5 convulsion for all groups.

Group	N	Primary Site	Secondary Site
Intact	12	14.6 (7-23)	9.0 (6-15)
Sham split	9	13.6 (5-24)	8.0 (4-14)
Total split	7	11.1 (7-14)	5.7 (4-9)*
Complete CC	9	14.2 (6-25)	9.8 (4-23)
Anterior CC	7	13.1(5-28)	6.6(3-10)
Posterior CC	7	16.6 (12-27)	9.1 (8-19)
Subcallosal	4	17.8 (9-24)	6.3 (2-12)
Anterior CC & AC	4	12.5 (9-19)	8.8 (6-10)

* = different from the intact and sham control, $P < 0.05$
N = the number of rats per group.

In my first split-brain study, using the amygdala as the kindled focus, I observed that complete section of all forebrain commissures, including the CC, dorsal and ventral hippocampal commissures (HC), anterior commissure (AC) and massa intermedia (Figure 2), resulted in lateralization of the amygdala after-discharge (AD) and development of hemiconvulsions. Surprising, however, such a section tended to facilitate the speed or rate of amygdala kindling (McIntyre 1975; McCaughran, Corcoran, and Wada 1978). The rate of kindling was measured as the number of stimulations necessary to develop the first clonic, rearing-and-falling, convulsion. Although total forebrain commissurotomy tended to facilitate amygdala kindling, the bisection of single commissures did not (see Table 1).

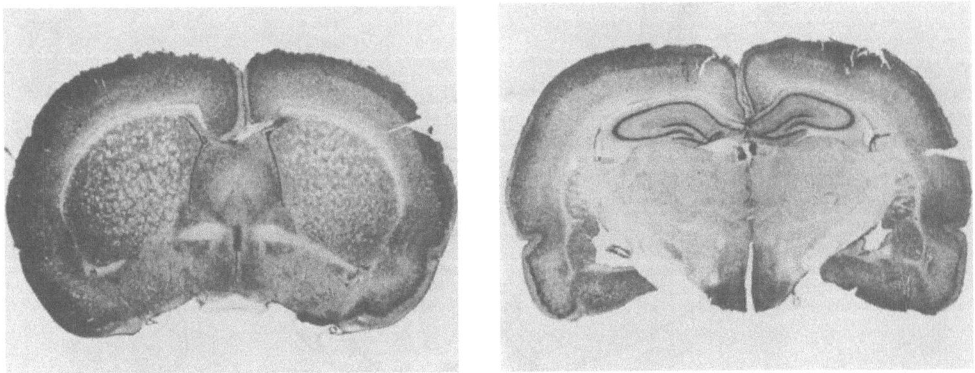

A. B.

Figure 2. Photomicrographs of cresyl violet stain coronal sections showing commissurotomy of the corpus callosum alone (A), or the total forebrain (B). Note the bisection in B extends through the massa intermedia to the third ventricle.

The form of the 'generalized' convulsion in commissurotomized rats, of course, was not the same as in intact rats. In intact rats, the stage-5 (Racine 1972) convulsion appeared as strong, symmetrical forelimb clonus, from a seated position, followed by rearing-and-falling. After falling, the rat regained its seated posture only to rear-and-fall again. On the other hand, with total forebrain commissurotomy, the rat also showed rearing-and-falling, but falling was heralded by a contraversive turn of the trunk with unilateral clonus of the forelimb contralateral to the kindled site. After falling, and without regaining a seated posture, the rat usually manifested clonus additionally in its contralateral hindlimb. The remainder of the rat's hemiconvulsion was characterized by an attempt to assume a righted posture using ipsilateral fore- and hindlimbs as support. It appeared that the non-seizing hemisphere of the commissurotomized rat was attempting to control the disruptive thrashing of the contralateral limbs by continuous, shuffling and hopping readjustments of the ipsilateral limbs. I suggest this was a normal, 'conscious-like' behavioral adjustment rather than a reflexive act since fully commissurotomized rats, unlike intact rats, do not experience amnesia for events before or during the convulsion (McIntyre 1978). Hence, the fully commissurotomized rat is capable of normal information acquisition, processing and retrieval during these strong hemiconvulsions.

In spite of the differences between bisected and intact rats indicated above, I have referred to the strong hemiconvulsions in bisected rats as generalized. By this I mean that the behavioral and electrographic progression of seizure leading to a convulsion in the total commissurotomized rat was essentially like the intact rat, except for elimination of the final projection of seizure across the midline. Indeed, we have observed that if commissurotomy was not complete, leaving only the anterior half of the CC intact, the convulsions always were bilateral, symmetrical and indistinguishable from those of an intact rat.

From the above experiment, it was also apparent that the forebrain commissures carrying amygdala electrographic seizures and those supporting the forelimb convulsive seizures were not identical. To achieve a completely lateralized electrographic seizure, total forebrain commissurotomy was required; bisection of the CC and AC together, however, could temporarily achieve this effect. On the other hand, lateralization of convulsive seizures initially required bisection of only the anterior CC. With sufficient repetitions of the kindling stimulus, however, anterior CC section alone could not contain the spread of the convulsive seizure and, eventually, bilateral asynchronous or synchronous convulsions were observed.

One completely unexpected observation was made in this experiment. The positive transfer in the rate of kindling from one amygdala to the other amygdala, normally seen in intact rats, was preserved or even enhanced in total forebrain commissurotomized rats (Table 1). These interamygdala transfer data (lowered scores at the secondary site) strongly suggested a substantial role in amygdala kindling for the brainstem. This proposal was consistent with the proposal of Wada and Sato (1974; 1975a).

Additional support for brainstem involvement in amygdala kindling can be derived from the following observations in the intact rat. Transfer between the two amygdalae in intact rats normally occurs with a positive saving of about 50%. This means that kindling the second amygdala requires only half of the number of stimulations to trigger a generalized convulsion as the first amygdala (McIntyre and Goddard 1973). However, this positive transfer is not a function of increased excitability in the second amygdala, since the AD triggered there, after primary amygdala kindling, is usually very brief and similar to the unstimulated or 'naive' amygdala. In addition, there is no facilitation in kindling rate from the initial AD triggered at the secondary site to the onset of behavioral automatisms (stages 1-2). The facilitation in secondary site amygdala kindling is the result of enhanced progression from stage-2 seizures to fully generalized convulsions (stage-5). The structures in the intact rat which mediate this late facilitation (positive transfer) are unknown. Considered with the results of the commissurotomy study described above, these data suggest that in the rat, but not the cat and baboon (Wada and Sato 1975b; Wada and Mizoguchi 1984), positive transfer might involve the brainstem (McIntyre 1975; McCaughran, Corcoran, and Wada 1977, 1978).

HIPPOCAMPAL KINDLING IN THE BISECTED BRAIN

Kindling from the dorsal hippocampus (DH) in rats provided a profile different from amygdala kindling. First, DH kindling in intact rats was much slower than amygdala kindling. Second, DH kindling, unlike amygdala kindling, was not facilitated by forebrain commissurotomy involving the CC and HC. In fact, it tended to be somewhat retarded. And third, DH kindling in intact rats produced positive transfer to the contralateral DH with an average saving of 95-100%. Here the first AD triggered in the secondary DH site was not at all similar to a 'naive' hippocampus, but appeared near-fully developed (McIntyre and Stuckey 1985). We observed that forebrain commissurotomy (involving the CC and HC) prior to primary site kindling eliminated nearly 80% of this positive transfer (Table 2). Thus DH transfer largely was a product of forebrain commissure participation (probably the HC). Could the 20% positive transfer in DH kindling which remained after commissurotomy be based on the same subcortical (brainstem?) mechanisms as inter-amygdala transfer? To date, this question remains unanswered.

Table 2. Primary and secondary site dorsal hippocampal kindling rates (mean + SEM) to the first stage-5 convulsion for all groups.

Group	N	Primary Site	Secondary Site
Sham split	10	38.2 ± 4.1	2.7 ± 0.5
Anterior CC	7	41.0 ± 6.9	5.7 ± 2.1
Total split (CC & HC)	10	51.0 ± 3.7	38.0 ± 4.9*

* = different from all other groups, $P < 0.001$
N = number of rats per group.

The observation that hippocampal kindling in intact rats produced nearly 100% transfer to the contralateral hippocampus begged the question as to whether the contralateral DH was fully kindled during primary site kindling, or whether the positive transfer to the secondary site merely reflected the ability of the secondary site to commissurally-trigger the previously kindled primary site. To answer this question, we kindled the primary DH and then commissurotomized the rats. The integrity of the primary site was then redetermined before initiating kindling of the contralateral DH. After total forebrain commissurotomy, stimulation of the primary site quickly regained its efficacy for triggering behavioral seizures. The latter, of course, now were hemiconvulsive. Then we kindled the secondary DH site. To our surprise, kindling of the secondary DH in the commissurotomized rats proceeded very quickly, much like intact rats (Table 3). It appeared conclusively, therefore, that primary site epileptogenesis, kindled in the DH of the rat, resulted in virtually complete epileptogenesis (a mirror focus?) in the homotopic DH (McIntyre and Edson 1987). We believe the pathway supporting the secondary epileptogenesis during primary kindling was the HC, and not the CC.

A similar observation was made by McCaughran, Corcoran, and Wada (1977) when commissurotomizing rats after amygdala kindling. They showed that the positive transfer in kindling between amygdalae was not altered by commissurotomy, even though the previously bilateral, symmetrical motor convulsions were now hemiconvulsive. Of course, those results were expected, since commissurotomy before amygdala kindling produced hemiconvulsions and did not alter positive transfer (McIntyre 1975; McCaughran, Corcoran, and Wada 1978).

Table 3. Rate of dorsal hippocampal kindling at the primary site before and after commissurotomy, and then at the secondary site.

Commissurotomy Groups	N	Primary Site	Primary Site Retest	Secondary Site
Sham split	9	20.0 ± 3.2	2.5 ± 0.8	1.2 ± 0.2
Partial split (CC & HC)	15	18.5 ± 2.7	2.1 ± 0.3	2.3 ± 0.4
Total split (CC & HC)	9	26.0 ± 3.5	3.7 ± 0.7	3.5 ± 0.7*

* = different from sham group, $P < 0.05$; secondary site scores of all groups are different from their respective primary site scores, $P < 0.001$.

STATUS EPILEPTICUS IN THE BISECTED BRAIN

We have also used kindling to develop a model of focally initiated status epilepticus (SE). This model was achieved by first kindling the amygdala (McIntyre, Nathanson, and Edson 1982) or the hippocampus (McIntyre, Stokes, and Edson 1986), and then exposing the kindled site to 60 min of continuous stimulation at an intensity 50-100% higher than the local AD threshold. The stimulation was interrupted for one minute every nine or ten minutes to record electrographic activity. Such a stimulation procedure initiated continuous behavioral and/or electrographic SE in approximately 50% of the kindled rats.

Developed in this manner, amygdala-triggered SE appeared in one of four basic forms, each with its own behavioral, electrographic and metabolic profile (McIntyre, Don, and Edson 1991). The mildest form produced low frequency electrographic seizure, no brain damage, and was non-convulsive. The next two forms were progressively more severe (called ambulatory and masticatory SE), produced increasingly higher frequencies of electrographic seizures, considerable brain damage involving limbic, thalamic and restricted neocortical structures, and were convulsive (similar to stage-2 kindled seizures). And lastly, far more severe by every measurement, the final form was fully generalized (similar to stage-3 to 5 seizures) and always lethal.

The course of SE induction from a kindled amygdala focus in rats was little affected by commissurotomy (McIntyre and Kelly, in progress). This was not surprising since commissurotomy had only a small facilitative effect on the rate of amygdala kindling (McIntyre 1975; McCaughran, Corcoran, and Wada 1978). However, commissurotomy did affect SE induction from a hippocampal focus. In this case, the effect was facilitative. This was shown by the greater proportion of bisected compared to intact rats which developed SE (McIntyre, Stokes, and Edson 1986).

In both amygdala and hippocampal-triggered SE models, the occasional (or frequent) spontaneous forelimb convulsions during SE were rendered either hemiconvulsive, or bilaterally asymmetric, by anterior CC sections. On the other hand, the electrographic seizures were lateralized only by more extensive sections including all of the HC and, in the case of the amygdala-triggered SE, the AC.

Using commissurally-intact rats, in both amygdala and hippocampal-kindled SE models, the brain damage following spontaneous recovery from ambulatory or masticatory SE involved the same neuroanatomic structures. Histological comparisons of the aftermath of SE from the two kindled SE models revealed differences only in the severity and laterality of the brain damage. SE triggered from the amygdala of intact rats resulted in *unilateral* damage to the basolateral amygdala and pyriform cortex in all rats, with additional bilateral damage to CA1 hippocampus in 50% of the rats. By contrast, SE triggered from the kindled hippocampus of intact rats produced bilateral damage in the hippocampus, and *bilateral* damage to the basolateral amygdala and pyriform cortex.

The difference between these two kindled SE models in laterality of temporal lobe brain damage provided additional support for the notion that previous hippocampal kindling alters temporal lobe circuits bilaterally, while amygdala kindling, to a minimal level of 6 stage-5 convulsions, alters the circuits unilaterally (McIntyre 1981; 1986; McIntyre and Kelly 1990a). I would predict that protracted kindling of the amygdala (perhaps 30 convulsions or more) would give rise to bilateral temporal lobe circuit alterations (e.g., McIntyre 1979), and the potential for bilateral SE pathology like hippocampal-kindled rats.

As expected, forebrain commissurotomy did not alter the lateralized amygdala-pyriform damage seen previously following SE in intact amygdala-kindled rats, but did lateralize the CA1 hippocampal damage. The lateralization of hippocampal damage, however, occurred only when the HC was completely bisected. If the HC bisection was incomplete, the CA1 damage, when present, was always bilateral (McIntyre and Kelly, in progress). In the case of hippocampal-kindled rats, forebrain commissurotomy was effective in lateralizing all SE-induced brain damage but, again, only if the entire HC was bisected. In these rats, the lateralized damage was independent of additional sectioning of the AC and/or CC. It appeared, therefore, that the *bilateral* damage normally inflicted on temporal lobe structures during SE triggered from a kindled hippocampal focus was produced by interhemispheric propagation through the HC, and without involvement of the AC and CC (McIntyre, Stokes, and Edson 1986). Speculatively, the SE triggered in these rats made use of the bilateral neural circuits developed during hippocampal kindling. The changed excitability of these altered circuits now could provide the enhanced bilateral, recurrent synaptic drive necessary to destroy both the local hippocampal cell fields and several efferent temporal lobe structures.

A final set of experiments involving commissurotomy examined the neural populations which trigger the generalized stage-5 convulsion. Our previous speculations (McIntyre and Racine 1986; McIntyre 1986; McIntyre and Kelly 1990a), based on several varieties of data, pointed to the pyriform and perirhinal cortex as pivotal for triggering limbic-kindled convulsions. The ultimate manifestation of convulsions, however, likely resulted from recruitment of other neural populations, for example, those of the frontal cortex, striatum and brainstem. To determine temporal lobe involvement in the triggering of stage-5 convulsions, first we kindled an amygdala focus in commissurotomized rats. Second, we subjected the rats to amygdala-triggered SE. Of course, the SE experience generally destroyed the ipsilateral pyriform cortex. Then third, we exposed the ipsilateral DH to the kindling procedure.

The outcome of the DH kindling procedure depended upon the degree of brain damage (both in the temporal lobe and hippocampus) caused previously by the SE experience, and the extent of the forebrain commissurotomy. As we saw previously, when the forebrain was either completely bisected, or partially bisected to include the anterior half of the CC and HC, amygdala-kindled convulsions were lateralized to contralateral limbs. In the present experiment, before amygdala-triggered SE, this was also true. After recovery from amygdala-triggered SE, those rats with minimal hippocampal damage showed a normal primary AD to hippocampal stimulation. This AD grew progressively during daily repetitions of the kindling stimulation. In those rats with *total* commissurotomy and a *completely* destroyed pyriform and perirhinal cortex, stage-5 convulsions could not be again developed. But in those rats with only *partial* pyriform cortex damage, hemiconvulsions contralateral to the stimulated hippocampus were again developed, and appeared identical to those triggered from the ipsilateral amygdala before SE.

On the other hand, the really interesting rats were those with bisection of only the anterior CC and HC plus complete loss of the pyriform and perirhinal cortex. These anterior bisected rats also developed hemiconvulsions to hippocampal kindling but, unlike the partially-damaged pyriform rats described above, these hemiconvulsions did not involve the contralateral limbs, but rather the ipsilateral limbs. This represented a shift in convulsion laterality from their previous (before SE) amygdala-triggered convulsions. The shift most likely resulted from the inability of the discharge, triggered in kindled hippocampus, to gain access ipsilaterally to convulsion-trigger mechanisms in the previously kindled temporal lobe (since the ipsilateral basal amygdala and pyriform and perirhinal cortex were now lost). In these rats, we believe that the kindled hippocampal discharge, through the remaining posterior HC, eventually gained access to the contralateral hippocampus. From there it invaded, and recruited, the contralateral amygdala-pyriform-perirhinal cortex, triggering a hemiconvulsion from that hemisphere. For these reasons, the hemiconvulsion appeared, with long latency, in the limbs ipsilateral to the kindled DH focus (McIntyre and Kelly 1990b).

In summary, the results of our commissurotomy studies in the kindled rat indicated that the anterior half of the CC was the preferred interhemispheric pathway carrying forelimb convulsions. This was true regardless of whether convulsions were kindled from the amygdala or hippocampus. With many repetitions of the kindling stimulus, however, these partial commissurotomies could not contain the seizures and, eventually, bilateral, usually asynchronous, convulsions were produced. Total forebrain commissurotomy was effective in containing both convulsive and electrographic seizures, but the latter with greater difficulty. With this in mind, the important experiments have not been conducted to determine the longevity of seizure lateralization following total forebrain commissurotomy. Based on the above results, one might predict that the intact brainstem of a total forebrain commissurotomized rat would eventually transfer seizures from the initiating hemisphere to the contralateral hemisphere. If this were to happen, one should record a hemiconvulsion in the contralateral limbs, followed with a long latency, by a second hemiconvulsion in the ipsilateral limbs. I describe this behavioral scenario with confidence because I have seen it numerous times. However, because ephaptic transmission and other possibilities exist, I cannot, in the rat, ascribe with certainty the seizure transfer mechanism to the brainstem.

REFERENCES

Bogen, J.E., and Vogel, P.J., 1963, Treatment of generalized seizures by cerebral commissurotomy, *Surg. Forum* 14:431.

Curtis, H.J., 1940, Intercortical connections of the corpus callosum indicated by evoked potentials, *J. Neurophysiol.* 3:404.

Erickson, T.C., 1940, Spread of epileptic discharge: An experimental study of the afterdischarge induced by electrical stimulation of the cerebral cortex, *Arch. Neurol. Psychiatry* 43:429.

Fox, C.A., Fisher, R.R., and Desalva, S.J., 1948, The distribution of the anterior commissure in the monkey (*Macaca mulatta*), *J. Comp. Neurol.* 89:245.

Frost, L.L., Baldwin, M., and Wood, C.D., 1958, Investigation of the primate amygdala: movements of the face and jaw, *Neurology* 8:543.

Goddard, G.V., McIntyre, D.C., and Leech, C.K., 1969, A permanent change in brain function resulting from daily electrical stimulation,*Exp. Neurol.* 25:295.

McCaughran, J.A., Jr., Corcoran, M.E., and Wada, J.A., 1977, Facilitation of secondary-site amygdaloid kindling following bisection of the corpus callosum and hippocampal commissure in rats, *Exp. Neurol.* 57:132.

McCaughran, J.A., Jr., Corcoran, M.E., and Wada, J.A., 1978, Role of the forebrain commissures in amygdaloid kindling, *Epilepsia* 19:19.

McIntyre, D.C., 1975, Split-brain rat: transfer and interference of amygdala kindled convulsions, *Can. J. Neurol. Sci.* 2:419.

McIntyre, D.C., 1978, Kindling and memory: the adrenal system and the bisected brain, *in* "Limbic Mechanisms," K.E. Livingston and O. Hornykiewicz, eds., Plenum Press, New York

McIntyre, D.C., 1979, Effects of focal vs. generalized kindled convulsions from anterior neocortex or amygdala on CER acquisition in rats, *Physiol. Behav.* 23:855.

McIntyre, D.C., 1981, Catecholamine involvement in amygdala kindling of the rat, *in* "Kindling 2," J.A. Wada, ed., Raven Press, New York

McIntyre, D.C., 1986, Kindling and the pyriform cortex, *in* "Kindling 3," J.A. Wada, ed., Raven Press, New York.

McIntyre, D.C. and Goddard, G.V., 1973, Transfer, interference and spontaneous recovery of convulsions kindled from the rat amygdala, *Electroenceph. Clin. Neurophysiol.* 35:533.

McIntyre, D.C.,and Stuckey, G.N., 1985, Dorsal hippocampal kindling and transfer in the split-brain rat, *Exp. Neurol.* 87:86.

McIntyre, D.C.,and Racine, R.J., 1986, Kindling mechanisms: current progress on an experimental epilepsy model, *Prog. Neurobiol.* 27:1.

McIntyre, D.C. and Edson, N., 1987, Facilitation of secondary site kindling in the dorsal hippocampus following forebrain bisection, *Exp. Neurol.* 96:569.

McIntyre, D.C. and Kelly, M.E., 1990a, Is the pyriform cortex important for limbic kindling, *in* "Kindling 4," J.A. Wada, ed., Plenum Press, New York

McIntyre, D.C. and Kelly, M.E., 1990b, Pyriform cortex involvement in limbic kindling, *Neurosci. Abstr.* 16:1107.

McIntyre, D.C., Nathanson, D., and Edson, N., 1982, A new model of partial status epilepticus based on kindling, *Brain Res.* 250:53.

McIntyre, D.C., Stokes, K.A., and Edson, N., 1986, Status epilepticus following stimulation of a kindled hippocampal focus in intact and commissurotomized rats, *Exp. Neurol.* 94:554.

McIntyre, D.C., Don, J.C., and Edson, N., 1991, Distribution of [^{14}C]2-deoxyglucose after various forms and durations of status epilepticus induced by stimulation of a kindled amygdala focus in rats, *Epilepsy Res.* 10:119.
Myers, R.E. and Sperry, R.W., 1953, Interocular transfer of a visual form discrimination habit in cats after section of the optic chiasma and corpus callosum, *Anat. Rec.* 115:351.
Niedermeyer, E., 1990, "The Epilepsies: Diagnosis and Management," Urban and Schwarzenberg, Baltimore
Pinel, J.P.J., Mucha, R.F., and Phillips, A.G., 1975, Spontaneous seizures generated in rats by kindling: a preliminary report, *Physiol. Psychol.* 3:127.
Poblete, R., Ruben, R.J., and Walker, A.E., 1959, Propagation of after discharge between temporal lobes, *J. Neurophysiol.* 22:538.
Racine, R.J., 1972, Modification of seizure activity by electrical stimulation. II. Motor seizure, *Electroenceph. Clin. Neurophysiol.* 32:281.
Racine, R.J., 1978, Kindling: the first decade, *Neurosurgery* 3:234.
Sato, M., Racine, R.J., and McIntyre, D.C., 1990, Kindling: basic mechanisms and clinical validity, *Electroenceph. Clin. Neurophysiol.* 76:459.
van Wagenen, W.P. and Herren, R.Y., 1940, Surgical division of commissural pathways in the corpus callosum; related to spread of epileptic attach, *Arch. Neurol. Psychiatry* 44:740.
Wada, J.A. and Sato, M., 1974, Generalized convulsive seizure induced by daily stimulation of the amygdala in cats: correlative electrographic and behavioral features, *Neurol. (Minneap.)* 24:565.
Wada, J.A. and Sato, M., 1975a, Effects of unilateral lesion in the midbrain reticular formation on kindled amygdaloid convulsions in cats,*Epilepsia* 16:693.
Wada, J.A. and Sato, M., 1975b, The generalized convulsive seizure state induced by daily electrical stimulation of the amygdala in split-brain cats,*Epilepsia* 16:417.
Wada, J.A. and Osawa, T., 1976, Spontaneous recurrent seizure state induced by daily electric amygdaloid stimulation in Senegalese baboons (Papio papio), *Neurol.* 26:273.
Wada, J.A. and Mizoguchi, T., 1984, Limbic kindling in the forebrain bisected photosensitive baboon, *Papio papio, Epilepsia* 25:278.
Wada, J.A., Sato, M., and Corcoran, M.E., 1974, Persistent seizure susceptibility and recurrent spontaneous seizures in kindled cats, *Epilepsia* 15:465.

CALLOSAL AND THALAMIC TRANSECTION: EFFECTS ON SPONTANEOUS AND PENTYLENETETRAZOL-INDUCED ABSENCE SEIZURES IN RATS

M.Vergnes, Ph.D.[1] and C.Marescaux, M.D.[2]

[1]U 398 INSERM, Centre de Neurochimie du CNRS
5 rue Blaise Pascal, 67084 Strasbourg Cedex, France

[2]Clinique Neurologique, CHU, 1 place de l'Hôpital
67091 Strasbourg Cedex, France

INTRODUCTION

In a strain of Wistar rats inbred in our laboratory, all animals are affected with generalized nonconvulsive epilepsy (absence or petit mal epilepsy). The seizures occur when animals are awake and inactive, with a frequency of about one per minute, and they disappear during activity and sleep. The behavioral characteristics during the seizures are immobility, reduced tone in neck muscles and usually rhythmic vibrissal twitches. In contrast with these discrete clinical signs, drastic EEG manifestations characterize the seizures: large 7-9 c/s spike and wave discharges (SWD), the duration of which may exceed one minute, with a mean duration of 20 s/min. in our recording conditions.

The SWD appear on a fast low voltage waking EEG or sometimes during transition between wake and sleep. No post-ictal depression or change of EEG activity ever appears as a consequence of a seizure. Drugs effective against absence seizures in humans suppress the SWD. EEG, behavioral and pharmacological studies demonstrate that spontaneous SWD in these Wistar rats fulfill all the requirements for an experimental model of absence epilepsy. The SWD are genetically determined with an autosomal dominant inheritance. A strain with SWD occurring in all animals, called "Genetic Absence Epilepsy Rats from Strasbourg" (GAERS) was inbred in our laboratory. Simultaneously a non epileptic control strain was outbred from the same Wistar colony (Vergnes et al. 1982, 1990).

Epilepsy and the Corpus Callosum II, Edited by A.G. Reeves
and D.W. Roberts, Plenum Press, New York, 1995

The SWD are always bilateral and synchronous all over the cortex of both hemispheres and in the lateral part of the thalamus, whereas they do not appear in limbic structures. As in other models of absence-epilepsy the thalamo-cortical circuitry is the main substrate of the SWD (Gloor 1988; Vergnes et al. 1987).

The corpus callosum (CC) is the major interhemispheric commissure, connecting preferentially homotopic cortical areas (Innocenti 1986), and it is the most likely pathway underlying bilateral spread and synchronization of cortical paroxystic activities and especially of SWD in absence-epilepsy. However subcortical pathways, such as intrathalamic connections, may contribute to bilateral synchronization of the thalamo-cortical SWD. In order to examine these questions, the CC and/or the midline thalamus were sectioned in GAERS. The effects of callosotomy were also examined on pentylenetetrazol (PTZ)-induced seizures which are commonly used as a pharmacological model of absence-seizures in rats (Marescaux et al. 1984).

METHODS

Male Wistar rats (350-450 g) from our epileptic strain (GAERS) and from the non-epileptic control strain were operated under pentobarbital anesthesia (40 mg/kg ip). The CC was transected through a midline slit in the skull with a surgical blade fixed in a stereotaxic holder and moved anteriorily along the midline of the brain from 2 to 12 mm anterior, and 4 to 4.5 mm ventral to the lambda. When bleeding had ceased, absorbable gelatin sponge was tucked into the medial slit and the skin was sutured. The midline thalamus was sectioned in addition to the CC by lowering the blade between AP = 4.5 and 7 to DV = 7 mm. The midline thalamic nuclei were lesioned electrolytically by a 2 mA current for 15 sec/site (AP = 4, 5, 6, and 7 mm; DV = 6-7mm). After one week recovery, four single contact electrodes were implanted bilaterally over the fronto-parietal cortex for EEG control. The animals recovered rapidly and survived without any apparent deficit.

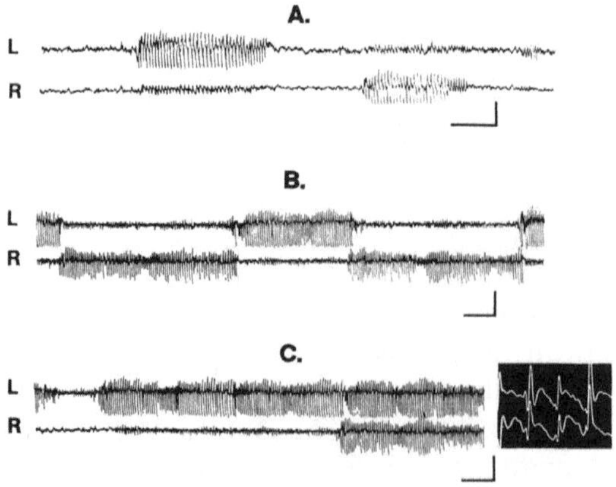

Figure 1. A, B, C. Three different patterns of SWD recorded in one rat with callosal transection. When SWD are bilateral, each single spike and wave complex occurs synchronously on both sides (C, see oscilloscopic recording). Calibrations 400 mV; 2 sec.

EEG was recorded repeatedly for several weeks in freely moving animals kept awake when necessary by gentle sensory stimulation. For each rat, the cumulated duration of SWD was measured on each hemisphere for 30 minute periods. Bilateral synchronization was appreciated visually on EEG and confirmed on oscilloscopic deviations.

PTZ was injected at three doses (20, 35 and 50 mg/kg ip). PTZ-induced seizures were continuously monitored by EEG and direct observation.

At the end of the experiment, the rats were killed by an overdose of pentobarbital and the brain was removed. Serial paraffin sections stained with cresyl violet were examined to ascertain the extent of the lesions.

RESULTS

In unlesioned GAERS the SWD are always bilateral and synchronous in both hemispheres. After transection of the CC (n = 12), the bilateral synchronism was abolished and three patterns emerged: (1) the SWD occurred unilaterally and independently on each hemisphere, and sometimes alternated from one side to the other (51% of the SWD); (2) SWD started on one side and then continued bilaterally after delays varying from 0.5 to several seconds (40% of the SWD); (3) some bilateral synchronous SWD still occurred in most animals (9% of the SWD) (Fig. 1). When the SWD were present bilaterally, each single spike-and-wave complex was synchronous on both hemispheres. The transection of the CC was complete in four rats and subtotal in eight rats. There was no relationship between the extent of the callosal section and the SWD patterns.

When the transection of the CC was associated with a midline cut through the thalamus (n = 5), the bilateral desynchronisation was more complete: the single spike and wave complexes occurred independently in bilateral SWD (Fig 2B). However the lesion of the midline thalamic nuclei (n = 4) separating the thalamus in two lateral parts, did not affect SWD. They occurred in the same amount as in intact animals and synchronously on both hemispheres (Fig. 2).

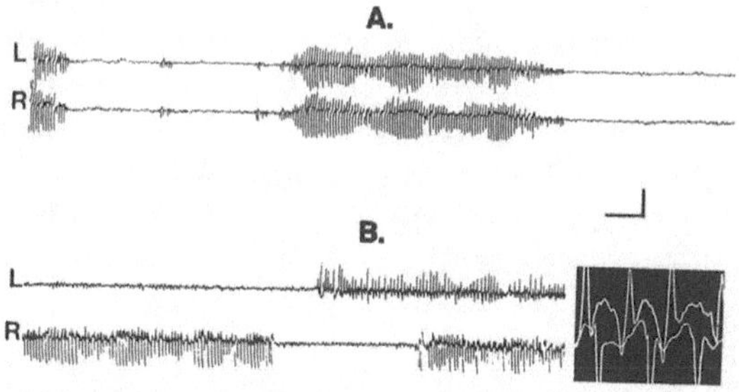

Figure 2. A. SWD recorded in a rat with medial thalamic lesions: SWD are bilateral and synchronous. B: SWD recorded in a rat with both corpus callosum and medial thalamus sections. SWD are completely desynchronized: when bilateral SWD occur, each spike and wave complex appears independently on both sides (see oscilloscopic recording). Calibrations 400 mV; 2 sec.

In spite of the desynchronisation of SWD after callosotomy, the total amount of seizures, measured on both hemispheres, was comparable to the values found in intact rats. In 12 unlesioned GAERS cumulated duration of SWD varied from 410 to 730 sec. per 30 min. (mean ± SEM = 548 ± 28 sec), whereas in callosotomized rats the mean SWD duration was 576 ± 50 sec. in the left cortex and 550 ± 43 sec. in the right cortex per 30 min.

In non-epileptic controls (n = 5), injection of a low dose of PTZ (20 mg/kg ip) induced within three to ten minutes bilateral and synchronous 5-7 c/s SWD lasting 1-5 sec. These SWD were associated with behavioral arrest and facial myoclonus. In 5 callosotomized rats these PTZ-evoked SWD occurred independently on each hemisphere (Fig. 3A).

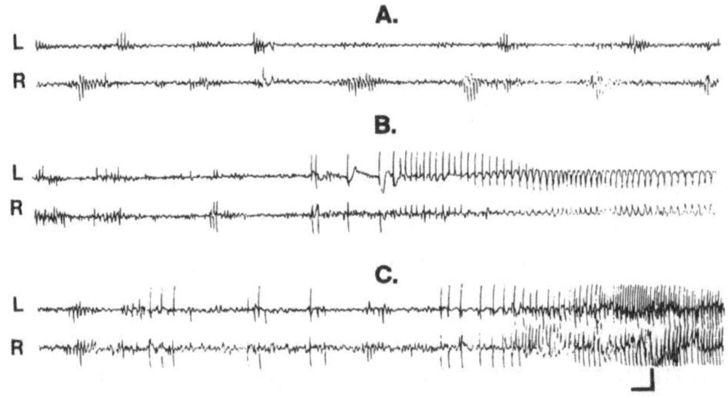

Figure 3. PTZ-induced seizures in callosotomized rats. A: PTZ 20 mg/kg, bilateral asynchronous SWD. B: PTZ 35 mg/kg, unilateral myoclonic seizure. C: PTZ 50 mg/kg, bilateral synchronous tonic-clonic seizure. Calibrations 400 mV; 1 sec.

The intermediate dose of PTZ (35 mg/kg) induced, after 5 to 25 minutes in controls, myoclonic seizures lasting 5 to 70 seconds. These seizures are characterized by bilateral and synchronous high amplitude spike-waves at 3-5 c/seconds, with rearing and bilateral symmetrical myoclonus of the forelimbs and the face. In callosotomized rats, these myoclonic seizures were no more symmetrical and were sometimes unilateral (Fig. 3B).

A high dose of PTZ (50 mg/kg) produced, in controls, within one to two minutes, bilateral and symmetrical tonic and tonic-clonic seizures with rearing and falling. The tonic phase was concomitant with high frequency large polyspikes. The clonic phase was accompanied by spikes and polyspikes and waves. The seizures were followed by slow and low amplitude activity with high voltage isolated bilateral spikes concomitant with body jerks. In callosotomized rats, the seizures became rapidly bilateral (Fig. 3C). After the tonic phase, when the clonus is fully developed, the spikes and waves were synchronized on both hemispheres and the clonic jerks were symmetrical. The post-ictal spikes remained bilateral and synchronous.

The latency and severity of PTZ-induced seizures were in no case decreased by callosotomy.

DISCUSSION

Our results confirm that the CC is the major pathway involved in bilateralization of spontaneous SWD in rats' absence epilepsy. However, the transection of the CC does not totally abolish bilateral synchronism of the SWD, suggesting that additional pathways are also involved in interhemispheric synchronization of SWD (Vergnes et al. 1989). According to our results, the midline thalamus is likely to fulfill this role: full interhemispheric dissociation of SWD is only achieved when the midline thalamus is sectioned in addition to the CC. However the midline thalamus plays a minor role in bilateral transfer, as destruction of this area does not affect occurrence of bilateral and synchronous SWD.

PTZ-elicited seizures are commonly considered as models of petit-mal seizures. The bilateral synchronization of SWD and of rhythmic spiking associated with myoclonic seizures induced by PTZ, is suppressed after callosotomy. However the spikes and polyspikes concomitant with convulsive seizures induced by high doses of PTZ remain bilateral and synchronous in animals with a full transection of the CC, suggesting involvement of brain-stem structures mediating the generalization of these tonic-clonic seizures.

In other animal models with bilateral and synchronous SWD, the CC appears to be the main structure ensuring bilateral synchronization. Section of the CC in rats, cats or primates abolished bilateral synchronization of discharges induced by epileptogenic substances (Marcus and Watson 1966, 1968; Mares et al. 1978). The CC is also involved in bilateral synchronization of discharges in PTZ-treated cats and primates (Marcus et al. 1969) and in the penicillin-induced generalized epilepsy in cats (Musgrave and Gloor 1980; Pellegrini et al. 1979). Finally, in the photosensitive baboon, CC section converts cortical SWD from synchronous to asynchronous ones (Naquet et al. 1975).

CONCLUSION

The typical SWD of petit mal epilepsy are built up in a cortico-thalamic substrate. In each hemisphere the thalamic relay nuclei and their cortical projection area are intimately associated in production of SWD, which are ipsilaterally synchronous in both structures. Bilateral spread and synchronization of the SWD occur through the CC with a minor participation of intrathalamic pathways. The SWD are initiated equally in each hemisphere in callosotomized animals, and their number and duration are similar to that of control rats. Therefore it appears clearly that callosotomy should have no beneficial effect in generalized non-convulsive absence epilepsy.

REFERENCES

Gloor P., 1988, Neurophysiological mechanism of generalized spike-and-wave discharge and its implication for understanding absence seizures, *in:* "Elements of Petit Mal Epilepsy," M.S. Myslobodsky, A.F. Mirsky, eds., Peter Lang, New York, pp. 159-209.

Innocenti, G., 1986, General organization of callosal connections in the cerebral cortex, *in*: "Cerebral Cortex, Vol. 5," E.G. Jones and A.Peters, eds., Plenum Press, New York, pp. 291-353.

Marcus, E.M. and Watson, C.W., 1966, Bilateral synchronous spike wave electrocorticographic patterns in the cat, *Arch. Neurol.* 14:601-610.

Marcus, E.M. and Watson, C.W., 1968, Symmetrical epileptogenic foci in monkey cerebral cortex, *Arch. Neurol.* 19:99-116.

Marcus, E.M., Watson, C.W., and Jacobson, S., 1969, Role of the corpus callosum in bilateral synchronous discharges induced by intravenous pentylenetetrazol, *Neurology* 19:309.

Mares, P., Filip, J., and Mares J., 1978, The role of the corpus callosum in the interhemispheric transmission of epileptic electrical acticity, *Physiol. Bohemoslov.* 27:23-29.

Marescaux, C., Micheletti, G., Vergnes, M., Depaulis, A., Rumbach L., and Warter, J.M., 1984, A model of chronic spontaneous petit mal-like seizures in the rat: comparison with pentylenetetrazol-induced seizures, *Epilepsia* 25:326-331.

Musgrave, J. and Gloor, P., 1980, The role of the corpus callosum in bilateral interhemispheric synchrony of spike and wave discharge in feline generalized penicillin epilepsy, *Epilepsia* 21:369-378.

Naquet, R., Catier, J., and Ménini, C., 1975, Neurophysiology of photically induced epilepsy in *Papio papio*, *in*: "Advances in Neurology, Vol. 10," B.S. Meldrum and C.D. Marsden, eds., Raven Press, New York, pp. 107-118.

Pellegrini, A., Musgrave, J., and Gloor, P., 1979, Role of afferent input of subcortical origin in the genesis of bilaterally synchronous epileptic discharges of feline generalized penicillin epilepsy, *Exp. Neurol.* 64:155-173.

Vergnes, M., Marescaux, C., Depaulis, A., Micheletti, G., and Warter, J.M., 1987, Spontaneous spike and wave discharges in thalamus and cortex in a rat model of genetic petit mal-like seizures, *Exp. Neurol.* 96:127-136.

Vergnes, M., Marescaux, Ch., Depaulis, A., Micheletti, G., and Warter, J.M., 1990, *in*: "Generalized Epilepsy. Neurobiological Approaches," M. Avoli, P. Gloor, G. Kostopoulos, and R. Naquet, eds., Birkhäuser, Boston, pp. 238-253.

Vergnes, M., Marescaux, C., Lannes, B., Depaulis, A., Micheletti, G., and Warter, J.M., 1989, Interhemispheric desynchronization of spontaneous spike-wave discharges by corpus callosum transection in rats with petit mal-like epilepsy, *Epilepsy Res.* 4:8-13.

Vergnes, M., Marescaux, C., Micheletti, G., Reis, J., Depaulis, A., Rumbach, L., and Warter, J.M., 1982, Spontaneous paroxysmal electroclinical patterns in rat: a model of generalized nonconvulsive epilepsy, *Neurosci. Lett.* 33:97-101.

ROLE OF THE CORPUS CALLOSUM IN THE PHOTOSENSITIVE EPILEPSY OF BABOONS

Ch. Menini, M.D.,[1] C. Silva-Barrat, M.D.,[1] H. Fukuda, M.D.,[2] R. Naquet, M.D.,[1] and J. Wada, M.D., Sc.D., F.R.C.P. [3]

[1]Institute A. Fessard
CNRS, 91190 Gif-sur-Yvette, France

[2]Department of Neurosurgery
Asahikawa Medical College

[3]Division of Neurosciences and Neurology
University of British Columbia
Vancouver, BC V6T2A1, Canada

SUMMARY

In predisposed baboons (Papio papio), intermittent light stimulation (ILS) at 25 Hz may induce cortical spikes which symmetrically occupy both fronto-rolandic cortical areas and are accompanied by myoclonic jerks. When ILS is pursued, spikes are transformed into spike-waves or polyspike-waves and can have the capacity to either self-perpetuate after cessation of ILS in an "after-discharge" or followed by a generalized tonic-clonic seizure. Electrographically, seizures start in the two fronto-rolandic areas simultaneously, and quickly diffuse to all brain structures except the rhinencephalon. Monocular as well as binocular ILS induces paroxysmal discharges (PDs) and seizures which, in both cases, occur bilaterally and symmetrically. In naturally photosensitive animals, PDs and seizures induced by ILS never start in the occipital areas.

Initially, corpus callosum sections were performed through a visual approach, occasionally causing a lesion of the motor cortex, one side. In this case, binocular photic stimulation of the animals induced seizures which remained localized in the non-lesioned hemisphere and were associated with a contralateral hemiclonus. Seizures were followed by an electrical silence of the same hemisphere and contralateral hemiparesis. When corpus callosum section could be performed without lesioning thc

Epilepsy and the Corpus Callosum II, Edited by A.G. Reeves
and D.W. Roberts, Plenum Press, New York, 1995

motor cortex, ILS-induced seizures could be clinically and electrically bilateral and symmetrical, but they also could start bilaterally and then develop into a unilateral seizure associated with a contralateral hemiclonus and a postictal hemiparesis. In a given animal, the unilateral seizure might sometimes develop on one side and other times on the other.

Subsequently, corpus callosum sections were performed through a stereotactic approach, avoiding cortical lesions. Such operated animals were tested in two different situations.

1) Visual activation of a single occipital lobe could be obtained by monocular ILS applied to a single retinal hemifield in paralyzed baboons. These animals had been previously injected with allylglycine, a drug which renders paralyzed baboons very photosensitive. In such a situation, PDs and seizures started in either hemisphere, depending on the retinal hemifield stimulated. However, electroclinical correlations could not be established in this preparation.

2) Similar single lobe activation could be obtained in non-paralyzed baboons after laser-destruction of the temporal retinal hemifield in one eye. Monocular stimulation of such a lasered eye activates only the nasal retinal hemifield and the visual pathways directed to the contralateral visual lobe. In this case, spike-waves and seizures occurred only in the activated hemisphere with contralateral hemiclonus and hemiparesis. Subsequent stimulation of the non-operated eye in the same animal produced spike-waves and seizures only in the opposite hemisphere when the stimulation took place during the post-ictal silent period in the first hemisphere. Otherwise, stimulation of the non-operated eye induced bilateral, asynchronous spike-waves and seizures. In all cases, the tonic phase of the seizure was always bilateral but asymmetrical, clinically and electrically.

These data support the following evidences:

a) PDs and seizures have a cortical, hemispheric origin;
b) the cortical visual afferents play a critical role for triggering epileptic phenomena in photosensitive baboons;
c) there is an independent epileptogenic excitability of each cerebral hemisphere;
d) the corpus callosum is the major, if not unique structure responsible for the bilateral synchronization of the cortical epileptic events;
e) bisynchronization can be suppressed and unilateral seizures can be obtained following callosal bisection; and
f) the tonic phase of the seizure is mediated, at least partly, through crossed (probably interreticular) pathways; its bilateralization is not suppressed by corpus callosum section.

INTRODUCTION

When baboons (Papio papio) are submitted to intermittent light stimulation (ILS) at 25 Hz, predisposed animals present myoclonic jerks involving initially the eyelids and then the face. When the photic stimulation is prolonged, myoclonic jerks may involve the whole body and, in the most sensitive animals, develop into a generalized tonic- clonic seizure.

EEG recordings have shown that, during ILS and after a variable latency period, spike-waves appear in the frontal regions and are synchronous with the myoclonic jerks. When generalized myoclonic jerks appear, cortical PDs increase in amplitude, invade other cortical territories and may then be followed by a fully generalized seizure starting in the rolandic area. PDs occur always bilaterally and synchronously in large cortical territories and are accompanied by bilateral and symmetrical clinical signs. Similarly, generalized seizures are always bilateral and synchronous. According to this brief description, the baboon's photosensitive epilepsy can be considered a "primary generalized epilepsy" sensitive to ILS (Gastaut 1973) and similar to that seen in epileptic patients without any identified neurological deficit and any obvious etiological factors.

Data collected over years have initially suggested and then demonstrated that PDs and seizures induced by ILS in photosensitive baboons originate in the motor cortex (Silva-Barrat and Menini 1990, Menini, et al. 1992a, 1992b):

1) ILS-induced spike-waves first appear in the fronto-rolandic cortical regions (areas 4 and 6) and from there radiate to other cortical territories and to subcortical structures such as the internal capsule, the pons, and the brain stem except the limbic structures (Fischer-Williams, et al. 1968). Spike-waves have never been observed to appear initially in the occipital or temporal regions or in deep nuclei.

2) Intralaminar recordings in the frontal cortex, and calculation of current source density have determined that neuronal generators responsible for ILS-induced PDs lie in the motor cortical area (Silva-Barrat, et al. 1988a). When a spike-wave occurs, the earliest, most activated zone is situated in the superficial layers of the motor cortex, suggesting that visual cortico-cortical afferents arriving at this level are essential for the production of PDs. A second activated zone is situated in the deep layers of the motor cortex, suggesting that pyramidal cells situated at this level are activated later by thalamo-cortical afferents.

3) Under the effects of ILS, fronto-rolandic cortical neurons are activated in bursts of action potentials which first appear in the absence of EEG abnormality (Menini, et al. 1981) PDs occur when the bursts have increased their internal frequency, both phenomena being synchronous. In other cerebral structures, such a bursting activity occurs after the appearance of the fronto-rolandic neuronal bursts. A study of multiunitary activities in different subcortical structures (Silva-Barrat, et al. 1986) has shown that the pontine and mesencephalic reticular formations are activated only after PDs have reached an amplitude greater than 200μV. Thalamic nuclei (VL, CM, LP) are activated only after PDs have reached an even greater amplitude (>400μV). This study has also shown that a powerful activation of some deep structures, mainly of the brain stem reticular formation, precedes the appearance of a generalized seizure. These observations suggest that the brain stem reticular formation is involved in the development of seizures, but only as a consequence of the production of PDs by the motor cortex.

4) Bilateral chronic infusions of GABA were performed into the cerebral cortex of naturally photosensitive baboons by using osmotic minipumps. These infusions produced a complete blockade of the epileptic EEG discharges and associated clinical signs induced by ILS, but only

when GABA was directed to the cortical motor and occipital areas (Brailowsky, et al. 1987). Similar GABA infusions into the premotor or prefrontal cortical areas did not significantly affect photosensitivity (Brailowsky, et al. 1989). Chronic bilateral GABA infusions at the reticular level have confirmed that this structure is involved in seizure mechanisms since their anticonvulsant effect against seizures is stronger than against PDs (Silva-Barrat, et al. 1988b).

Photosensitive epilepsy of baboons is characterized by ILS-induced spike-waves and seizure discharges which start bilaterally and synchronously in both fronto-rolandic areas. These bilateral PDs are easily induced when stimulation is applied to both eyes and they can also occur when only one eye is stimulated. Since PDs are generated in the motor cortex, their constantly observed bilateralization and synchronization can only depend on bifrontal interhemispheric connections, either directly through the corpus callosum or indirectly through a brain stem structure such as the brain stem reticular formation which is reciprocally interconnected with large cortical territories.

This chapter is devoted to review a series of data which altogether demonstrate that the corpus callosum is the major, if not unique structure responsible for the bilateral synchronization of spike-waves as well as of seizures, with the exception of the tonic phase of seizures.

EFFECTS OF CALLOSOTOMY

In non-operated baboons, PDs and seizures always occur symmetrically and synchronously on both hemispheres, even when they are induced by monocular ILS (frame 1 in Fig. 1). Sectioning the corpus callosum decreases the synchronization of ILS-induced seizures in the two hemispheres. In an initial experimental series, callosotomy was performed under visual control and callosotomized baboons were subsequently submitted to fullfield binocular or monocular stimulation (Catier et al. 1970, Naquet et al. 1972). In this situation, a given baboon submitted to ILS could show seizure discharges which one day started and terminated in the two hemispheres simultaneously, the day after evolved unilaterally, and another day predominated in one hemisphere and had independent termination in the two hemispheres. These variations occurred in spite of the fact that visual stimulation was applied to both eyes and that visual afferents activated the two hemispheres symmetrically. When they occurred, lateralized discharges were accompanied by a convulsion of the contralateral hemibody, and followed by an electrical silence in the predominantly discharging hemisphere, with a transitory paresis of the contralateral hemibody. Similar observations could be made during binocular or monocular stimulation (frame 2 in Fig. 1).

In a more recent experimental series, callosotomy was performed by using a stereotactic approach (see Naquet and Wada 1992), and operated baboons were subsequently submitted to discrete visual stimulation in order to see whether this stimulation could induce localized or generalized seizures (Fukuda et al. 1988, 1989). It was possible to apply focal photic stimulation precisely limited to the fovea and to one retinal hemifield of one eye in baboons which were previously and temporarily paralyzed and artificially ventilated. Given that paralysis markedly diminishes the natural photosensibility of baboons (Naquet and Menini 1972), it was necessary to systemically administer allyglycine, a drug which blocks GABA synthesis and permits the appearance of ILS-induced PDs even in paralyzed animals (Menini et al. 1977). The necessary paralysis of the animals made it impossible to establish electro-clinical correlations. Due to the anatomical disposition of visual pathways, the discrete stimulation of the nasal field of the retina of one eye activates mainly visual fibers directed to the contralateral occipital lobe. In callosotomized baboons, stimulation of the nasal retinal field of one eye induced EEG seizure discharges which appeared and always predominated in the hemisphere contralateral to the stimulated eye, i.e., in the hemisphere receiving activated visual afferents. Electrical silence followed the seizure discharge only in this hemisphere. After such a seizure, similar stimulation of the same eye was unable to induce another seizure for at least one hour, while, for stimulation of the other eye, the delay was only 20 minutes (Fukuda et al. 1988). These data demonstrate that, in the absence of the corpus callosum, the excitability of each hemisphere is independent one from the other, which is never the case in animals without lesions.

EFFECTS OF CALLOSOTOMY PLUS MOTOR CORTICAL LESION

In the initial experiments reported above, section of the corpus callosum was sometimes associated with a significant lesion of the motor cortex on one side, as a consequence of the visual approach for the surgery (Catier et al. 1970, Naquet et al. 1972). In these cases, ILS-induced epileptic EEG manifestations always occurred in the non-lesioned "healthy" hemisphere, with clinical signs localized in the opposite hemibody. The seizures were followed by an electrical depression involving the same hemisphere and also by a transient paresis of the contralateral hemibody. This observation confirms that the motor cortex is important for the generation of seizures since a single hemisphere, while receiving activated visual afferents, is unable to develop PDs and seizures when the motor area is destroyed.

EFFECTS OF CALLOSOTOMY PLUS RETINA COAGULATION

In order to better assess the role of the corpus callosum in the bilateral synchronization of ILS-induced PDs and seizures, corpus callosum sections were performed in animals submitted to retinal lesion (Fukuda et al. 1989). In these animals, the temporal half of the macula and retina of one eye was lasered. After this operation, monocular non-focused stimulation of the operated eye activated only the nasal retina hemifield and visual afferents reaching the contralateral visual cortex even if the animals were moving, which procedure allowed electroclinical correlations during seizures. In non-callosotomized baboons, monocular stimulation of the operated eye always induced bilaterally synchronous PDs and bilaterally synchronous seizures (frame 3 in Fig. 1). In split brain baboons, the same stimulation induced strictly unilateral PDs associated with myoclonus of the contralateral hemibody. ILS-induced electrographic seizures occurred uniquely in the hemisphere receiving

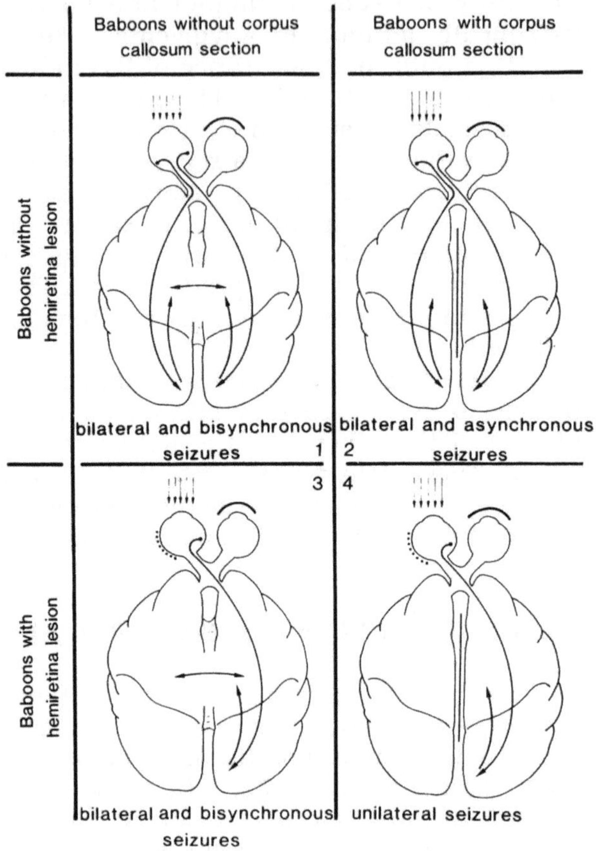

Figure 1. Schematic representation of the effects of monocular intermittent light stimulation (ILS) in four baboon preparations:

1. In non-operated photosensitive animals, monocular ILS activates visual afferents for both hemispheres which, in their turn, activate the generators of PDs located in both motor cortical areas. Interhemispheric connections are functional; in these conditions, PDs and seizures occur bilaterally and bisynchronously.
2. In baboons with corpus callosum section, both hemispheres are activated by visual afferents, but interhemispheric connections are not functional; in these conditions, PDs and seizures occur bilaterally but asynchronously.
3. In baboons with destruction of the temporal retinal hemifield of one eye, monocular stimulation of the operated eye activates visual afferents for only the contralateral hemisphere; however, since interhemispheric connections are functional, PDs and seizures occur bilaterally and bisynchronously.
4. In baboons with the combination of corpus callosum section and temporal retinal destruction, monocular stimulation of the operated eye activates visual afferents for only the contralateral hemisphere; since interhemispheric connections are not functional, PDs and seizures develop unilaterally and are restricted to the hemisphere receiving activated visual afferents; in such a preparation, stimulation of the non-operated eye produces effects similar to Case 2.

visual afferents (frame 4 in Fig. 1). In that case, the tonic phase of the seizure was associated with a bilateral tonic spasm; the clonic phase of the seizure concerned only the hemibody contralateral to the discharging hemisphere; the postictal silent period was also restricted to the discharging hemisphere with a paralysis of the contralateral hemibody, while normal EEG activity was observed in the other hemisphere. Immediately after such a unilateral seizure, photic stimulation of the opposite, non-operated eye, while activating visual afferents for both hemispheres, might induce a seizure in the contralateral hemisphere, even during the postictal silent period of the preceding seizure.

ROLE OF THE OCCIPITAL CORTEX

In photosensitive baboons, the occipital cortex is not a generator of PDs or seizures since ILS-induced occipital spike-waves have never been observed in the absence of fronto-rolandic PDs, except when animals were injected with a convulsant dose of allylglycine. Moreover, intracortical single unit recordings in the occipital cortex (Menini, et al. 1981, Silva-Barrat, et al. 1986) have shown that the occipital neuronal activity is correlated with visual stimuli and not with frontal PDs, and the same is true when a generalized, tonic-clonic seizure occurs.

Conversely, numerous experimental data demonstrate that the occipital cortex plays a determinant role as a relay for ILS-activated visual afferents triggering the fronto-rolandic PDs and the seizures (Menini, et al. 1989), since these afferents reach primarily the specific visual cortex. Bilateral ablation of the occipital lobes makes the ILS-induced electrographic and clinical epileptic manifestations disappear completely and definitively (Wada et al. 1973). Similarly, depression of the occipital activity by bilateral chronic infusion of GABA (Brailowsky et al. 1987, 1988) blocks ILS-induced epileptic manifestations as efficiently as did GABA infusions into the motor cortex. A supplementary evidence for the participation of visual cortical afferents in triggering epileptic manifestations in baboons comes from the comparison of fullfield and hemifield monocular visual stimulations in callosotomized and non-callosotomized baboons, as reviewed in the present paper and schematized in Figure 1. Since no long and direct occipito-frontal connections exist in primates, the ILS-activated afferents reach the motor cortex indirectly and activate the neuronal generators of PDs which are located there. A possible participation of some subcortical visual relay structures such as the superior colliculus (Wada et al. 1972) or the pulvinar can be eliminated since their unilateral or bilateral destruction had only transitory effects on the photosensitive reactivity of baboons and did not influence the bilateral expression of epileptic manifestations.

ROLE OF THE RETICULAR FORMATION

Very soon after the discovery of photosensitive epilepsy of baboons, the involvement of the reticular formation in seizure mechanisms was hypothesized, since lowering in state of vigilance increases the photosensitive reactivity (Balzamo 1968). Moreover, seizure onset is preceded and accompanied by vegetative manifestations such as blood pressure increase and regional blood volume changes (Ancri et al. 1981). These blood volume changes are characterized by a clear reduction in some peripheral (nasal and hepatic) regions and by a substantial increase in the brain, indicating the early involvement of the cerebrovascular centers and of the brain stem reticular formation where these centers are situated.

A multiunitary analysis performed during the initiation of grand mal seizures (Silva-Barrat et al. 1986) showed that, at the beginning of the EEG tonic phase of the seizure, most of the deep structures are strongly activated, the reticular pontine and mesencephalic formations showing the highest activation level. This activation occurs a few seconds before the seizure onset and correlates with the aforementioned vegetative modifications. Concomitantly with the tonic phase of the seizure, the EEG cortical spike-waves disappear, the frontal neuronal bursting activity becomes disorganized, and the extracellular calcium concentration in the fronto-rolandic cortex is strongly reduced (Pumain et al., 1985). From these data, the strong activation of brain stem reticular structures was interpreted as resulting from a loss of the control normally exerted by the cortex on the deep structures.

Chronic local GABA infusions were performed into the brain stem reticular formation (magnocellular nucleus) in order to temporarily depress the reticular activity and observe the effects on photosensitive reactivity (Silva-Barrat et al. 1988b). Even though they were not performed in callosotomized baboons, these experiments pointed to the secondary involvement of reticular mechanisms in photosensitive seizures, since local GABA infusions into the reticular structures had anticonvulsant effects against seizures significantly stronger than against cortical PDs: GABA infusions partially blocked the appearance of seizures while they only diminished but did not suppress the frequency of cortical PDs. Furthermore, the cortical PDs and the few seizures which were induced by ILS during intrareticular GABA infusion still appeared bilaterally and synchronously, thus indicating that their bilateralization is not influenced by the level of reticular activity.

These data:

- demonstrate the involvement of the brain stem reticular formation in seizure mechanisms;
- are compatible with other data such as vascular changes observed during seizures;
- are compatible with the observation of a bilateral tonic spasm during the hemispheric seizures induced in callosotomized baboons;
- strongly suggest that the triggering of a seizure, while involving reticular mechanisms, is dependent on cortical paroxysmal phenomena;
- suggest also that the bilateralization of seizures does not depend on the level of reticular activity.

CONCLUSIONS

In non-callosotomized baboons, ILS always induces bilateral and synchronous PDs and seizures, whatever the visual stimulation mode (binocular, fullfield monocular, hemifield monocular).

In callosotomized baboons, the bilateralization and synchronization of PDs and seizures varies depending on the stimulation mode utilized:

- When baboons are submitted to binocular or fullfield monocular stimulations, i.e., stimulation modes activating visual afferents directed to both cerebral hemispheres, they can present unilateral as well as bilateral seizures, these bilateral seizures being sometimes asynchronous and sometimes synchronous in both hemispheres;
- When baboons are submitted to hemifield monocular stimulation, i.e., a stimulation mode activating visual afferents directed to only one cerebral hemisphere, they present only unilateral PDs and seizures limited to this hemisphere.

The activation of the brain stem reticular formation which is a consequence of both the photic stimulation and the appearance of cortical spike-waves, seems responsible for the bilateral tonic spasm which accompanies bilateral as well as unilateral seizures, and also probably for the vegetative manifestations which occur during seizures. Otherwise, the bilateralization of cortical epileptic manifestations is not influenced by changes in brain stem reticular activity: A depression of this activity, as caused by bilateral chronic GABA infusion, while blocking the seizure appearance and diminishing the cortical PD frequency, does not prevent the appearance of bilateral and synchronous spike-waves.

All these data and considerations constitute an evidence that interhemispheric synchronization of PDs and seizures depends mainly on transcallosal interconnections rather than on brain stem structures. It is suggested that callosotomy could be considered a possible surgical therapy against some convulsive generalized seizures.

REFERENCES

Ancri, D., Naquet, R., Menini, Ch., Meldrum B.S., Stutzman, J.M., and Basset, J.Y., 1981, Cerebral and extra-cerebral blood volume in generalized seizures in the baboon Papio papio, *Electroenceph. Clin. Neurophysiol.* 51:91-103.

Balzamo, E., 1968, Etude polygraphique du sommeil nocturne du Papio papio, babouin du Sénégal., *Thèse*, Marseille.

Brailowsky, S., Menini, Ch., Silva-Barrat, C., and Naquet, R., 1987, Epileptogenic gamma-aminobutyric acid-withdrawal syndrome after chronic, intracortical infusion in baboons, *Neurosci. Lett.* 74:75-80.

Brailowsky, S., Silva-Barrat, C., Menini, Ch., Riche, D., and Naquet, R., 1989, Effects of localized, chronic GABA infusions into different cortical areas of the photo-sensitive baboon, Papio papio, *Electroenceph. Clin. Neurophysiol.* 72:147-156.

Catier, J., Choux, M., Cordeau, J.P., Dimov, S., Riche, D., Everhard, A., et Naquet, R., 1970, Résultats préliminaires des effets électrographiques de la section du corps calleux chez le Papio papio photosensible, *Rev. Neurol.* (Paris), 122:521-522.

Fischer-Williams, M., Poncet, M., Riche, D., and Naquet, R., 1968, Light-induced epilepsy in the baboon Papio papio: cortical and depth recordings, *Electroenceph. Clin. Neurophysiol.* 25:557-569.

Fukuda, H., Valin, A., Bryere, P., Riche, D., Wada, J.A., and Naquet, R., 1988, Role of forebrain commissure and hemispheric independence in photosensitive response of epileptic baboon Papio papio, *Electroenceph. Clin. Neurophysiol.* 69:363-370.

Fukuda, H., Valin, A., Menini, Ch., Boscher, C., De La Sayette, V., Riche, D., Kunimoto, M., Wada, J.A., and Naquet, R., 1989, Effect of macular and peripheral retina coagulation on photosensitive epilepsy in the forebrain bisected baboon Papio papio, *Epilepsia* 30:623-630.

Gastaut, H., 1973, "Dictionnaire de l'Epilepsie," OMS, Geneve, 80 p.

Menini, Ch., Brailowsky, S., Silva-Barrat, C., Riche D., and Naquet, R., 1992a, Cortical GABAergic control of epilepsy in photosensitive baboons, *in:* "Frontal Lobe Seizures and Epilepsies," P. Chauvel, A.V. Delgado-Escueta, E. Halgren, and J. Bancaud, eds., Advances in Neurology, Raven Press, New York, 57:571-578. (Revue).

Menini, Ch. and Silva-Barrat, C., 1989, Role of visual afferents in the photosensitive epilepsy of baboons Papio papio, *in:* "Reflex Seizures and Reflex Epilepsies," A. Beaumanoir, H. Gastaut, and R. Naquet eds., Ed. Médecine & Hygiène, Genève, pp. 39-48.

Menini, Ch., Silva-Barrat, C., and Naquet, R., 1992b, The Epileptic baboon, Hypothetical neuronal network and action of anticonvulsant drugs, *in:* "Drugs for Control of Epilepsy; Actions on Neuronal Networks Involved in Seizure Disorders," C.L. Faingold and G.H. Fromm, eds., CRC Press (Boca Raton), pp. 325-347.

Menini, Ch., Silva-Comte, C., Stutzmann, J.M., and Dimov, S., 1981, Cortical unit discharges during photic intermittent stimulation in the Papio papio. Relationships with paroxysmal fronto-rolandic activity, *Electroenceph. Clin. Neurophysiol.* 52:42-49.

Menini, Ch., Stutzmann, J.M., Laurent, H., et Valin, A., 1977, Les crises induites -ou non- par la stimulation lumineuse intermittente chez le Papio papio après injection d'allylglycine, *Rev. EEG Neurophysiol.* 7:232-238.

Naquet, R. et Menini, Ch., 1972 Suppl., La photosensibilité et pharmacologique de ses mécanismes, *Electroenceph. Clin. Neurophysiol.* 31:13-26.

Naquet, R., Menini, Ch., and Catier, J., 1972, Photically induced epilepsy in Papio papio. The initiation of discharges and the role of the frontal cortex and of the corpus callosum, *in:* "Synchronization of the EEG in the Epilepsies," M.A.B. Brazier and H. Petsche, eds., Springer Verlag, Vienna pp. 347-367.

Naquet, R. and Wada, J., 1992, Role of the corpus callosum in photosensitive seizures of epileptic baboon Papio papio, *Adv. Neurol.* 57:579-587.

Pumain, R., Menini, Ch., Heinemann, U., Louvel, Jr., and Silva-Barrat, C., 1985, Chemical synaptic transmission is not necessary for epileptic seizures to persist in baboon Papio papio, *Exp. Neurol.* 89:250-258.

Silva-Barrat, C., Brainlowsky, S., Levesque, G., and Menini, Ch., 1988a, Epileptic discharges induced by intermittent light stimulation in photosensitive baboons: a current source density study, *Epilepsy Research* 2:1-8.

Silva-Barrat, C., Brainlowsky, S., Riche, D., and Menini, Ch., 1988b, Anticonvulsant effects of localized chronic infusions of GABA in cortical and reticular structures of baboons, *Exp. Neurol.* 101:418-427.

Silva-Barrat, C. and Menini, Ch., 1990, Photosensitive epilepsy of the baboon: a generalized epilepsy with a motor cortical origin, *in:* "Generalized Epilepsy; Neurobiological Approaches," M. Avoli, P. Gloor, G. Kostopoulos, and R. Naquet, eds., Birkhäuser, Boston, pp. 286-297.

Silva-Barrat, C., Menini, Ch., Bryère, P., and Naquet, R., 1986, Multiunitary activity analysis of cortical and subcortical structures in paroxysmal discharges and grand mal seizures in photosensitive baboons, *Electroenceph. Clin. Neurophysiol.* 64:455-468.

Wada, J.A., Catier, J., Charmasson, G., Menini, Ch., and Naquet, R., 1973, Further examination of neural mechanisms underlying photosensitivity in the epileptic senegalese baboon Papio papio, *Electroenceph. Clin. Neurophysiol.* 34:786.

Wada, J.A., Terao, A., and Booker, H.E., 1972, Longitudinal correlative analysis of epileptic baboon Papio papio, *Neurology* (Minnea.) 22:1272-1285.

III

CLINICAL EPILEPSY

SOME HISTORICAL ASPECTS OF CALLOSOTOMY FOR EPILEPSY

Joseph E. Bogen, M.D.

Clinical Professor of Neurological Surgery
University of Southern California
Los Angeles, CA 90033

INTRODUCTION

The large size, central location and widespread connections of the corpus callosum stimulated, for centuries, investigations which were motivated as much by scientific curiosity as by therapeutic considerations. Callosal physiology has more recently been important to surgeons concerned primarily with other structures, including those neighboring the third ventricle, which can be approached through the corpus callosum. But the principal medical interest has been in the role of the corpus callosum in the generalization of seizures. At the end of the 19th century and on two other, separate occasions in the 20th century, callosotomy was promoted as a treatment for seizure disorders. Before 1900, less obviously around 1940, and most clearly in the 1960's, these therapeutic proposals were stimulated by animal experimentation. One theme of this chapter is the reciprocal interaction of surgical therapy and laboratory experimentation: in particular, the most recent therapeutic use of callosotomy has been accompanied by widespread physiologic and psychologic interest in this conspicuous brain structure.

In this chapter we consider first a brief history of studies of the corpus callosum. Then follows a chronological account of interest in callosotomy as a treatment for epilepsy. I will conclude with a few personal observations.

STUDIES OF CALLOSAL FUNCTION

Studies of the corpus callosum were first undertaken by the Humoral Anatomists. These were the writers of antiquity whose concepts of brain function emphasized the contents of the brain cavities and the flow of various fluids such as air, phlegm, CSF, blood, etc. For them, the corpus

Epilepsy and the Corpus Callosum II, Edited by A.G. Reeves
and D.W. Roberts, Plenum Press, New York, 1995

callosum seemed largely a supporting structure. This view persisted for a millennium. Even that original Renaissance genius, Vesalius, believed that the corpus callosum served mainly as a mechanical support, maintaining the integrity of the various cavities. In 1543 he wrote:

> *There is a part [whose] external surface is gleaming white and harder than the substance on the remaining surface of the brain. It was for this reason that the ancient Greeks called this part "tyloeides" ["callosus" in Latin] and, following their example, in my discourse I have always referred to this part as the corpus callosum. If you look at the right and left brain . . . and also if you compare the front and rear, the corpus callosum is observed to be in the middle of the brain; . . . Indeed, it relates the right side of the cerebrum to the left; then it produces and supports the septum of the right and left ventricles; finally, through that septum it supports and props the [fornix] so that it may not collapse and, to the great detriment of all the functions of the cerebrum, crush the cavity common to the two [lateral] ventricles of the cerebrum.* (Clarke and O'Malley 1968, p. 597.)

In the 17th century the "traffic anatomists" took a major step forward. It was at about the time of Willis (1664) that anatomists began thinking more in terms of a traffic or communication between the more solid parts of the brain. This view became quite explicit in the statement of Viq d'Azyr who wrote in 1784:

> *It seems to me that the commissures are intended to establish sympathetic communications between different parts of the brain, just as the nerves do between different organs and the brain itself. . . .* (Clarke and O'Malley 1968, p. 592.)

For over two centuries, beliefs about callosal function consisted almost solely of inferences from its central location, widespread connections, and large size (larger than all those descending and ascending tracts, taken together, which connect the cerebrum with the outside world). Willis, de la Peyronie, and Lancisi, among others, thought the corpus callosum a likely candidate for "the seat of the soul;" or they used some other expression intended to cover that highest or ultimate liaison which brings coherent, vital unity to a complex assemblage.

The observations of the early anatomists have been often supported by subsequent anatomic observations, including the large number of callosal fibers, at least 600 million of them.[1] Because these fibers interconnect so much of the cerebral cortex, especially that cortex considered associative, it has often been suggested that they serve some of the "highest," most educable, and characteristically human functions of the cerebrum (Bremer 1958).

Inference of function from observable structure is time-honored and productive. However, such inference has its limitations. The physiological evidence has only partially sustained anatomic inference. We now

[1]The $2x10^8$ estimate of Tomasch (1954) was based on light microscopy. According to G. Innocenti (personal communication in 1991), electron microscopy will at least triple the old estimate.

know from various observations (notably the split-brain) that the corpus callosum is indeed an important integrative structure; we also know it is neither sufficient nor indispensable, providing only one of a number of integrative mechanisms.

That the corpus callosum is not the exclusive "seat of the soul" is evident from the apparent normality in social situations of patients who have had complete callosotomies. That it is an important integrating mechanism is clear from the peculiarities of such patients. These include, among other things, a unilateral tactile *anomia*, a left *hemialexia*, and a unilateral *apraxia*. That is, for the right-hander with complete callosotomy, there is an inability to name aloud objects felt with the left hand, an inability to read aloud written material presented solely to the left half-field of vision, and an inability to execute with the left hand actions verbally named or described by the examiner. The apraxia usually recedes in a few months, whereas the hemialexia and unilateral anomia persist for years.

Such deficits are now readily demonstrable in individuals who have had surgical section of the corpus callosum. But these deficits were first recognized in patients with vascular disease causing very complex and evolving syndromes.

In the closing decades of the 19th century (or more broadly construed, in the period between the American Civil War and the First World War) there emerged that group of neurologists whose discoveries and formulations are still at the core of current clinical knowledge. Among them were several, including Wernicke, Liepmann, Dejerine, and Goldstein, who interpreted various neurological symptoms as resulting from disconnection, including interruption of information flow through the corpus callosum.

The concept of apraxia was developed by Liepmann expressly to describe a patient who could carry out commands with one of his hands but not with the other. Liepmann and Maas (1908) described a right-hander whose callosal lesion caused a left apraxia as well as a left-hand *agraphia* -- an inability to write -- in the absence of aphasia. These disabilities have subsequently been observed many times. Unilateral apraxia and the unilateral agraphia are not always present, and they may subside when a stroke victim progressively recovers, but they remain among the cardinal signs of callosal interruption.

Liepmann considered the corpus callosum instrumental in most left-hand responses to verbal command: the verbal instruction was comprehended only by the left hemisphere and the left hand followed instructions which were delivered not by a directly descending pathway which we now call "ipsilateral control," but by a route involving callosal interhemispheric transfer from left to right and then by way of what we now call "contralateral control," that is, by right hemisphere control of the left hand. Necessarily then, callosal interruption would result in an inability to follow verbal commands with the left hand although there would be no loss of comprehension (as expected from a left hemisphere lesion). And there would be no weakness or incoordination of the left hand (as could result from a right hemisphere lesion). This view was largely ignored or rejected (particularly in the English-speaking countries) for nearly half a century, although it is now thought to be essentially correct [reasons for this rejection are considered in detail in Bogen 1993].

Correspondingly, we now recognize the notion of spatial or pictorial instructions understood by the right hemisphere and requiring callosally mediated interhemispheric communication for correct right-hand execution. This right-to-left aspect of callosal function was not part of Liepmann's original callosal concept although, in retrospect, it seems a natural corollary. We also understand that most right hemispheres have some verbal comprehension, hence may respond, without callosal participation, with either left or right hand movements (Bogen 1993).

Familiarity with the symptoms following callosal section is essential for surgeons who employ this procedure, whether to treat epilepsy or to gain access to other structures. Partial section produces fractions of the full callosal syndrome, depending upon the part sectioned; in particular, most of the well recognized callosal symptoms can be avoided by sparing the splenium (Bogen 1987).

CORPUS CALLOSUM AND EPILEPSY

The famous experiments of Fritsch and Hitzig (1870) showed that there was a limited region of the cerebral cortex (the "motor cortex") whose electrical stimulation resulted in movements of the contralateral limbs. In two of their dogs, removal of the electrically excitable cortex resulted in subsequent incoordination of the contralateral limbs. In another two dogs, tetanization caused "epileptic attacks" beginning first in the contralateral limbs and then becoming generalized. The generalization of the "epileptic attacks" was interpretable, in retrospect, as due to transmission across the corpus callosum.

A next step was taken by Bubnoff and Heidenhain (1881). They stimulated the white matter exposed by ablation of the motor cortex. This stimulation could produce convulsive movements in the unparalyzed ipsilateral limbs. They concluded that excitation had spread across the corpus callosum to involve the motor cortex of the opposite hemisphere. A few years later, Horsley (1886) found that cutting the corpus callosum could prevent the spread of seizures.

In succeeding decades, the possible role of the corpus callosum in seizure spread was studied in experimental animals by many investigators. When Spiegel (1931) reviewed the physiology of epilepsy, he emphasized the common finding that generalized epileptic convulsions could occur after the corpus callosum had been severed. His own experiments included section of all crossing fibers down to the rhombencephalon and "even after this operation, we could observe that general clonic convulsions developed following one-sided cortical stimulation." He also described experiments with a sagittal section of the rhombencephalon in the midline, again with generalized convulsions being possible from stimulation of one hemisphere. Evidently generalization can occur through fibers which cross the midline at several levels.

The multiplicity of routes for seizure spread rendered unattractive for decades the idea of cutting the corpus callosum for epilepsy, particularly in view of the technical difficulties which this would involve. The feasibility of cutting the corpus callosum was urged by Dandy (1936) who used this approach to reach midline tumors. But neither he nor others of the time recommended callosotomy for epilepsy.

The first callosotomies for epilepsy were reported in Volume 44 of the *Archives of Neurology and Psychiatry* by Van Wagenen and Herren (1940); their first case was operated on 6 February 1939. The rationale they gave for this procedure was their observation of two cases in which callosal tumor had lessened seizure frequency and two cases of vascular insult which stopped the seizures altogether. The paper contained no references, hence it is not clear if the authors knew of Speigel's report or the material which he summarized. Nor is it clear if they were aware of the work of Erickson. Erickson (1940) had done monkey experiments showing a role of the corpus callosum in seizure spread, his paper having appeared in the preceding volume, Volume 43 of the *Archives*. One might assume that Erickson's animal experiments at the Montreal Neurological Institute were known to Van Wagenen since the neuroscience community was quite small in those days. Moreover, both attended meetings of the American Neurologic Association.

However, Robert Joynt (1985), who subsequently arrived at the University of Rochester, has emphasized that the Van Wagenen series was undertaken "solely on clinical observations [which were personally made] by the two authors." Joynt gave special attention to the summary of the Van Wagenen-Herren paper. Parts of the summary are well worth repeating here because their conclusions have remained largely correct after more than fifty years.

Van Wagenen and Herren summarized their first ten experiences with callosotomy as follows:

> *The impression is gained from this small number of observations that the type of case in which section of commissural fibers in the corpus callosum is most favorable is the one in which a large cortical or subcortical scar exists, as in cases 1 and 10. Whether section of various commissural pathways to prevent the spread of an epileptic wave is indicated for patients having multiple irritable foci is a matter for future study . . . the observation on patient's having jacksonian seizures on the right side after section of the corpus callosum on one occasion and on the left side on another suggests that there are at least bilateral foci from which seizures may originate (case 7).*
>
> *The inhibitory effect of the cortex of one hemisphere on the activity of the other must also be considered seriously . . . it may be that in certain instances the cortex of one hemisphere may inhibit abnormal activity of an abnormal zone and that section of commissural pathways is contraindicated.*
>
> *Section of the commissural pathways contained in the corpus callosum may be carried out without any untoward effect on the patient. Such a section may serve to limit the spread of an epileptic wave to the opposite hemisphere. When such limitation occurs, the patients do not seem to lose consciousness or have generalized convulsions.*

Although Van Wagenen and Herren seemed happy with their results, no one else took up the operation. This may have been in part because of the onset and continuation for five years of World War II, or, long-term outcomes may have been unfavorable and, although unpublished, well known (Smith 1991). Another factor was the growing conviction, reaching a peak in the 1950s, that the reticular formation and its rostral projections were of particular importance in seizure spread. As Penfield and Jasper (1954) stated in one of the great classics of epileptology:

> *It seems reasonable to assume, therefore, that generalization of the motor seizure does not take place by spread of excitation through cortical circuits. It must spread through the more closely interrelated neuronal network of the higher brain stem, in a contrencephalic system with symmetrical functional relationships to both sides of the body* (Penfield and Jasper 1954).

The irrelevance of callosal transmission of seizure activity beginning unilaterally, was suggested by the occasional recurrence of generalized convulsions in humans with hemispherectomy (White 1961). Indeed, in a few experiments generalized convulsions could be produced in dogs subsequent to bilateral decortication (van Harreveld 1947).

In spite of the foregoing, a number of other considerations discussed in the next section, led us to expect improvement, in properly chosen cases, of seizure disorders otherwise untreatable. Briefly, our reintroduction in 1962 of callosotomy, together with anterior and hippocampal commissurotomy, as a treatment for medically intractable epilepsy was stimulated, indeed made possible, by animal experimentation involving similar procedures in cats and monkeys (for example, Myers and Sperry 1953; Sperry 1961).

Because of the remarkable improvement of both our first patient (Bogen and Vogel 1962) and our second patient (Bogen et al. 1965) we continued to offer the operation. A few years later we briefly reported our rewarding results in nine (of ten total) cases each having at least two years follow-up, concluding:

> *It thus appears that the combination of cerebral commissurotomy plus postoperative medication has limited propagation of seizure activity from a cortical focus* (Bogen, Sperry, and Vogel 1969).

The improved status of these patients made possible their participation in a long and still continuing series of neuropsychological investigations (Sperry, Gazzaniga and Bogen 1969; Sperry 1974; Bogen 1993; Zaidel 1994).

Our reports were followed by a few others (Luessenhop et al. 1970). However, wider acceptance of this procedure was only made possible by the sustained effort during the decade of the 1970s by Donald Wilson and co-workers, particularly Alex Reeves, at Dartmouth. By 1982, at the Dartmouth conference organized by Wilson and Reeves, callosotomy for epilepsy was reported from six more clinical centers. There had also been a burgeoning of experimental studies, many of them presented at that conference. Both the clinical and experimental contributions were subsequently included in a book edited by Reeves (1985).

During the 1980s, callosotomy, either complete or (more often) partial, became an established procedure. In the words of Spencer et al. (1987):

Over the past decade, corpus callosum section has become a widely accepted, relatively safe, clearly needed, broadly practiced, and continually evolving addition to the medical treatment of certain types of severe and uncontrolled seizures in certain patients who are not candidates for resective procedures.

Over the time span of about thirty years, during which callosotomy came to be more favorably viewed as a treatment for epilepsy, increased familiarity with both the anatomy and with the physiology of the corpus callosum has encouraged the use of partial callosotomy as an approach to the third ventricle and other midline structures (Apuzzo 1987).

SOME PERSONAL COMMENTS

In 1959, I wrote an essay entitled, "A Rationale for Splitting the Brain in Cases of Epilepsy." I took it to Roger Sperry whom I had known for several years at Cal Tech and who was, I felt, the neuroscientist of our time. He offered two suggestions: "Change the title" and "Look up those papers by Akelaitis." My subsequent review of the Van Wagenen-Akelaitis series is summarized in the appendices. Meanwhile, there was also the question of who would be a suitable candidate for such a procedure.

I first met Bill Jenkins in the summer of 1960 when he was brought to the emergency room in status epilepticus; I was the neurology resident then on call. The heterogeneity as well as the intractability and severity of his multicentric seizure disorder became clearer to me over the next months. Both in the clinic and in the hospital I witnessed psychomotor spells, sudden tonic falls and unilateral jerking as well as generalized convulsions. In late 1960, I wrote to Maitland Baldwin, then Chief of Neurosurgery at the National Institutes of Health in Bethesda. A few months later, Bill was admitted to the NIH epilepsy service where he spent six weeks. he was sent home in the spring of 1961, having been informed that there was no treatment, standard or innovative, then available for his problem.

Bill and his wife, Fern, were then told of Van Wagenen's partial results, mainly with partial sections of the corpus callosum. I suggested that a complete section might help. Their enthusiasm encouraged me to approach Prof. Vogel, knowing of his experience with removal of callosal AVM's. He suggested that we practice a half-dozen times in the morgue. By the end of the summer (during which I was again on the neurosurgery service) the procedure seemed reasonably in hand. There was further delay during which Bill underwent testing in Sperry's laboratory, mainly with Michael Gazzaniga who had arrived in September as a beginning graduate student in psychobiology. During this delay we also had an opportunity to keep a reasonably complete record of his many seizures (Bogen 1985).

It was during this period of preoperative testing that Bill said, "You know, even if it doesn't help my seizures, if you learn something it will be more worthwhile than anything I've been able to do for years." He was operated upon in February1962.

It seems to me, in retrospect, that if there had been a research committee at our hospital whose multi-member approval was required, the procedure would never have been done. At that time, a department chairman could decide on his own, similar, I expect, to the situation at Rochester in the late 30s.

From the start, our procedure included not only complete callosotomy (requiring two skull openings) but also section of the anterior commissure, accessed in most cases by entering the third ventricle through its roof. We chose to do as complete a section as possible for two reasons: 1) monkeys so operated were without neurological disability and participated well in demanding psychological testing (Sperry 1961), and 2) if a complete neocommissural section failed in the ideal case (an intelligent, personable individual with a supportive family whose multicentric seizure disorder could hardly have been much worse) then we would be through. Fortunately, it succeeded.

The completeness of our procedures (Bogen et al. 1988) was attained without the operating microscope (which I first used in 1970), the good light the scope provides, the bipolar cautery, osmotic diuretics, modern neuroanesthesia, and a variety of instruments only subsequently available. It is a tribute to Vogel's operative skills, including his sense of tolerable retraction and his remarkable vision at the usual operating distances. (He never did take up the microscope.) And how impressive it is that Van Wagenen worked under even less auspicious conditions!

Our next major step was to do a callosotomy (and anterior commissurotomy) which spared the splenium, whose section we understood by then to be the main source of disconnection symptoms. We operated upon two patients whose seizures caused life-threatening psychomotor behavior and whose bitemporal foci appeared to be independent. Their seizure disorders were markedly improved (one subsequently obtained a steady job for the first time) and they had no discernible disconnection symptoms (Gordon et al. 1971). In the words of Wada (1987), this report, ". . . revitalized our interest in re-examining brain bisection as a possible new treatment modality . . . "

By now, sparing both splenium and anterior commissure has become commonplace, particularly because a section restricted to the anterior two-thirds to three-fourths of the corpus callosum can alleviate drop attacks; and drop attacks are, in the opinion of many, the prime indication for callosotomy. That drop attacks could be eliminated by callosotomy never occurred to us, even by 1974 when we summarized our criteria for operation (Bogen and Vogel 1975). This was in spite of the fact that for Bill Jenkins, the commissurotomy eliminated his drop attacks as well as (except for two occasions in ten years) his generalized convulsions. We were still influenced to some extent by the concept of "centrencephalic seizures," our theoretical views preventing us from recognizing a fact in front of our eyes.

The idea that extremely rapid generalization of seizures required a centrencephalon weighed even more heavily with others than with us, and was probably responsible in part for the disbelief with which our reports were received. In addition, our work was done at a medical school (Loma Linda University) better known in those days for training medical missionaries than for scientific advances. Not only was the procedure at odds with a well-known theory, it was worse! Had not this approach already been tried and failed? When I wrote to Frank Smith, the Chief of Neurosurgery at Rochester, asking for as much information as he could provide, his reply was quite short, including that "Dr. Van Wagenen always was sorry about what he did to those patients." For over a decade there were persons in Boston who referred to us as "the West Coast butchers." Without the excellent work of Wilson and Reeves, it is quite possible that our efforts (as well as of others, like Wada) would not have been widely accepted.

History teaches us much. Among other lessons, we know that no matter how useful a therapeutic technique, the odds are high that it will eventually be outmoded. Meanwhile, however, callosotomy has illustrated to notable degree the interplay among social, scientific and clinical concerns.

SUMMARY OF APPENDICES I - III

The original attempt to control medically intractable epilepsy by commissural section was undertaken by W. P. Van Wagenen; his first operation was on 11 February 1939. It was incumbent upon us, when we reintroduced this approach in 1962, to be as familiar as possible with his results.

When one undertakes an innovation, familiarity with what has gone before is essential to avoid so far as possible any waste of time and energy and, in the case of medical innovation, unnecessary risk to the patients. However, a totally thorough review of the literature while clearly desirable, is often not available at the time that, in Van Wagenen's words "an adequate and justifiable opportunity" presents itself.

In Appendix I are listed all of the two dozen reports of which I am now aware, of the Van Wagenen patients. At the time of our first operation (on WJ on 13 February 1962) we had reviewed less than half of these: they are (in Appendix I) A, B, C, 1, 2, 3, 4, 6, 7, and 8.

Our 1961 summary in tabular form is included here as Appendix II. In the succeeding year (before operating on NG on 5 September 1963) we added to our collection references D, E, F, G, H, and 5. Our in-house summary, prepared in 1963 on the basis of the sixteen references then available, is included here as Appendix III.

APPENDIX I
Publications From the Van Wagenen Series of Callosotomies for Epilepsy

A. Van Wagenen, W.P. and Herren, R.Y., 1940, Surgical division of commissural pathways in the corpus callosum. Relation to spread of an epileptic attack, *Arch. Neurol. Psychiat.* 44:740-759.

B. Akelaitis, A.J.E., 1942, Psychobiological studies following section of the corpus callosum. A preliminary report, *Amer. J. Psychiac.* 47:519-543.

1. Smith, K.U. and Akelaitis, A.J., 1942, Studies on the corpus callosum. I. Laterality in behavior and bilateral motor organization in man before and after section of the corpus callosum, *Arch. Neurol. Psychiat.* 47:519-543.
2. Akelaitis, A.J., 1941, Studies on the corpus callosum. II. The higher visual functions in each homonymous field following complete section of the corpus callosum, *Arch. Neurol. Psychiat*. 45:788-796.
3. Akelaitis, A.J., Risteen, W.A., Herren, R.Y., and Van Wagenen, W.P., 1942, Studies on the corpus callosum. III. A contribution to the study of dyspraxia in epileptics

following partial and complete section of the corpus callosum, *Arch. Neurol. Psychiat.* 47:1971-1008.

4. Akelaitis, A.J., 1944-45, Studies on the corpus callosum. IV. Diagonostic dyspraxia in epileptics following partial and complete section of the corpus callosum, *Amer. J. Psychiat.* 101:594-599.
5. Akelaitis, A.J., 1942, Studies on the corpus callosum. V. Homonymous defects for color, object and letter recognition (homonymous hemiamblyopia) before and after section of the corpus callosum, *Arch. Neurol. Psychiat.* 48:108-118.
6. Akelaitis, A.J., 1942, Studies on the corpus callosum. VI. Orientation (temporal-spatial gnosis) following section of the corpus callosum, *Arch. Neurol. Psychiat.* 48:914-937.
7. Akelaitis, A.J., 1943, Studies on the corpus callosum. VII. Study of language functions (tactile and visual lexia and graphia) unilaterally following section of the corpus callosum, *J. Neuropath. Exper. Neurol.* 2:226-262.
8. Akelaitis, A.J., 1941, Studies on the corpus callosum. VIII. The effects of partial and complete section of the corpus callosum on psychopathic epileptics, *Amer. J. Psychiat.* 98:409-414.
9. Akelaitis, A.J., Risteen, W.A., and Van Wagenen, W.F., 1943, Studies on the corpus callosum. IX. Relationship of the grasp reflex to section of the corpus callosum, *Arch Neurol. Psychiat.* 49:820-825.
10. Akelaitis, A.J., 1944, Studies on the corpus callosum. X Reaction time and speed of motor co-ordination in man before and after section of the corpus callosum. Cited in *J. Neurosurg.* 1:102.

C. Akelaitis, A.J., 1944, A study of gnosis, praxis and language following section of the corpus callosum and anterior commissure, *J. Neurosurg.* 1:94-102.

D. Parsons, F.H., 1943, Eight cases of section of corpus callosum in individuals with a history of epileptic seizures: psychological tests, *J. Gen. Psychol.* 19:227-241.

E. Parsons, F.H., 1940, Psychological tests of patients one year after section of corpus callosum, *Psychol. Bull.* 37:498.

F. Bridgman, C.S. and Smith, K.U., 1945, Bilateral neural integration in visual perception after section of the corpus callosum, *J. Comp. Neurol.* 83:57-68.

G. Hursh, J.B., 1945, Origin of the spike and wave pattern of petit mal epilepsy. An electroencephalographic study, *Arch. Neurol. Psychiat.* 53:274-82.

H. Smith, K.U., 1951, Learning and the associative pathways of the human cerebrral cortex, *Science* 114:117-120.

I. Smith, K.U., 1952, Experimental analysis of the associative mechanism of the human brain in learning functions, *J. Comp. Physiol. Psychol.* 45:66-72.

J. Akelaitis, A.J.E., 1940, A study of gnosis, praxis and language following partial and complete section of the corpus callosum and anterior commissure, *Trans. Amer. Neurol. Assoc.* 66:182-185.

K. Akelaitis, A.J., Risteen, W.A., Herren, R.Y., and Van Wagenen, W.P., 1941, Studies on the corpus callosum. III. A contribution to the study of dyspraxia in epileptics following partial and complete section of the corpus callosum, *Trans. Amer. Neurol. Assoc.* 67:75-78.

L. Smith, K.U., 1947, Bilateral integrative action of the cerebral cortex in man in verbal association and sensori-motor coordination, *J. Exper. Psychol.* 37:367-376.

M. Goldstein, M.N. and Joynt, R.J., 1969, Long-term followup of a callosal-sectioned patient, *Arch. Neurol.* 20:96-102.

N. Goldstein, M.N., Joynt, R.J., and Hartley, R.B., 1975, The long term effects of callosal sectioning, *Arch. Neurol.* 32:52-53.

APPENDIX II
Extent of Section and Complications in the Van Wagenen Operations

Patient	Sources		Post-Op Complication
EC 43	A	mid 1/2	"uneventful"
EJB 24	A, III	mid 1/2 remainder cc	"surprisingly smooth" "sick" 2 days
GB 14	A	complete cc	"uneventful"
EE 12	III, 1	complete cc	R hemiparesis & confusion
LC 16	III, VI	nearly complete	L hemiparesis & died 10 days post-op (hematoma)
GE 35	I	complete cc A.C.	"uneventful"
EK 30	I, III, VI	complete cc	catatonic mutism 2 wks.
HK 15	A, B, III	ant 1/2	"convalescence was slow"; died during 2nd op. (too rapid infusion?)
EL 22	III	mid 1/2	apathetic & confused for 1 week
AM 25	A, III	complete cc	"stormy post-op course" delirium for 2 weeks
LM 14	VI	ant 1/2	"uneventful"
RM 10	A, VI	mid 1/3	"exceptionally smooth"
WM 18	I, III	nearly complete	"uneventful" L hand weak & astereognosis 1 wk.
CN 36	A, III	ant 2/3	confused 2 days, transient L hemiparesis
FN	VI	complete cc	"never disoriented"
FO'B	III	complete cc	confused 1 wk. (10 days amnesia)
CP 46	C	ant 1/2, cc	local anesthesia, cooperative but disoriented -- confused 3 wks.
FP 17	III	nearly complete	"uneventful" transient R paresis (slight)
MLP 10	I	nearly complete	R motor defect & astereognosis 4 wks.
CR	VI	nearly complete	uneventful
FR 21	A, II, IV	mid 1/3 completed cc	akinetic 2 days, spoke well in 1 wk., L. hemiparesis
FR 26	I, VI	agenesis	uneventful "some muteness 1st wk. caused by headache"
GR 33	A	mid-third	"satisfactory"
CS 166	I, II, III, VI	complete cc	L hemiparesis & sensory loss 3 wks.
WS 26	A, B A, VI	 cc	L hemiparesis & "apathy"; flap re-elevated "psychomotor retardation"
JW 17	VI	ant 1/2	none mentioned
SL 21	I	ant 1/2	"uneventful"

APPENDIX III, PART 1

The Twenty-Eight Patients Operated by Dr. Van Wagenen and Referred to in Various Papers by Akelaitis, et al.

(Papers with post-operative seizure data are underlined)

3	A & B	1		
4	8	11	EB	A, B, 1, 3, 6, 7, 8, D (called EC in A and EVB in 1 where the age is given incorrectly
8	5	6	EJB	A, B, 1, 2, 3, 6, 7, 8, D (called EB in 1)
	10	8	GB	A, B, 1, 2, 6, 7, D
17		1	EC	1, 2, 3, 6, 7 (these initials erroneously used in A for EB)
12		19	LC	3,6
18		3	GE	1, 2, 3, 6, 7, C (1st op "complete," 2nd op A.C.)
15		9	EH	1, 3, 6, 7
10		7	EK	1, 2, 3, 5, 6, 7 (2 year follow-up)
1	6		HK	A, B, 2, 3, 6, 7, D (died during 2nd op as flap turned)
2			EL	3, 6, 7, H
		14	SL	1
5	9		AM	A, B, 3, 6, 7, D, H (per 7, died in status)
			LM	6
	4		RM	A, B, 6, 7, D
6		16	WM	1, 3, 6, 7
3	7	10	CN	A, B, 1, 3, 6, 7, 8, D
			FN	6, 7
9		4	FO'B	1, 3, 4, 5, 6, 7
			CP	C (ant 1/2, right fornix, bilat. frontal leucotomy)
6		13	FP	1, 3, 6, 7, 8
14		15	MLP	1, 3, 3, 6, 7
			CR	6,7
		17	mFR	6, 1
13	3	5	fFR	A, B, 1, 2, 3, 4, 6, 7, D
	11		GR	A, B, 6 (called GMR in B and in 6)
16		2	CS	1, 2, 3, 6, 7
11	2	12	WS	A, B, 1, 3, 6, 7
			JW	6, 7

APPENDIX III, PART 2
Dr. Van Wagenen's Results
(34 Operations on 28 Patients)

Of operations with partial section (12), there were seven poor results (EB, EJB, HK, EL, RM, fFR, and WS) and five without data (LM, CP, CR, GR, and JW).

Of operations with splenial tip preserved (11), there were two good results (EH and MFR), three fair results (WM, CN, and MLP), two poor results (EJB, and WS, both originally partial sections) and four without data (LC, SL, AM, FP).

Of operations with complete section of callosum (8) there were two good results (EK, FN), three fair results (EC, fFR, who was a previous partial section, FO'B), two poor results (GB, and GE who had A.C. section later), and one without data (CS).

1. Of 28 patients, 10 had improvement.

2. Eliminating those with only partial section (8): of 20 patients, 10 had improvement. (Assumes lack of improvement in cases with insufficient data.)
3. Eliminating further those with inadequate data (including 2 deaths): of 14 patients, 10 had improvement, of whom 5 had occasional grand mal; and 5 were free of grand mal, although 4 continued to have focal seizures (EH, EK, EN, and FN).
4. Of the 14 patients with follow-up and near-complete operation: there were 4 injured as adults with 2 good, 1 fair, and 1 poor result; 3 injured in teens with 1 good, 1 fair, and 1 poor result; 7 injured in childhood with 2 good, 3 fair, and 2 poor results.
   There were 6 with head injuries, of whom 3 had good, 2 fair, and 1 poor result; and 3 with infantile hemiplegia, of whom there were 1 good, 1 fair, and 1 poor result.
   Of the 9 cases with definite focal changes, there were 5 good, 2 fair, and 2 poor results. Of the 5 cases without a focus, there were 3 fair and 2 poor results.
5. The highest proportion of poor results was in those having only a partial section. Of these, 3 had further section but only 1 was improved.

CONCLUSIONS

1. Unilateral seizures are quite common after callosal section, either on one side (GB, EC, LC, EH, EK, AM, MLP, FN, GR) or both sides (CN, fFR, WS).
2. There is at least a 50 percent chance of improvement following a nearly-complete operation. Of those surviving a nearly-complete operation and followed up to two years, two-thirds were improved including one-third who were free of generalized convulsions.
3. Chances for benefit are possibly better if the responsible lesion is acquired in later life, and definitely better if manifested as focal rather than diffuse disease.

REFERENCES

(In addition to these selected references and those in Appendix I, a 600 item bibliography can be found in Bogen 1993)

Apuzzo, M.L.J., ed. 1987, "Surgery of the Third Ventricle," Williams & Wilkins, Baltimore.

Bogen, J.E., 1985, Concluding overview, *in:* "Epilepsy and the Corpus Callosum," A.G. Reeves, ed., Plenum, New York.

Bogen, J.E., 1987 Physiological consequences of complete or partial commissural section, *in:* "Surgery of the Third Ventricle," M.L.J. Apuzzo, ed., Williams & Wilkins, Baltimore.

Bogen, J.E., 1993, The callosal syndromes, *in:* "Clinical Neuropsychology, 3rd Ed.," K. Heilman and E. Valenstein, eds., Oxford University Press, New York.

Bogen, J.E., Fisher, E.D., and Vogel, P.J., 1965, Cerebral commissuroty: a second case report, *J. Am. Med. Assn.* 194:1328-1329.

Bogen, J.E., Schultz, D.H., and Vogel, P.J., 1988, Completeness of callosotomy shown by magnetic resonance imaging in the long term, *Arch. Neurology* 45:1203-1205.

Bogen, J.E., Sperry, R.W., and Vogel, P.J., 1969, Commissural section and propagation of seizures, *in:* "Basic Mechanisms of the Epilepsies, H.H. Jasper, A.A. Ward, Jr., and A. Pope, Little, Brown and Co., Boston.
Bogen, J.E. and Vogel, P.J., 1962, Cerebral commissurotomy in man: preliminary case report, *Bull. Los Angeles Neurol. Soc.* 27:169-172.
Bogen, J.E. and Vogel, P.J., 1975, Neurologic status in the long term following cerebral commissurotomy, *in:* "Les Syndromes de Disconnexion Calleuse Chez L'Homme," F. Michel and B. Schott, eds., Hôpital Neurologique, Lyon.
Bremer, F., 1958, Physiology of the corpus callosum, *Res. Publ. Assoc. Res. Nerv. Ment. Dis.* 36:424-448.
Bubnoff, N. and Heidenhain, R., 1881, Ueber erregungs und hemmungsorgänge innerhalf der motorischen hirncentren, *Arch. Ges. Physiol.* 26:137-200. Translated into English by G. von Bonin and W.S. McCulloch, *in:* "The Precentral Motor Cortex, Ed. 2," P.C. Bucy, ed., University of Illinois Press, Urbana, Ill., Ch. VII, pp. 173-210.
Clarke, E. and O'Malley, C.D., 1968, "The Human Brain and Spinal Cord," University of California Press, Berkeley, CA.
Dandy, W.E., 1936, Operative experience in cases of pineal tumor, *Arch. Surg.* 33:19-46.
Erickson, T.C., 1940, Spread of the epileptic discharge, *Arch. Neurol. Psychiat.* 43:429-452.
Fritsch, G. and Hitzig, E., 1870, On the electrical excitability of the cerebrum, *Arch. f. Anat., Physiol., und Wissenschaftl. Mediz.*, pp. 300-332 (English translation in Von Bonin, 1960).
Gordon, H.W., Bogen, J.E., and Sperry, R.W., 1971, Absence of deconnexion syndrome in two patients with partial section of the neocommissures, *Brain* 94:328-336.
Horsley, V., 1886, Abstracts of Brown Lectures: Epilepsy. *The Lancet*, 2:1211-1213.
Joynt, R.J., 1985, History of forebrain commissurotomy, *in:* "Epilepsy and the Corpus Callosum," A.G. Reeves, ed., Plenum Press, New York.
Liepmann, H. and Maas, O., 1908, Fall von linksseitiger agraphie und apraxie bei rechtsseitiger Lähmung, *J. Psychol. Neurol.* 10:214-227.
Luessenhop, A.J., de la Cruz, T.C., and Fenichel, G.M., 1970, Surgical disconnection of the cerebral hemispheres for intractable seizures; results in infancy and childhood, *JAMA* 213:1630-1636.
Myers, R.E. and Sperry, R.W., 1953, Interocular transfer of visual form discrimination habit in cats after section of the optic chiasm and corpus callosum, *Anat. Record.* 115:351-352.
Penfield, W. and Jasper, H., 1954, "Epilepsy and the Functional Anatomy of the Human Brain, Little, Brown and Company, Boston.
Reeves, A.G., ed., 1985, "Epilepsy and the Corpus Callosum," Plenum, New York.
Smith, F.P., 1991, Callosotomy for seizures, *J. Neurosurg.* 75:668-669.
Spencer, S.S., Gates, J.R., Reeves, A.G., Spencer, D.D., Maxwell, R.E., and Roberts, D., 1987, Corpus callosum section, *in:* "Surgical Treatment of the Epilepsies," J. Engel, ed., Raven Press, New York.
Sperry, R.W., 1961, Cerebral organization and behavior, *Science,* 133:1749-1757.
Sperry, R.W., 1974, Lateral specialization in the surgically separated hemispheres, *in:* "Neuroscience 3rd Study Prog.," F.O. Schmitt and F.G. Worden, eds., MIT Press, Cambridge, MA.
Sperry, R.W., Gazzaniga, M.S., and Bogen, J.E., 1969, Interhemispheric relationships: the neocortical commissures; syndromes of hemisphere disconnection, *in:* "Handbook of Clinical Neurology, Vol. IV," P.M. Vinken and G.W. Bruyn, eds., pp. 272-290, North-Holland Publishing Co., Amsterdam.
Spiegel, E., 1931, The central mechanism of generalized epileptic fits, *Amer. J. Psychiat.* 88:595-609.
Van Harreveld, A., 1947, On the mechanism and localization of the symptoms of electroshock and electronarcosis, *J. Neuropath. Exper. Neurol.* 6:177-184.
Van Wagenen, W.P. and Herren, R.Y., 1940, Surgical division of the commissural pathways in the corpus callosum, *Arch Neurol. Psychiat.* 44:740-759.
Von Bonin, G., 1960, "Some Papers on the Cerebral Cortex," Charles C. Thomas, Springfield.
Wada, J.A., 1987, Anterior 2/3 callosal bisection: comparative observations in animals and man, *in:* "Fundamental Mechanisms of Human Brain Functions," J. Engel, G.A. Ojemann, H.O. Luders, and P.D. Williamson, eds., Raven Press, New York.

White, H.H., 1961, Cerebral hemispherectomy in the treatment of infantile hemiplegia, *Confin. Neurol.* 21:1-150.
Willis, T., 1664, "Cerebri Anatome," Translated *in:* "The Human Brain and Spinal Cord," 1968, E. Clarke, and O'Malley, C.D., University of California Press, Berkeley and Los Angeles.
Wilson, D.H., Reeves, A., Gazzaniga, M., and Culver, C., 1977, Cerebral commissurotomy for control of intractable seizures, *Neurology* 27:708-715.
Zaidel, E., 1994, Interhemispheric transfer in the split brain: long-term status following complete cerebral commissurotomy, *in:* "Human Laterality," R.H. Davidson and K. Hugdahl, eds., MIT Press, Cambridge.
Zaidel, E., Zaidel, D.W., and Bogen, J.E., 1990, Testing the commisurotomy patient, in: "Neuromethods, Vol. 17: Neuropsychology," A.A. Boulton, G.B. Baker, and M. Hiscock, eds., Humana Press, Clifton, NJ.

CORPUS CALLOSUM SECTION: PREOPERATIVE EVALUATION

Peter D. Williamson, M.D.

Section of Neurology
Dartmouth-Hitchcock Medical Center
One Medical Center Drive
Lebanon, NH 03756-0001

INTRODUCTION

Successful division of the corpus callosum (CC) in humans was initially described by Dandy in 1931. Nine years later, Van Wagenen and Herren (1940) reported the use of CC section in an attempt to control the spread of epileptic seizures. Van Wagenen may have operated on as many as 27 patients during less than a year (Williamson 1985). Details concerning preoperative evaluation and postoperative outcome in these patients are sparse, with most information being anecdotal. Van Wagenen, after a flurry of activity, apparently abandoned CC section, as the procedure no longer interested him. During the 1960s, Bogen and colleagues (Bogen et al. 1962, 1963, 1965, 1969, 1975) began to use CC section in medically refractory epileptic patients. Few details of outcome in terms of seizure control were provided, but the Bogen series generated considerable interest in the neuropsychological consequences of dividing the CC (Sperry 1968, 1970). Neurosurgical complications ultimately led Bogen to discontinue the procedure.

In 1970, Luessenhop (1970) provided some clinical detail and follow-up on four hemiplegic pediatric patients who had complete CC sections. Excellent results were achieved in three of the patients, but the unfavorable result in the fourth patient apparently discouraged additional CC sections. During 1971, the late Don Wilson began using section of the CC in medically refractory epileptic patients at Dartmouth-Hitchcock Medical Center (Wilson 1975). Reports from the Dartmouth series provided many of the initial clinical details about preoperative findings and postoperative outcome (Wilson 1977, 1982). Besides introducing more objectivity, Wilson and his colleagues, after observing a number of postoperative complications, decided to simplify and modify the procedure. Initially, division of the central commissures included complete section of

Epilepsy and the Corpus Callosum II, Edited by A.G. Reeves
and D.W. Roberts, Plenum Press, New York, 1995

the CC, hippocampal commissure, anterior commissure and division of one fornix (Wilson 1977). In the modified procedure, using microsurgical techniques, only the CC and hippocampal commissure were sectioned without entering the ventricles. This modification resulted in fewer complications without reduced efficacy. Performing the procedure in two stages, separated by months, further reduced morbidity (Roberts 1985, 1991).

The reports from the Dartmouth series describing beneficial results with increased safety prompted other epilepsy centers to use corpus callosum section in selected patients with intractable epilepsy. At that time, the late 1970s and early 1980s, the problem was selection. Although Wilson et al. (1975, 1977, 1982) had provided considerable information about the types of patients undergoing surgery at Dartmouth, selection criteria remained rather vague. Nine years ago, at the first symposium on Epilepsy and the Corpus Callosum (Reeves 1985), I reviewed the available literature and listed some preliminary selection criteria (Williamson 1985). Since that time, a number of studies and reviews aimed at improving selection criteria have been published (e.g., Rayport et al. 1983, Geoffroy et al. 1983, Blume 1984, Bouvier et al. 1983, Gates et al. 1984, Gates et al. 1987, Spencer et al. 1988, Purves et al. 1983, Murro et al. 1988, Roberts and Reeves 1987, Marino 1985, Spencer SS 1988, Wylie 1988, Sass et al. 1988, Spencer et al. 1987, Goodman et al. 1985, Maxwell et al. 1987, Purves 1991, Nordgren et al. 1991). Despite advances in understanding, confusion regarding selection criteria persists. In a recent review, Purves (1991) states, "The selection of patients who will have a good outcome from this procedure (corpus callosum section) remains a difficult problem, as does the decision of how much of the corpus callosum should be divided."

PREOPERATIVE EVALUATION

General Considerations

When evaluating patients for possible section of the CC, there are a number of general considerations that must be taken into account. While most of these will be clarified during the overall evaluation, they are discussed here for both convenience and emphasis.

Medical intractability must be assessed in all patients being considered for epilepsy surgery. This usually is not a difficult problem with candidates being considered for CC section, as most will have severe epilepsy and will have been tried on all available antiepileptic drugs alone and in combination, often at toxic levels. Nevertheless, a historical survey of antiepileptic drug usage, toxicity, response and serum levels is necessary, as even some patients with severe Lennox-Gastaut syndrome can be controlled with medications (Niedermeyer 1990). If there is a "gap" in the antiepileptic drug history, untried appropriate drugs or combinations should be used before proceeding with the presurgical evaluation. On the other hand, there is little to be gained by delaying possible beneficial surgery while evaluating the effects of various second-line antiepileptic drugs. After obtaining a thorough antiepileptic drug history, clinical judgement combined with common sense should suffice.

The impact of seizures on a patient's life must be assessed. This can be a difficult issue when considering any type of epilepsy surgery. Certainly, when dealing with a healthy, active, normally intelligent teenager, whose two to three complex partial seizures per week on optimal therapy are handicapping his education, mobility and socialization, and who probably has unilateral mesial temporal lobe seizure origin, the answer is obvious. The chance of eliminating seizures and thereby producing enormous benefit is very high.

Patients being considered for CC section, however, rarely present with such obvious answers. Often, the effect of seizures on family and caregivers must also be considered. Corpus callosum section is rarely curative and persistent postoperative seizures will remain a handicap in most patients. How reduction in seizure frequency or intensity or elimination of certain seizure types will affect the patient and the family must be addressed during initial contact and repeatedly throughout the presurgical evaluation.

Corpus callosum surgery is usually a palliative procedure. If there is a possibility that resective surgery might eliminate seizures without producing unacceptable neurological deficits, then this must take precedence over callosal section. The current state of neuroimaging technology makes it unlikely that a localized potentially epileptogenic lesion would not be detected. However, subtle lesions such as small neoplasms or cortical dysplasias are just beginning to be appreciated on MRI (Kuzniecky et al. 1991), and they can still be missed. Localized EEG abnormalities or clinical seizure characteristics can also suggest localized seizure origin. While these issues will be discussed further, the possibility of seizure origin from a resectable region of brain should be considered throughout the evaluation, regardless of the initial impression. We have been fooled more than once in the Yale program.

There are a number of other general considerations that relate to epilepsy surgery, such as the possible presence of progressive neurological or medical disease, and the presence of psychiatric conditions, and these must be determined for each patient (Williamson 1991). The presence of such conditions and their nature can be absolute or relative contraindications to epilepsy surgery, depending upon specific circumstances in each patient.

History

A careful history is the foundation of good medicine. A detailed comprehensive history is particularly relevant in evaluating patients being considered for epilepsy surgery, as it can provide important clues regarding etiology, seizure frequency and progression, seizure types and severity, precipitating factors, and familial predisposition. Only the patient can describe subjective aspects of seizures. Since warnings or auras provide clues to region of seizure origin and are highly suggestive of localized seizure onset, detailed descriptions of initial seizure symptoms can have a direct bearing on eventual evaluation strategies. Patients rarely are able to provide accurate descriptions of the objective manifestations of seizures, therefore, at least one family member should always be present during the initial interview. Family members should be carefully questioned regarding seizure type and whether or not there are distinctly different types, severity and frequency of seizures, and any other pertinent features such as initial seizure type and precipitating factors. Family members, particularly the mother, can provide important information about potentially significant events that occurred early in the patient's life such as head injuries, febrile seizures, birth difficulties and CNS infections. A careful psychosocial history, obtained from both the patient and family, is an essential part of the initial evaluation, as this information will be crucial in determining impact of seizures on patient and family, and will have a direct bearing on goals and expectations.

Physical Examination

Most patients being evaluated for callosal section will have been referred from other physicians and will have undergone numerous prior examinations. Nevertheless, a thorough general physical and neurological examination must be conducted on all patients to search for conditions that might affect prognosis or alter evaluation strategies. A careful dermatological examination can reveal stigmata of the phakomatoses. Subtle hemiatrophies and hemipareses have important localizing significance. When an obvious lateralized weakness is present, its degree and extent, and associated neurological deficits, such as visual field defects and impaired cortical sensation, must be carefully assessed, as these findings will relate to the type of surgical treatment offered.

Neuroimaging

The introduction of modern neuroimaging has changed the practice of neurology and neurosurgery. Many lesions and abnormalities previously undetectable are now easily diagnosed. Subtle abnormalities such as hippocampal sclerosis and some cortical dysplasias are beginning to be appreciated (Berkovic et al. 1991, Kuzniecky et al. 1991). Positron emission tomography has revealed different subtypes of some of the intractable epilepsies, findings that might have prognostic significance (Chugani and Engel 1986).

In patients being evaluated for callosal section, neuroimaging can demonstrate unsuspected, potentially resectable epileptogenic lesions. Two patients from the Yale program had circumscribed lesions detected after undergoing callosal section (Williamson et al. 1992) and additional patients being considered for CC section have had lesions detected that altered the course of their therapy. Neuroimaging in patients with incomplete infantile hemiplegia will demonstrate the location and extent of the lesion and help direct surgical strategies. Finally, postoperative MRI will reveal the extent or completeness of the CC section (Gates et al. 1986).

Neuropsychological Testing

Neuropsychological testing has become an integral part of the presurgical evaluation (Rausch 1987). As discussed below, general intelligence level may be a factor in predicting outcome after CC section.

Most important, detailed preoperative testing of all cognitive function provides the essential baseline data for eventual postoperative comparisons. Although most reports describe minimal long term psychological deficits (Gazzaniga et al. 1975, Ferral et al. 1983, Purves et al. 1983, Beniak et al. 1985, Wylie 1988), there is a growing body of evidence that this might not be true, at least in some patients. Persistent disconnection syndromes and declines in memory and language have been observed and can result in significant handicaps (Wilson et al. 1977, Sussman et al. 1983, Rayport et al. 1983, Ferguson et al. 1985, Sass et al. 1988, Woodhurst et al. 1989). Sass et al. (1988) examined pre- and postoperative neuropsychological and neurological examination results and found that some declines could be predicted. For example, when intracarotid Amytal testing demonstrated crossed dominance, patients appeared to be at risk for developing postoperative language deficits.

CCTV/EEG Monitoring

It might be argued that the time and expense generated by prolonged CCTV/EEG monitoring is not justified in some patients being evaluated for CC section, but I would disagree with that attitude. There are still too many inconsistencies and uncertainties surrounding CC section. The value of CCTV/EEG monitoring to document ictal and interictal EEG patterns and seizure types has been well established (Gumnit 1987, Gotman 1985). While descriptions of seizure types by family members are helpful, they are not always reliable, as we have repeatedly observed. This is particularly true when patients have multiple seizure types, as many CC section candidates will have (Niedermeyer 1990). Since different seizure types appear to respond differently to surgical disconnection (Purves et al. 1983, Gates 1984, Spencer et al. 1987, Spencer et al. 1988, Purves 1991, Nordgren et al. 1991), accurate documentation of all seizure types in each patient has obvious prognostic significance. Furthermore, preoperative documentation of seizure types is essential for meaningful comparison with postoperative seizures. Although postoperative seizure documentation should be as thorough as preoperative, this has not been consistent in our program. If some of the uncertainties surrounding CC section are ever to be resolved, CCTV/EEG monitoring before and after surgery will have to be done consistently on large numbers of patients.

Data Summation

When the evaluation has progressed to this point, a number of the questions discussed above can usually be answered satisfactorily. Familiarity gained with the patient and family will allow assessment of the impact of seizures on the patient's and family's lifestyle. Reasonable expectations and possible complications can be thoroughly reviewed. If, during the course of the evaluation the possibility of a resectable focus has been raised, then this issue needs to be carefully reviewed among the medical team involved with the case, and discussed with the patient and family. This issue has further implications in terms of possible intracranial electrode placement and monitoring (Williamson 1991).

PATIENT SELECTION CRITERIA

Criteria for patient selection have improved during the past eleven years, but not dramatically. Syndromic classification identified several groups of patients that appeared to benefit from CC section (Williamson 1985), but this system was too restrictive in that some patients could not be classified, and syndromes are seldom pure. The major advance in selection criteria relates to specific types of seizures that are most likely to respond favorably to the procedure (Purves 1991). Other factors, such as general intelligence, laterality of disease, and specific neuropsychological deficits all need to be considered (Spencer 1987). In practice, no single system of selection criteria is mutually exclusive of the other, as they all must be taken into account for each patient. Despite the advances in understanding the criteria for selection, a number of uncertainties and inconsistencies persist. In the discussion that follows, general selection criteria will be examined first, followed by a review of specific patient categories or syndromes. These systems are interrelated, so some redundancies cannot be avoided.

General Selection Criteria

1. Seizure type: Many patients, particularly those with Lennox-Gastaut syndrome (Niedermeyer 1990), will have more than one seizure type and the different types respond differently to CC section. There is agreement that secondarily generalized seizures (tonic, atonic, tonic-clonic) respond best to CC section (Purves et al. 1983, Rayport et al. 1983, Gates et al. 1984, Spencer 1988), with some programs selecting only patients with these seizure types (Rayport et al. 1983, Gates et al. 1984). Recurrent bouts of convulsive status epilepticus are reduced or eliminated by partial or complete CC section (Purves et al. 1983, Geoffroy et al. 1983).

Complex partial seizures generally respond less favorably (Spencer et al. 1988). Intuitively, partial or complete CC section would not seem to have a place in the treatment of temporal lobe epilepsy yet, though lacking in detail, descriptions from Bogan's (1975) earlier series suggested that some patients may have had temporal lobe type seizures. Bilateral independent mesial temporal seizure origin was documented during intracranial recording in one of our patients who subsequently had an excellent response from an anterior two-thirds CC section. Another of our patients with similar intracranial EEG findings did poorly following anterior two-thirds CC section. The latter patient had an unsuspected parietal lobe neoplasm, the resection of which eliminated seizures (Williamson et al. 1992). Roberts and Reeves (1987) reported that CC section eliminated partial seizures in 40% of patients, with most of the remaining patients experiencing a reduction in severity or frequency of this seizure type. This study, as is true of most others, suffers from the use of complex partial seizure in the generic sense. It is well known that complex partial seizures can originate from a variety of locations in the cerebral cortex (Williamson 1985, 1987, Waterman et al 1986, Swartz et al 1988, Williamson 1992 a, b). As discussed below, it may be that only certain types of complex partial seizures predictably respond to CC section.

Although simple partial seizures usually do not respond favorably to CC section, there are exceptions. Patients with infantile hemiplegia presumably establish aberrant neural connections, and because of this, simple partial seizures in such patients can respond favorably to CC section. We reported a patient with infantile hemiplegia who developed simple partial status epilepticus on an almost daily basis, manifested by aphemia, facial twitching, and hemiparesthesias (Williamson et al. 1985). Two-stage complete CC section abolished all seizure activity except rare episodes of paresthesia involving only the cheek. Presumably seizure discharges from the atrophic hemisphere were producing most clinical manifestations by spreading to the good hemisphere via the CC.

Spencer et al. (1984) have presented evidence that occasionally new or more intense partial seizures occur after CC section and postulate this is due to reduced inhibition . This apparently occurs almost twice as frequently in children as adults (Spencer 1988, Epilepsia suppl). Gates et al. (1987) reported that new or intense partial seizures are actually fragments of previous generalized tonic-clonic seizures, a view shared by the Dartmouth group (Roberts 1991). We have recently observed that some patients whose families describe persistent generalized tonic-clonic seizures after CC section are actually having hemiconvulsions. Furthermore, these hemiconvulsions often occur on different sides during different seizures. Consciousness is maintained during the hemiconvulsions, but language is lost ictally and transiently postictally when the seizure involves the language dominant hemisphere. Corresponding Todd's hemiparesis is also seen. When this occurs in children, maintenance of contact during the hemiconvulsions is often a frightening experience, whereas older patients are more accepting and appreciate having some control during seizures. Therefore, division of the CC may split the generalized convulsion into two halves. Whether or not CC section actually produces more intense focal seizures needs to be reexamined.

2. Cognitive function: Mental subnormality has been related to poor outcome following CC section with a lower IQ limit of 70 the suggested cut-off point (Wilson 1982, Blume1984, Purves et al. 1983, Murro et al. 1988). Other series have reported poor results in patients with IQ below 50 (Geoffroy et al. 1983, Spencer et al. 1988). Low IQ, particularly below 50, would imply diffuse bilateral brain disease with probable multifocal seizure origin, a condition associated with poor outcome (Purves et al. 1983, Spencer et al. 1988). Purves et al. (1983), however, reported excellent results in two patients with infantile hemiplegia and very low IQs, and Geoffroy et al. (1983) observed some improvement following complete CC section in severely retarded children.

In conclusion, although a low IQ is a poor prognostic factor in many patients, there will be exceptions. Certainly, a severely retarded child subject to frequent falls and recurrent bouts of status epilepticus might benefit from a CC section. Whether or not it is justified to use CC section to achieve modest improvements in severely retarded patients is a moral and ethical issue that must be decided individually with consideration given to available resources.

3. Structural lesions: When structural lesions are detected in patients with intractable epilepsy, the relationship of the lesion to the seizures must be determined. Circumscribed mass lesions that involve the cerebral cortex are almost always the cause of epilepsy and, unless there are extremely noncongruent findings, they should be resected, if possible (Boon et al. 1991). Large unilateral atrophic lesions are usually associated with obvious neurological deficits. Most such patients being considered for CC section will be included in the infantile hemiplegia category discussed below. However, lateralized disease of less severe proportions as determined by EEG, neurological examination, or neuroimaging is generally associated with a good outcome following CC section (Spencer et al. 1988).

Epileptic Syndromes and Specific Patient Categories

1. Infantile hemiplegia: During the 1982 conference of the Corpus Callosum and Epilepsy, I concluded that patients with infantile hemiplegia and intractable seizures were excellent candidates for CC section (Williamson 1985). This conclusion was based upon our experience and that of others (Luessenhop 1972, Goodman et al. 1985, Wilson 1975). Since then, Rasmussen (1983) reported that a modified hemispherectomy, in which the occipital and frontal lobes were disconnected but left in place, avoided the delayed complications of total hemispherectomy. Comparing CC section and "functional hemispherectomy" in patients with infantile hemiplegia, it appears that the latter procedure offers the best chance of complete freedom from seizures (Goodman 1985, Tinuper et al. 1988). However, hemispherectomy should only be performed on patients who have no useful finger control. Intact visual fields and preserved cortical sensation are additional factors to be considered. If there is any doubt, CC section is the procedure of choice, as hemispherectomy can always be performed later.

If patients with less than a complete infantile hemiplegia are candidates for CC section, how far can this concept be extended? During the 1982 symposium, I included a rather imprecise category of "forme-fruste" infantile hemiplegia to designate patients with lateralized brain disease of probable early onset who did not have obvious hemiplegia (Williamson 1985). As noted above, this concept has been further extended to include any evidence of lateralized brain disease as being associated with a good prognosis (Purves 1983, Spencer et al. 1988).

2. Rasmussen's syndrome: Rasmussen's syndrome is a form of unilateral chronic encephalitis associated with refractory seizures and progressive mental and neurological deterioration (Rasmussen 1978). It usually presents with epilepsy partialis continua in an otherwise healthy child. During the early part of the disease in which focal motor seizures predominate, there is no reason to believe that CC section would be beneficial. There are two situations where CC section might be useful in Rasmussen's syndrome. The first is when frequent generalized seizures or recurrent convulsive status epilepticus occur. The second is when the disease stops progressing before complete hemiplegia has developed, yet generalized seizures persist. When the disease has progressed to the point that there is no useful finger movement, a functional hemispherectomy is the preferred treatment, provided that the good hemisphere can support language (Rasmussen 1987).

3. Lennox-Gastaut syndrome: The Lennox-Gastaut syndrome encompasses a variety of conditions that are linked by similar EEG findings and seizure types (Engel 1989, Niedermeyer 1990). About two-thirds have no recognized cause (idiopathic Lennox-Gastaut), while the remainder (symptomatic) are caused by different metabolic, congenital or structural abnormalities (Niedermeyer 1990). Although the diagnosis often has to be inferred, unquestionably the idiopathic Lennox-Gastaut syndrome is the most common condition associated with CC section (Rayport 1983, Gates 1984, Williamson 1985). This is not surprising since tonic, atonic and generalized tonic-clonic seizures are characteristic of this syndrome (Niedermeyer 1990) and are the seizure types that respond most favorably to CC section.

Although surgical results in patients with Lennox-Gastaut syndrome are generally favorable, results have been uneven. Functional imaging of patients with cryptogenic Lennox-Gastaut syndrome has revealed diffuse, multifocal, focal and lateralized abnormalities in different patients (Chugani et al. 1987). Perhaps these subvarieties of the syndrome could explain some of the postoperative inconsistencies, but at present this is purely speculative.

4. Frontal lobe epilepsy: A variety of different seizure types have been associated with origin in the frontal lobes (Williamson et al. 1987, Swartz et al. 1987). These include asymmetrical tonic seizures, clonic seizures, secondarily generalized tonic-clonic seizures, and a special type of complex partial seizures (Williamson et al. 1985, Waterman et al. 1987). Focal clonic and asymmetrical tonic seizures are usually simple partial seizures and a favorable response to CC section would not be expected. Secondarily generalized seizures of frontal lobe origin would be expected to respond favorably to CC section and there is evidence that they do (Purves 1991). The special type of frontal lobe complex partial seizures can be difficult to lateralize even with intracranial recording (Williamson 1985). There is accumulating evidence that excellent results can be achieved with CC section in patients with frontal lobe complex partial seizures (Bouvier et al. 1983, Purves 1983, Gates et al. 1984, Williamson 1985, St Hilaire 1987, Murro et al. 1988).

5. Bilateral multifocal disease: There is agreement that patients with bilateral multifocal disease do not do well (Purves 1983, Spencer 1988). Yet there are exceptions (Garcia-Flores 1987). In addition, as noted above, some patients with Lennox-Gastaut syndrome have bilateral, probably multifocal disease. Whether or not these specific patients respond poorly to CC section is not known. Clearly more information is needed.

SUMMARY AND CONCLUSIONS

This communication was designed to present a general overview of the preoperative evaluation of patients for CC section. While the preoperative evaluation for CC section does not differ significantly from the preoperative evaluation of patients being considered for resective surgery, the patient population differs as do the goals of surgery. Most patients undergoing preoperative evaluation for CC section will have severe medically refractory epilepsy with frequent injury-inducing falls and will be mildly to moderately subnormal mentally. While the elimination of seizures has been reported, most patients continue to have seizures following CC section. The goals and expectations, as well as the uncertainties, must be discussed openly with the patient and family initially and throughout the evaluation. Patients and their families are often desperate for help and will grasp at any straws of hope. Because of this, extra care, thought and time must go into counseling in order to define realistic goals and avoid the development of false expectations.

The goal of the preoperative evaluation is to collect as much pertinent information as possible using available technology and specially trained personnel. This information must be compared with currently understood selection criteria and prognostic factors in order to decide whether or not CC section offers a reasonable opportunity to improve the quality of life of each patient. If CC section is done, the data from the preoperative evaluation will also be used for comparison with postoperative results. It is only through comparison of the data from thorough pre- and postoperative evaluations that the uncertainties surrounding CC section can be resolved.

REFERENCES

Beniak, T.E., Gates, J.R., Risse, G.L., 1985, Comparison of selected neuropsychological test variables pre- and post-operatively on patients subjected to corpus callosotomy, *Epilepsia* 26:534.

Berkovic, S.F., Andermann, F., Olivier, A., Ethier, R., Melanson, D., Rohitaille, Y., Kuzniecky, R., Peters, T., and Feindel, W., 1991, Hippocampal sclerosis in temporal lobe epilepsy demontrated by magnetic resonance imaging, *Ann. Neurol.* 29:175-182.

Blume, W.T., 1984, Corpus callosum section for seizure control: rationale and review of experimental and clinical data, *Cleve. Clin. Quar.* 5:319-332.

Bogen, J.E. and Vogel, P.J., 1962, Cerebral commissurotomy in man. Preliminary case report. *Bull. Los Angeles Neurol. Soc.* 27:169-172.

Bogen, J.E. and Vogel, P.J., 1963, Treatment of generalized seizures by cerebral commissurotomy, *Surg. Forum* 14:431-433.

Bogen, J.,1965, Cerebral commissurotomy. A second case report, *JAMA* 194:1328-1329.

Bogen, J.E., Sperry, R.W., and Vogel, P.J., 1969, Commissural section and propagation of seizures, *in:* "Basic Mechanisms of the Epilepsies," H. Jasper, ed. Little, Brown, Boston, pp 439-440.

Bogen, J.E., Fisher, E.D., and Vogel, P.J., 1975, Neurologic status in the long term following complete cerebral commissurotomy, *in:* "Les Syndromes de Disconnexion Calleuse chez l'Homme," F. Michel and B. Schott eds., Hopital Neurologique, Lyon, pp. 227-251.

Boon, P.A., Williamson, P.D., Fried, I., Spencer, D.D., Novelly, R.A., Spencer, S.S., and Mattson, R.H., 1991, Intracranial, intraaxial, space-occupying lesions in patients with intractable partial seizures: an anatomo-clinical, neuropsychological and surgical correlation, *Epilepsia* 32:467-476.

Bouvier, G., Mercier, C., Saint-Hilaire J.M., Giard, N., Labrecque, R., and Weique, R.A., 1983, Anterior callosotomy and chronic depth electrode recording in the surgical management of some intractable seizures, *Appl. Neurophysiol.* 46:52-56.

Chugani, H.T. and Engel, J., 1986, PET in intractable epilepsy, *in:* "Intractable Epilepsy: Experimental and Clinical Aspects," D. Schmidt and P.L. Morselli, eds., Raven Press, New York, pp. 119-128.

Chugani, H.T., Mazziotta, J.C., Engel, J., Phelps, M.E., 1987, The Lennox-Gastaut syndrome: metabolic subtypes determined by 2-deoxy 2[18F]fluro-D-glucose positron emission tomography, *Ann. Neurol.* 21:4-13.

Dandy, W.E., 1931, Congenital cerebral cysts of the cavum septum pellneidi (fifth ventricle) and cavum vergael (sixth ventricle): diagnosis and treatment, *Arch. Neurol. Psychiatr.* 25:44-66.

Engel, J., 1989, The Lennox-Gastaut syndrome, *in:* "Seizures and Epilepsy," J. Engel ed., F.A. Davis, Philadelphia, pp. 203-207.

Ferguson, S.M., Rayport, M., and Currie, W.S., 1985, Neuropsychiatric observations on behavioral consequences of corpus callosum section for seizure control, *in:* "Epilepsy and the Corpus Callosum," A.G. Reeves, ed., Plenum Press, New York, pp. 501-514.

Ferrell, R.B., Culver, C.M., and Tucker, G.J., 1983, Psychosocial and cognitive function after commissurotomy for intractable seizures, *J. Neurosurg.* 58:374-380.

Garcia-Flores, E., 1987, Corpus callosum section for patients with intractable epilepsy, *Appl. Neurophysiol.* 50:390-397.

Gates, J.R., Leppik, I.E., Yap, J., and Gumnit, R.J., 1984, Corpus callosotomy: clinical and electroencephalographic effects, *Epilepsia* 25:308-316.

Gates, J.R., Mireles, R., Maxwell, R., Sharbrough, F., and Forbes, G., 1986, Magnetic resonance imaging, electroencephalogram and selected neuropsychologic testing in staged corpus callosotomy, *Arch. Neurol.* 43:1188-1191.

Gates, J.R., Rosenfeld, W.E., Maxwell, R.E., and Lyons, R.E., 1987, Response of multiple seizure types to corpus callosum section, *Epilepsia* 28:28-34.

Gazzaniga, M.S., Risse, G.L., Springer, S.P., Clark, E., and Wilson, D.H., 1975, Psychologic and neurologic consequences of partial and complete cerebral commissurotomy, *Neurology* 25:10-15.

Geoffroy, G., Lassonde, M., Delisle, F., and Decarie, M.,1983, Corpus callosotomy for control of intractable epilepsy in children, *Neurology* 33:891-897.

Goodman, R.N., Williamson, P.D., Reeves, A.G., Spencer, S, Spencer, D.D., Mattson, R.H., and Roberts, D.W., 1985, Interhemisphere commissurotomy for congenital hemiplegics with intractable epilepsy, *Neurology* 35:1351-1354.

Gotman, J., Ives, J.R., and Gloor, P., eds., 1985, "Long-term Monitoring in Epilepsy," Elsevier, Amsterdam.

Gumnit, R.J., ed., 1987, "Intensive Neurodiagnostic Monitoring," Raven Press, New York.

Kuzniecky, R., Garcia, J.H., Faugut, E., and Morawetz, R.B., 1991, Cortical dysphasia in temporal lobe epilepsy: magnetic resonance imaging correlations, *Ann. Neurol.* 29:293-298.

Luessenhop, A.J., Dela Cruz, T.E., and Fenichel, G.M., 1970, Surgical disconnection of the cerebral hemispheres for intractable seizures, *JAMA* 213:1630.

Marino, R., 1985, Selective partial micro surgical callosotomy for intractable multiform seizures: criteria for clinical selection and results, *Appl. Neurophysiol.* 48:404-407.

Maxwell, R.E., Gates, J.R., and Gumnit, R.J., 1987, Corpus callosotomy at the University of Minnesota, *in:* "Surgical Treatment of the Epilepsies," J. Engel, ed., Raven Press, New York pp. 659-666.

Murro, A.M., Flanigin, H.F., Gallagher, B.B., King, DW, and Smith J.R. , 1988, Corpus callosotomy for the treatment of intractable epilepsy, *Epil. Res.* 2:44-50.

Niedermeyer, E., 1990, Lennox Gastaut syndrome, *in:* "The Epilepsies: Diagnosis and Management," E. Niedermeyer, ed., Urban and Schwarzenberg, Bulimor-Munich, pp. 130-135.

Nordgren, R.E., Reeves, A.G., Viguera, A.C., and Roberts, D.W., 1991, Corpus callosotomy for intractable seizures in the pediatric age group, *Arch. Neurol.* 48:364-372.

Purves, S.J., Wada, J.A., Woodhurst, W.B., Moyes, P.D., Strauss, E., Kosaka, B., and Li, D., 1983, Results of anterior corpus callosum section in 24 patients with medically intractable seizures, *Neurology* 38:1194-1201.

Purves, S.J., 1991, Selection of patients for corpus callosum section, *in:* "Surgery For Epilepsy," S.S. Spencer and D.S. Spencer, eds., Blackwell Scientific Pubs, Boston, pp. 69-85.

Rayport, M., Ferguson, S.M., Corrie, W.S.D., 1983, Outcomes and indications of corpus callosum section for intractable seizure control, *Appl. Neurophysiol.* 46:47-51.

Rasmussen, T., 1978, Further observations on the syndrome of chronic encephalitis and epilepsy, *Appl. Neurophysiol.* 41:1-12.

Rasmussen, T., 1983, Hemispherectomy for seizures revisited, *Can. J. Neurol. Sci.* 10:71-78.

Rasmussen, T., 1987, Commentary: extratemporal cortical excisions and hemispherectomy, *in:* "Surgical Treatment of the Epilepsies," J. Engel, ed, Raven Press, New York, pp. 417-424.
Rausch, R.. 1987, Psychological evaluation, *in:* "Surgical Treatment of the Epilepsies," J. Engel. ed., Raven Press, New York, pp. 181-201.
Reeves, A.G., ed., 1985, "Epilepsy and the Corpus Callosum," Plenum Press, New York.
Roberts, D.W., 1985, Corpus callosotomy: surgical technique, *in:* "Epilepsy and the Corpus Callosum," A.G. Reeves ed., Plenum Press, New York, pp. 259-267.
Roberts, D.W. and Reeves, A., 1987, Effect of commissurotomy on complex partial epilepsy in patients without a resectable seizure focus, *Appl. Neurophysiol.* 50:398-400.
Roberts, D.W., 1991, Corpus callosum section, *in:* "Surgery For Epilepsy," S.S. Spencer, and D.S. Spencer, eds., Blackwell Scientific Pubs., Boston, pp. 168-178.
Saint-Hilaire, J.M., Giard, N., Bouvier, G., and Labreque, R., 1985, Anterior callosotomy in frontal lobe epilepsies, *in:* "Epilepsy and the Corpus Callosum," A.G. Reeves ed., Plenum Press, New York, pp. 303-314.
Sass, K.J., Spencer, D.D., Spencer, S.S., Novelly, R.A., Williamson, P.D., and Mattson, R.H., 1988, Corpus callosotomy for epilepsy. II. Neurologic and neuropsychological outcome, *Neurology* 38:24-28.
Swartz, B. and Delgado-Escueta, A.V., 1987, Complex partial seizures of extratemporal origin: "the evidence for," *in:* "The Epileptic Focus," H.G. Wieser, E. J. Speckman and J. Engel, eds., John Libbey, London-Paris, pp. 137-174.
Spencer, S.S., 1988, Corpus callosum section and other disconnection procedures for medically intractable epilepsy, *Epilepsia* 29 (Suppl 2):585-599.
Spencer, S.S., Gates, J.R., Reeves, A.G., Spencer, D.D., Maxwell, R.E., and Roberts, D., 1987, Corpus callosum section, *in:* "Surgical Treatment of the Epilepsies," J. Engel, ed., Raven Press, New York, pp. 425-444.
Spencer, S.S., Spencer, D.D., Glaser, G.H., Williamson, P.D., and Mattson, R.H., 1984, More intense focal seizure types after callosal section: the role of inhibition, *Ann. Neurol.* 16:686-693.
Spencer, S.S., Spencer, D.D., Williamson, P.D., Sass, K., Novelly, R.A., and Mattson, H., 1988, Corpus callosotomy for epilepsy. I. Seizure effects, *Neurology* 28:19-24.
Sperry, R.W., 1968, Hemisphere disconnection and unity in conscious awareness, *Am. Psychol.* 23:723-733.
Sperry, R.W., Vogel, P.J., and Bogen, J.E., 1970, Syndrome of hemisphere disconnection, *in:* "Proc 2nd Pan-Am Cong Neurol, P. Baily and R. Frol, eds., Puerto Rico, pp. 195-200.
Sussman, N.M., Gur, R.C., Gur, R.E., and O'Connor, M., 1983, Mutism as a consequence of callosotomy, *J. Neurosurg.* 59:514-519.
Tinuper, P., Andermann, F., Villemerre, J.G., Rasmussen, T., and Quesney, L., 1988, Functional hemispherectomy for treatment of epilepsy associated with hemiplegia: rationale, indications, results and comparisons with callosotomy, *Ann. Neurol.* 24:27-34.
Van Wagenen, W.P. and Herren, R.Y., 1940, Surgical division of commissural pathways in the corpus callosum: relation to spread of an epileptic attack, *Arch. Neurol. Psychiat.* 44:740-759.
Waterman, K., Purves, S.J., Kosaka, B., Strauss, E., and Wada, J.A., 1987, An epileptic syndrome caused by mesial frontal lobe seizure foci, *Neurology* 37:577-582.
Williamson, P.D., 1985, Corpus callosum section for intractable epilepsy, *in:* "Epilepsy and the Corpus Callosum," A.G. Reeves, ed., Plenum Press, New York, pp. 243-257.
Williamson, P.D., 1991, Evaluation of patients for epilepsy surgery at the West Haven VA/Yale-New Haven Epilepsy Unit, *in:* "Surgery for Epilepsy," S.S. Spencer and D.D. Spencer, eds., Blackwell Scientific Pubs, Boston, pp. 36-52.
Williamson, P.D., Boon, P.A., Thadani, V.M., Darcey, T.M., Spencer, D.D., Spencer, S.S., Novelly, R.A., and Mattson, R.H., 1992, Parietal lobe epilepsy: diagnostic considerations and results of surgery, *Ann. Neurol.* 31:193-201.
Williamson, P.D., Spencer, D.D., Spencer, S.S., Novelly, R., and Mattson, R.H., 1985, Episodic aphemia and epileptic focus in nondominant hemisphere: relieved by section of corpus callosum, *Neurology* 35:1069-1071.
Williamson, P.D., Spencer, D.D., Spencer, S.S., Novelly, R.A., and Mattson, R.H., 1985, Complex partial seizures of frontal lobe origin, *Ann. Neurol.* 18:497-504.

Williamson, P.D., Thadani, V.M., Darcey, T.M., Spencer, D.D., Spencer, S.S., and Mattson, R.H., 1992, Occipital lobe epilepsy: clinical characteristics, seizure spread patterns, and results of surgery, *Ann. Neurol.* 31:3-13.

Williamson, P.D., Wieser, H.G., and Delgado-Escueta, A.V., 1987, Clinical characteristics of partial seizures, *in:* "Surgical Treatment of the Epilepsies," J. Engel, ed., Raven Press, New York, pp. 101-120.

Wilson, D.H., Culver, C., Waddinton, M., and Gazzinaga, M., 1975, Disconnection of the cerebral hemispheres: an alternative to hemispherectomy for the control of intractable seizures, *Neurology* 25:1149-1153.

Wilson, D.H., Reeves, A., Gazzaniga, M., and Culver, C., 1977, Cerebral commissurotomy for control in intractable seizures, *Neurology* 27:708-715.

Wilson, D.H., Reeves, A., and Gazzaniga, M., 1982, Central commissurotomy for intractable generalized epilepsy: series two, *Neurology* 32:687-697.

Woodhurst, W.B., Wada, J.W., and Purves, S.J., 1989, Speech disturbances after anterior callosal section, *Epilepsia* 30:677-8.

Wylie, E., 1988, Corpus callosotomy for intractable generalized epilepsy, *J. Pediatrics* 113:255-261.

EEG SELECTION FOR CORPUS CALLOSOTOMY

John R. Gates, M.D.

The Minnesota Epilepsy Group, P.A.
310 N. Smith Ave., Suite 300
St. Paul, MN 55102

INTRODUCTION

A consistent ethos has characterized the approach to callosotomy since the beginning. As reported in several previous studies, the effect of corpus callosum division on the appearance of electroencephalographic epileptiform discharges is quite dramatic.[1-10] In 1940 when VanWagenen and Herren[11] published their first ten cases of corpus callosum division based upon purely clinical grounds, they stated "as a rule, consciousness is not lost when the spread of the epileptic wave is not great or when it is limited to one cerebral cortex." Consequently, the concept of a dramatic electrophysiologic change as a consequence of corpus callosum division has been pervasive since the beginning of this procedure. Is this effect, however, more apparent than real?

HISTORY

The electroencephalographic effects of division of the corpus callosum were first reported by Hursh in 1945[12] when he reviewed two cases of corpus callosotomy from the VanWagenen and Herren series. Both cases were partial callosotomy and, in both, there was persistence of bilateral synchronous discharges documented despite the apparent seizure improvement post-surgically. In 1962, Bogen[13] and his colleagues began to report the results of their series of 25 cases. The electroencephalographic consequences were reported in an abstract in 1965.[14] In that paper, multiple epileptogenic foci on the EEG were demonstrated preoperatively. In some, there was a demonstration of a progression from focal to generalized epileptiform disturbance following the operation which was essentially a variably complete commussurotomy. New abnormalities appeared which gradually were resolved with the two

Epilepsy and the Corpus Callosum II, Edited by A.G. Reeves
and D.W. Roberts, Plenum Press, New York, 1995

hemispheres "tending to follow separate courses of subsistence." The final comment was that in several cases EEGs had become essentially normal with the patient continuing on medication.

In 1970, Luessenhop[15] reported a series of four patients and commented specifically that the best results were obtained in children with clear cut hemispheric lateralization as evidenced by the characteristics of seizures and the EEG. Poor results in this small group were associated with EEGs demonstrating independent bilaterally spiking foci. In 1976, Amacher[16] reported the results of four patients ages 14 to 20. In all cases, significant improvement was observed for a variably complete commissurotomy and major motor seizures. Three cases had an improvement in atonic seizures and two had apparent improvement in partial seizures. All cases had electroencephalograms which demonstrated diffuse and multifocal seizure discharges confined to one hemisphere beginning unilaterally and spreading to the opposite hemisphere. In 1975, Wilson published the first of the Dartmouth series[17] and this was the first comprehensive effort to document the efficacy of corpus callosum section on seizure control. The surface recorded EEG data documented transformation of previously generalized epileptiform bursts into more lateralized bursts with cases that included bilateral independent foci. The procedure was performed with considerable variability as to the extent of the division of the commissural pathways, though in the last 12 cases as summarized by Reeves and O'Leary[18] complete section of the corpus callosum and hippocampal commissure was apparently consistently effected.

RECENT STUDIES

The more recent series from Yale, Minnesota, and Dartmouth[1-10, 19] have attempted to standardize to a certain extent the procedures employed for callosotomy. The Yale series of stage callosotomy has demonstrated dramatic reduction in generalized epileptiform discharges postoperatively even with partial callosotomy. The Minnesota experience further supports the observation that callosotomy can be effective in reducing generalized epileptiform discharges. In more recent times, several other centers have become involved in the callosotomy experience including Brazil;[19,21] the two centers in Montreal;[22,23,24] Memphis;[25] Augusta, Georgia;[26] Italy;[27,28] Osaka, Japan;[29] Mexico.[30] Each of these publications has focused more on the seizure outcome and is discussed elsewhere in this volume.

The Montreal Neurological Institute series[22] attempted to identify EEG patterns preoperatively that could predict postoperative outcome. Their preoperative EEG studies included tracings after reduction or cessation of antiepileptic medication with recordings during sleep and, if possible, recording of ictal events. However, it is not stated whether all the patients had recorded ictal events for analysis. Consequently, comments related to EEG are restricted to interictal EEG patterns. They divided the 43 patients into four groups according to the amount of generalized spike or sharp and slow wave discharges, lateralization of the electroencephalographic abnormality and the presence of interictal focal epileptiform discharges. There was a trend for lateralization of the interictal EEG pattern in the drop attacks and generalized spike and wave groups to distinguish the eventual outcome.

Marino et al. from Brazil[21] in Sao Paulo also reported 35 patients with medically refractory multiform seizures submitted to corpus callosum division guided by intraoperative electroencephalographic and electrocorticographic monitoring. As in the Montreal series[22], interictal EEGs were performed preoperatively demonstrating bilateral synchronous, slow spike and wave paroxysms with or without focal accentuation intraoperatively. Scalp and electrocorticograph recordings were concomitantly obtained indicating the sites of abnormal electrical activity. Apparently hyperpnea or brevital activation was utilized to observe abnormal activity intraoperatively. Only part of the frontal corpus callosum was sectioned to interrupt the bilateral synchrony of the epileptic discharges and 28 of the patients who had adequate long-term follow-up had significant improvement in seizure frequency. There are no specific comments regarding preoperative EEG patterns or intraoperative pattern predictive of the eventual outcome.

EEG PATTERNS AND PREOPERATIVE PREDICTION

One of the greatest frustrations in the electrodiagnostic evaluation for callosotomy is the considerable variation in selection criteria from various centers. At one end of the spectrum, patients are selected based upon a history of tonic or atonic seizures and a very active multifocal and/or generalized epileptiform interictal abnormality. At the other end, multiple ictal events are recorded in an attempt to identify a characteristic EEG pattern.

At The Minnesota Epilepsy Group of United and Children's Hospitals, the preoperative electrodiagnostic evaluation requires the recording of ictal events to precisely identify the seizure types and the electroencephalographic correlate.[10] A history of drop seizures may not prove, on monitoring, to be tonic or atonic seizures for which the callosotomy procedure has been demonstrated by multiple studies to be reasonably successful. It is possible, for example, that a patient with a unilateral complex partial seizure can fall with propagation of the epileptic discharge to the other hemisphere or the patient might have pure tonic-clonic events with falls. Consequently, at least ten ictal events are recorded and it is preferred to try to record these patients in a modified sailing sling suspended from the ceiling to document the capacity for falling and injury without sustaining such injuries. All patients selected for callosotomy at The Minnesota Epilepsy Group have documented secondarily generalized epileptiform discharges that accompany the clinical seizures with atonic or tonic seizures resulting in the falls and injury. Patients are excluded in which a single epileptiform focus can be identified in a surgically resectable location that could clearly account for the secondarily generalized events. Interictally, all patients had active multifocal epileptiform abnormalities. One publication, Fiol and Gates[1] suggested that an epileptiform fast ictal expression with tonic or atonic seizures was more likely to respond to the surgery than a slow spike and wave, irregular spike and wave, or regular spike and wave pattern.

The data has recently been updated by The Minnesota Epilepsy Group.[31,32] Fifty-nine patients, ages three through ten, who had partial, complete, or two-stage complete corpus callosum division were reviewed retrospectively to identify predictive factors for outcome. Two specific factors were reviewed: seizure type and ictal EEG pattern.

The 37 patients, ages 10-49 (mean 21.8 years), had had uncontrolled tonic, atonic, or tonic clonic seizures. Other seizure types were invariably present in these patients. Electrodecremental epileptiform fast and slow spike and wave were the ictal EEG patterns seen in these patients. Satisfactory outcome was defined as an 80 percent reduction in primary seizure type. Follow-up was 0.5-10 years with a mean of 3.35 years.

Seventy-five percent of patients achieved a satisfactory outcome. Seventy-five percent of patients with tonic seizures had a successful outcome compared to 25 percent with tonic clonic seizures ($P < .05$). There was a trend toward more successful outcome associated with the electrodecremental epileptiform fast ictal pattern ($P < .10$). However, when the patients with tonic seizures and the electrodecremental epileptiform fast pattern were combined, 92 percent had a successful outcome compared to 33 percent of patients not demonstrating both features ($P < .001$). Consequently, in the greater than age ten group, tonic seizures associated with an ictal epileptiform fast EEG pattern appears to predict a more successful outcome.

When this data was reviewed in the 22 patients ages 3-9 (mean 5.8), however, 53 percent who had tonic seizures and 50 percent who had atonic seizures as the primary seizure type demonstrated a satisfactory outcome. When these patients were analyzed relative to an enhanced outcome based upon an electrodecremental epileptiform fast versus a slow spike and wave ictal pattern, an improved outcome was not demonstrated. Similarly, those displaying primarily tonic seizures and electrodecremental epileptiform fast activity in combination did not demonstrate a significantly better outcome (56 percent) than those without these features (50 percent). Consequently, in the less than ten age group, neither tonic nor atonic seizure type, ictal EEG pattern, nor these factors in combination predicted a satisfactory outcome. Nonetheless, 50 percent achieved an 80 percent reduction in seizure frequency postoperatively.

POSTOPERATIVE EFFECTS

Torres and French[33] were the first to report an acute effect of resection of the corpus callosum upon epileptiform activity from the EEG recorded intraoperatively. They demonstrated resolution of both apparently independent foci and secondarily generalized epileptiform discharges in the spared hemisphere immediately after resection of the corpus callosum in preparation for a subsequent hemispherectomy.

A recent publication reviewed the value of intraoperative EEG changes during corpus callosotomy in predicting surgical results.[34] The intraoperative transformation of generalized epileptiform discharges to lateralized epileptiform activity during the course of corpus callosum resection for intractable epilepsy was reviewed in 37 patients. The percentage decrease in atonic-tonic seizures at a mean follow-up of 26 months, range 12-86 months, was documented. Twenty-seven patients

(73%) had intraoperative interictal discharges and 21 (78%) showed varying degrees of lateralization of the generalized epileptiform discharges during the corpus callosum division which was two-thirds to complete. All patients experienced a greater than 80% reduction in atonic or tonic seizures with falls. The group with the largest decrease in generalized epileptiform discharges had the greatest decrease in seizures (95.5%).

Those patients without the changes in the generalized epileptiform discharges had an 88% decrease in seizures (14 patients) with mild or moderate decreases in the discharges. There was no statistically significant correlation between the decrease in the generalized epileptiform discharges and seizure frequency after operation. Although lateralization of epileptiform discharges after corpus callosum resection was evident in 78% of patients, the degree of lateralization did not correlate with the degree of reduction in tonic-atonic seizures.

A previous study[9] does demonstrate this dramatic change in the EEG from corpus callosum division intraoperatively. It is quite striking and intuitively difficult to conclude that such a dramatic change does not appear to correlate with subsequent seizure outcome.

Another curious observation of postoperative effects of callosum section upon the interhemispheric EEG coherence was made by Montplaisir et al.[35] from Montreal documenting a dramatic postoperative change in a patient subjected to anterior section and posterior section; both studies documenting a dramatic generalized decrease in EEG coherence in most frequency bands and over most brain regions samples. There was, however, no statement regarding the significance of this observation as a prognostic factor.

A paper from Montreal discussed frontal lobe seizures in which frontal topectomies were combined with anterior callosotomy.[35] Twenty-five patients had implanted depth electrodes using the technique described by Talairach.[37] Thirteen of twenty-three patients implanted had a frontal corticectomy and an anterior callosotomy. They concluded, "The addition of anterior callosotomy to frontal lobe corticectomy was of no value." Overall Engel Class I outcomes were on the order of 30%.

A final electrophysiologic discussion revolves around the role of the corpus callosum for secondary bilateral synchrony. Spencer, et al.[38] have previously suggested that the corpus callosum in the human is the exclusive pathway for secondary bilateral synchrony and that in patients subjected to complete callosotomy, ictal expression postoperatively no longer allows secondary bilateral synchrony. Gates, et al., in a recent publication,[39] documented persistence of secondary bilateral synchrony postoperatively suggesting that the corpus callosum is not the exclusive pathway for such ictal electroencephalographic expression.

CONCLUSIONS

1. The dramatic improvement in apparently secondarily generalized seizures documented in many patients argues persuasively that the corpus callosum must be a significant pathway for bilateral synchrony but not the exclusive pathway.
2. Satisfactory outcomes have been documented in cases with bilateral independent epileptic foci with capacity for secondary generalization

as well as patients with more unilateral epileptogenesis.

3. The extent of corpus callosum division required to effect a satisfactory outcome is being elucidated as suggested from the Oguni, et al. study from Montreal.[22] Two-thirds division appears to be more effective in seizure improvement than one-half though, as evidenced from the Minnesota study, the difference between two-thirds and complete is unclear. Further studies are required.
4. A pre-operative ictal Electrodecremental epileptiform fast EEG pattern for patients ≥ age ten with tonic seizures appears to predict a better surgical outcome.
5. More uniformity is required in patient evaluation, follow-up in reporting as well as surgical technique for corpus callosum division. This will enable a more precise assessment of outcome. Surface EEG recording of ictal events with 24-hour, long-term monitoring for epilepsy appears to be sufficient for patient selection in most cases. Prospective studies need to be undertaken to compare the utility of an anterior 2/3 callosal section as the initial phase for treatment for patients with partial seizures with rapid secondary generalization which are of frontal origin.

The electrophysiological and clinical outcome of two-thirds or more division of the corpus callosum is both apparent and real in those patients appropriately selected.

REFERENCES

1. M.D. Fiol, and J.R. Gates, EEG studies and corpus callosotomy results, *Electroencephalogr. Clin. Neurophysiol.* 58(2):34 (1984).
2. J.R. Gates, EEG selection for corpus callosotomy, *in:* "Surgical Treatment of Epilepsy," (Epilepsy Res. Suppl. 5), W.H. Theodore, ed., Elsevier Science Publishers, B.V. (1992).
3. J.R. Gates, I. Leppik, J. Yap, and R. Gumnit, Corpus callosotomy clinical and electroencephalographic effects, *Epilepsia* (1984).
4. D.H. Wilson, C. Culver, M. Waddington, and M. Gazzaniga, Disconnection of the cerebral hemispheres: an alternative to hemispherectomy for the control of intractable seizures, *Neurology* 25:1149-1153 (1978).
5. D.H. Wilson, A. Reeves, and M.S. Gazzaniga, Division of the corpus callosum for uncontrollable epilepsy, *Neurology* 28:649-653 (1978).
6. D.H. Wilson, A.G. Reeves, and M.S. Gazzaniga, "Central" commissurotomy for intractable generalized epilepsy: Series two, *Neurology* 32:687-697 (1982).
7. D.H. Wilson, A. Reeves, M. Gazzaniga, and C. Culver, Cerebral commissurotomy for control of intractable seizures, *Neurology* 27:708-715 (1977).
8. S.S. Spencer, D.D. Spencer, P.D. Williamson, R.H. Mattson, and K.J. Sass, Corpus callosotomy for epilepsy: I. Seizure effects, *Neurology* 38:19-24 (1988).
9. S.S. Spencer, J.R. Gates, A.G. Reeves, D.D. Spencer, R. Maxwell, and D. Roberts, Corpus callosum section for uncontrolled epilepsy, *in:* Surgical Treatment of the Epilepsies," J. Engel, ed., Raven Press, New York (1986).
10. J.R. Gates, J.A. Wada, A.G. Reeves, M. Lassonde, I. Papo, S.S. Spencer, F. Andermann, G. Risse, F.J. Ritter, and S.J. Purves, Reevaluation of corpus callosotomy, *in:* "Surgical Treatment of the Epilepsies," Second Edition, J. Engel, ed., Raven Press, New York (1993).
11. W.P. Van Wagenen and R.U. Herren, Surgical division of commissural pathways in the corpus callosum, *Acta. Neurol. Psychiatr.* 44:740-759 (1940).
12. J.B. Hursh, Origin of the spike and wave pattern of petit mal epilepsy: an electroencephalographic study, *Arch. Neurol. Psychiatr.* 45:274-282 (1945).

13. J.E. Bogen and P.J. Vogel, Cerebral commissurotomy in man: preliminary case report, *Bull. LA Neurol. Soc.* 27:169-172 (1962).
14. J.E. Bogen, E.D. Fisher, and P.H. Vogel, Cerebral commissurotomy: a second case report, *JAMA* 194:1328-1329 (1965).
15. A.J. Luessenhop, T.C. de la Cruz, and G.M. Genichel, Surgical disconnection of the cerebral hemispheres for intractable seizures: results in infancy and childhood, *JAMA* 213:1630-1636 (1970).
16. A.L. Amacher, Midline commissurotomy for the treatment of some cases of intractable epilepsy, *Child Brain* 2:54-58 (1976).
17. D.H. Wilson, C. Culver, M. Waddington, and M. Gazzaniga, Disconnection of the cerebral hemispheres, *Neurol.* 25:1149-1153 (1975).
18. A.G. Reeves and P.M. O'Leary, Total corpus callosotomy for control of medically intractable epilepsy, *in:* "Epilepsy and the Corpus Callosum," A.G. Reeves, ed., Plenum Press, New York pp. 269-280 (1985).
19. J.R. Gates, F.J. Ritter, P.C. Ragazzo, A.G. Reeves, and R.E. Nordgren, Corpus callosum section in children: seizure responses, *J. Epilepsy* 3(Suppl.):271-278 (1990).
20. R. Nordgren, A.G. Reeves, A. Viguera, and D.W. Roberts, Corpus callosotomy for intractable seizures in the pediatric age group, *Arch. Neurol.* 48(4):364-372 (1991).
21. R. Marino, J. Radvany, F.R. Huck, C.H.P. DeCamargo, and G. Gronich, Selective electroencephalograph-guided microsurgical callosotomy for refractory generalized epilepsy, *Surg. Neurol.* 34:219-228 (1990).
22. H. Oguni, A. Olivier, F. Andermann, and J. Comair, Anterior callosotomy in the treatment of medically intractable epilepsies: a study of 43 patients with a mean follow-up of 39 months, *Ann. Neurol.* 30(3):357-364 (1991).
23. G. Geoffroy, M. Lassonde, F. Delisle, and M. Decarie, Corpus callosotomy for control of intractable epilepsy in children, *Neurol.* 33:891-897 (1983).
24. G. Geoffroy, M. Lassonde, H. Sauerwein, and M. Decarie, Effectiveness of corpus callosotomy for control of intractable epilepsy in children, *in:* "Two Hemispheres, One Brain," F. Lepore, M. Ptito, H.H. Jasper, eds., Alan Liss, New York, pp. 361-368 (1986).
25. K.S. Fuiks, A.R. Wyler, B.P. Hermann, and G. Somes, Seizure outcome from anterior and complete corpus callosotomy, *J. Neurosurg.* 74:573-578 (1991).
26. J.R. Smith, H.F. Flanigin, D.W. King, B.B. Gallagher, D.W. Loring, K.J. Meador, A.M. Murro, and G.P. Lee, Surgical management of epilepsy, *South. Med. J.* 82(6):736-742 (1989).
27. I. Papo, A. Quattrini, L. Provinciali, F. Rychlicki, M. Del Pesce, A. Paggi, F. Ortenzi, A. Recchioni, and B. Censori, Callosotomy for the treatment of drug resistant generalized seizures, *Acta Neurochirurgica Suppl.* 50:134-135 (1990).
28. I. Papo, A. Quattrini, L. Provinciali, F. Rychlicki, A. Paggi, M. Del Pesce, and A. Ortenzi, Callosotomy for the management of intractable non-focal epilepsy: a preliminary personal assessment, *Acta Neurochirurgica* 96:46-53 (1989).
29. M. Goto, J. Okuda, Y. Ikejiri, T. Nishikawa, M. Hirose, H. Tanabe, Y. Nii, S. Nakatani, and J. Shiraiski, Somatosensory evoked potentials to median nerve stimulation after partial section of the corpus callosum, *J. Neurol.* 238:161-167 (1991).
30. E. Garcia-Flores, Corpus callosum section for patients with intractable epilepsy, *Appl. Neurophysiol.* 50:390-397 (1987).
31. W. Courtney, J. Gates, F. Ritter, M.D. Frost, et al., Prediction of seizure outcome following corpus callosotomy in patients ten years or older, *Epilepsia* 34(S6):43 (1993).
32. J. Gates, W. Courtney, F. Ritter, et al., Prediction of seizure outcome following corpus callosotomy among young children, *Epilepsia* 34(S6):111 (1993).
33. F. Torres and L.A. French, Acute effect of section of the corpus callosum upon "independent" epileptiform activity, *Acta Neurol. Scan.* 49:47-62 (1973).
34. M.E. Fiol, J.R. Gates, R. Mireles, R.E. Maxwell, and D.M. Erickson, Value of intraoperative EEG changes during corpus callosotomy in predicting surgical results, *Epilepsia* 34(1):74-78 (1993).
35. J. Montplaisir, R. Nielsen, J. Cote, D. Boivin, I. Rouleau, and G. Lapierre, Interhemispheric EEG coherence before and after partial callosotomy, *Clinical Electroencephalography* 21:1 (1990).

36. P. Chauvel, A.V. Delgado-Escueta, et al., eds., Frontal lobe seizures and epilepsy, indications for cortectomies or callosotomies, *in:* "Advances in Neurology, Vol. 57, " Raven Press, New York (1992).
37. J. Talairach and J. Bancaud, eds., Stereotaxic approach to epilepsy in progress, *in:* "Neurological Surgery, Vol. 5," Karger, New York, pp. 297-354.
38. S.S. Spencer, D.D. Spencer, G.H. Glaser, P.D. Williamson, and R.H. Mattson, More intense focal seizure types after callosal section: the role of inhibition, Ann. Neurol. 16(6):686-693 (1984).
39. L. DePaola, J.R. Gates, F.J. Ritter, M.B. Dunn, M.D. Frost, and P.E. Penovich, Generalized ictal electroencephalographic features after total corpus callosotomy. Abstract presented at Eastern Association of Electroencephalographers 53rd Annual Ski Meeting, La Sapiniere, Val David, P.Q., February 18-20 (1993).

SEIZURE TYPES: RESULTS OF PARTIAL AND COMPLETE CALLOSOTOMY IN ADULTS

Susan S. Spencer, M.D. and Dennis D. Spencer, M.D.

Yale University School of Medicine
New Haven, CT 06510

INTRODUCTION

Corpus callosum section was initially conceptualized as a means of preventing seizure generalization by interrupting the major interhemispheric commissural systems (Van Wagenen and Herren 1940). Section of the anterior commissure, hippocampal commissure, corpus callosum, and one fornix were performed, and effective, but accompanied by an unacceptably high complication rate. Successive modification resulted in a procedure that now involves section of only the corpus callosum and sometimes the hippocampal commissure. The pioneering work of Wilson and his colleagues established that the desirable reduction in generalized seizures was realized despite the persistence of some commissural connections (Wilson et al. 1977, 1978, 1982).

Subsequently, staged operations were instituted, again to reduce post-operative morbidity. The excellent response of generalized seizures in some of the patients without completion of the second stage, as well as additional observations in animal models, culminated in the use of only anterior corpus callosum section at some centers (Oguni et al. 1991; Purves et al. 1988; Huck et al. 1980). Their reports indicate equivalent response rates of generalized seizures with reduced morbidity and neuropsychological sequelae. On this basis it would seem appropriate to restrict operations to the anterior two-thirds of the corpus callosum, but confusing statements with respect to desired results and degree of section have obstructed definitive analysis of reported data.

While a decrease or elimination of all seizure types was sought by those who initially applied this technique to uncontrolled epilepsy, others selected patients with drop attacks, falls, and atonic seizures as the most appropriate candidates. Gates et al. began the trend towards classifying specific seizure types and their responses, an important but not uniformly

Epilepsy and the Corpus Callosum II, Edited by A.G. Reeves
and D.W. Roberts, Plenum Press, New York, 1995

utilized gauge of outcome (Gates et al. 1987). The variable categorization, length of follow-up, and age of the patient populations at different centers has also confounded interpretation of the reported literature with respect to the effects of the varieties of corpus callosum section on specific seizure types. As a result, continued controversy exists where it should not.

Sufficient published information is now available to approach specific questions, including the differential effects of corpus callosum section on adults and children, the longevity and evolution of seizure responses, the contrasting effects of partial and total section performed as single procedures or in tandem, and the response of partial seizures by themselves to these operations. We have analyzed the literature with regard to these questions. For these analyses, we considered adults greater than 16 years of age, generalized seizures to include myoclonic, atonic, tonic, absence, and generalized tonic clonic types, and partial seizures to include simple and complex partial types. We used a combination of new data from 50 patients at Yale and published reports of adults from other centers (Garcia-Flores 1987; Purves et al. 1988; Papo et al. 1989; Marino 1985; Rapaport and Lerman 1988; Fuiks et al. 1991; Wilson et al. 1977, 1978, 1982; Gates et al. 1987; Oguni et al. 1991; Van Wagenen and Herren 1940; Bogen and Vogel 1962; Huck et al. 1980; Murro et al. 1988; Spencer et al. 1988). This combined information provides a total of 213 adults with partial corpus callosum section and 53 adults with total corpus callosum section from whom our conclusions regarding seizure responses in adults are drawn. These are contrasted with results in 64 children reported in the literature of whom 30 underwent total, and 34 underwent anterior corpus callosum section (Garcia-Flores 1987; Purves et al. 1988; Papo et al. 1989; Marino 1985, Rapaport and Lerman 1988; Luessenhop 1970; Lussenhop et al. 1970; Geoffroy et al. 1983; Wilson et al. 1977, 1978, 1982; Van Wagenen and Herren 1940; Amacher 1976; Murro et al. 1988; Spencer et al. 1988). On the basis of these results, we believe that some order to the persistent confusion regarding patient selection for callosotomy is possible.

EFFECTS OF PARTIAL AND TOTAL CORPUS CALLOSUM SECTION ON DIFFERENT SEIZURE TYPES IN ADULTS (TABLE 1)

Mixed ages of patients have traditionally been reported with respect to seizure effects of callosotomy, and this practice may have obscured important differences between children and adults with respect to responses of specific seizure types. We categorized all published reports of adults after partial (n=213) and total (n=53) corpus callosum section with respect to response of specific seizure types into two categories for ease of comparison and interpretation: cured or marked diminution (>80% reduction in seizure frequency), and no worthwhile improvement (<80% decrease in seizure frequency). The best response after anterior corpus callosum section in adults was seen for atonic seizures (69% cured or markedly decreased), but tonic clonic seizures responded nearly as well (62%). Tonic seizures (51%) and absence seizures (34%) had less favorable responses as did partial seizures (35%). The favorable response of partial seizures as often as some types of generalized seizures is unexplained and remarkable.

Table 1. Response of various seizure types to anterior and total callosotomy in adults (N=263)

Seizure Type	Cured or Markedly Diminished		No Worthwhile Improvement	
	Anterior	Total	Anterior	Total
partial	43 (35%)	16 (42%)	76 (64%)	22 (58%)
tonic clonic	90 (62%)	27 (73%)	56 (38%)	10 (27%)
tonic	18 (51%)	3 (60%)	17 (49%)	2 (40%)
atonic	27 (69%)	7 (78%)	12 (31%)	2 (22%)
drops	52 (62%)	13 (68%)	32 (38%)	6 (32%)a
absence	17 (34%)	2 (40%)	33 (66%)	3 (60%)
myoclonic	12 (60%)	1 (100%)	8 (40%)	0

Collected reports of response to total corpus callosum section in adults reveal an approximately 10% higher response rate for all of the seizure types, as compared to anterior section results. Atonic seizures continue to show the highest rate of improvement (78%), while tonic clonic (73%) and tonic (60%) seizures assume second and third places again. Partial seizure response rose also, to 42%, nearly equivalent to that of absence seizures (40%).

Several interesting observations can be made with respect to these statistics regarding the effect of anterior and total corpus callosum section on various seizure types in adults. First, a marked difference in the response of the various generalized motor seizure types (tonic, atonic, tonic clonic) is not observed in adults after anterior or total section; accordingly, selection of adult patients for this operation on the basis of atonic seizures and drop attacks only is not appropriate. Counseling should reflect the expected improvement in tonic clonic and tonic seizures also. Secondly, collected results of improvement after anterior or total corpus callosum section demonstrate higher response rates after total section.

EEG changes support the clinical observations in adults (Spencer et al. 1993). We recently analyzed ictal EEG patterns before and after anterior callosotomy in 18 predominantly adult patients, and before and after total callosotomy in ten predominantly adult patients. Bilaterally synchronous seizure onset recorded before surgery was lateralized in 5/11 anterior section patients and all five total section patients. Seven of 18 anterior section patients and 5/10 total section patients had more precise localization of seizure onset also; this localization was posterior only after total section, but frontal after either anterior or total sections. Thus, EEG data during seizures supports the conclusion that the corpus callosum may indeed be the sole pathway for production of bilateral synchronous seizure discharge in patients with secondarily generalized seizures. This observation stands in contrast to the situation with regard to interictal spikes and effects of callosotomy, since interictal bisynchronous spikes are often decreased but hardly ever eliminated (Spencer et al. 1985).

CALLOSOTOMY IN ADULTS VERSUS CHILDREN

We looked then at collective reports of response of various seizure types to anterior and total corpus callosum section in children under age 16 as a separate group and compared these statistics to those in adults noted above. Children may have a different spectrum of pathology and may thus show a different pattern of improvement. A summary of this analysis is given in Table 2. In children under 16, atonic seizures responded much better than any other generalized seizure type to anterior corpus callosum section (80% versus 25-30% for absence, tonic, tonic clonic). Total corpus callosum section in children also produced improved response rates, but these increases were much more profound for tonic clonic, tonic, and absence seizures, all of which rose to 80% or more excellent response. Thus, by grouping patients of different ages, the contrasting responses in children and adults have been diluted in prior reports. Centers with a greater representation of one age group found different results from others, generating a perpetual and artificial controversy. At least from these published collected reports, it can be said that adults and children both achieve approximately 80% cure or marked decrease in atonic, tonic, tonic clonic, and absence seizures after total corpus callosum section, but that adults and children differ strikingly in their response pattern to partial section. Anterior section is most effective for atonic seizures in children, but less effective for other generalized seizure types than in adults.

Table 2. Comparison of seizure effects after anterior and total callosotomy in children (64 reported patients).

Seizure Type	Cured or Markedly Diminished		No Worthwhile Improvement	
	Anterior	Total	Anterior	Total
partial	4 (36%)	4 (44%)	7 (64%)	5 (56%)
tonic clonic	7 (25%)	19 (79%)	21 (75%)	5 (21%)
tonic	3 (30%)	3 (100%)	7 (70%)	0
atonic	4 (80%)	7 (70%)	1 (20%)	3 (30%)
drops	1 (25%)	0	3 (75%)	0
absence	1 (25%)	5 (83%)	3 (75%)	1 (17%)
myoclonic	3 (75%)	2 (40%)	1 (25%)	3 (60%)

These observations do not mitigate the beneficial effects of callosotomy on generalized seizures, but merely place them in perspective in different age groups and for different procedures. It may be appropriate to indicate to adults the expected response of all generalized seizure types to anterior section, and imply a minimal but definite gain in response when total section is performed. In children, on the other hand, atonic seizures can be expected to respond dramatically to anterior section, but other generalized types do not appear to have this similar response, and total section may be indicated more often. It is important to recognize, however, that the collective results of responses to anterior and total corpus callosum section in any groups of patients do not necessarily address the utility of completing an anterior section when it has failed to achieve the desired result. This is because patient selection for anterior or total procedures as reported in the literature may have been biased to certain seizure types, syndromes, ages, or extent of seizure activity.

THE EFFECT OF COMPLETION OF FAILED ANTERIOR CORPUS CALLOSUM SECTION

As noted, the higher response rate after total section in various seizure types in different age groups does not necessarily mean that an improvement should be expected if a failed anterior section is completed. Recent reports raise doubts as to such incremental effects of completion (Purves et al. 1988; Fuiks et al. 1991), although documentation of improvement with completion of corpus callosum section is available in scattered publications (Van Wagenen and Herren 1940; Papo et al. 1989; Spencer et al. 1987).

We recently analyzed 14 patients at our center who had corpus callosum section completed after anterior 2/3 section failed to achieve the desired response of generalized seizures (Spencer et al. 1993). These patients were predominantly adults. In contrast with previous reports, we found a definite improvement in response of generalized seizures when a failed anterior section was completed. Of ten patients with persistent tonic clonic seizures in this group, six were cured or had marked diminution in those seizures. Two of four patients with tonic, and two of five patients with atonic seizures, were also dramatically improved by completion of the section. Partial seizures did not change appreciably (two of 14 improved). In a comparison of patients who responded to the anterior 2/3 section (n=25) and those who required the completion (n=14), the presence of more than two seizure types, a non-focal ictal EEG, and a non-focal imaging analysis was predictive of failure of the anterior section to control generalized seizure types.

This observation suggests that more diffuse cerebral damage associated with uncontrolled generalized seizures is more likely to require total corpus callosum section, at least in adults. The reported difference between children and adults implies that such information may be particularly important in children, since they seem to require total section more often, but no definitive data is available to specifically address the application of our predictive factors for total corpus callosum section to younger age groups.

PERSISTENCE OF CALLOSOTOMY SEIZURE EFFECTS

The response of both generalized and partial seizures to corpus callosum section of any degree appears to be long lasting (Spencer et al. 1992). Based on a recent review, in fact, continued improvement is to be expected. Of 25 predominantly adult patients at Yale with more than two year follow-up after callosotomy, 52% had remission of all generalized seizures at one year, while 70% were in total remission at latest follow-up. Sixty eight percent of these patients had some decrease in partial seizures after one year and 80% had a decrease at latest follow-up (compared to preoperatively). Very few of these patients changed status for the worse with respect to generalized or partial seizure frequency with ensuing years.

PARTIAL SEIZURES AFTER CORPUS CALLOSUM SECTION

A variety of alterations in partial seizures besides the above documented decrease in frequency have also been reported (Spencer et al. 1984). New varieties of partial seizures may appear; these occurred in six of 41 patients at Yale after partial section and five of 20 after total section. These new partial seizures may be mild, but can be rather intense and include alternating hemiclonic events with a prominent Todd's paralysis. We observed the latter in five out of 41 patients after partial section and three out of 20 after total section. The relative rarity of these events has so far eluded our ability to predict their occurrence. The argument continues as to whether these new partial seizures are "aborted" generalized seizures, despite their sometimes entirely new symptomatology and occasionally a frequency that is greater than all seizure types combined prior to corpus callosum section. Nevertheless, so far their phenomenology has defied etiologic interpretation. Partial seizures may also become increasingly frequent, which occurs in 16-18% of collected adults after partial or total corpus callosum section. The need for patient counseling with regard to these unpredictable events is obvious. Further studies should be done to elucidate their pathophysiology and predict their occurrence.

CONCLUSION

By assessing the response of specific seizure types to both partial and total corpus callosum section and stratifying these results into different age groups, important information emerges with regard to patient counseling and expected responses. Collected reports suggest similar efficacy of anterior or total section for partial (35% versus 42%) or atonic (69% versus 78%) seizures in adults. This similarity is also present in children (36% versus 44% for partial seizures, 80% versus 70% for atonic seizures). Absence, tonic clonic, and tonic seizures responded much better to total than anterior section in children (83% versus 25%, 79% versus 25%, 100% versus 30%). This incremental response is not nearly as striking in adults (40% versus 34%, 73% versus 62%, 60% versus 51%). These observations may explain what has been a persistent area of confusion over which seizures are most appropriately treated by callosotomy. Completion of section appears to produce incremental improvements which are long standing.

REFERENCES

Amacher, A.L., 1976, Mid-line commissurotomy for the treatment of some cases of intractable epilepsy, *Child's Brain* 2:54-8.

Avila, J.O., Radvany, J., Huck, F.R., deCamargo, C.H.P., Marino R., Ragazzo, P.C., and Riva, D., 1980, Anterior callosotomy as a substitute for hemispherectomy, *Acta Neurochir. (suppl) (Wein)* 30:137-143.

Bogen, J.E. and Vogel, P.J., 1962, Cerebral commissurotomy in man: preliminary case report, *Bull. Los Ang. Neurol. Soc.* 27:169-172.

Fuiks, K.S., Wyler, A.R., Hermann, B.P., and Somes, G., 1991, Seizure outcome from anterior and complete corpus callosotomy, *J. Neurosurg.* 74:573-578.

Garcia-Flores, E., 1987, Corpus callosum section for patients with intractable epilepsy, *Appl. Neurophysiol.* 50:390-397.

Gates, J., Rosenfeld, W., Maxwell, R.E., and Lyons, R., 1987, Response of multiple seizure types to callosotomy, *Epilepsia* 28:28-34.

Geoffroy, G., Lassonde, M., Delisle, F., and Décarie, M., 1983, Corpus callosotomy for control of intractable epilepsy in children,*Neurology (Cleveland)* 33:891-897.

Huck, F.R., Radvany, J., Avila, J.O., deCamargo, C.H.P., Marino, R., Ragazzo, P.C., Riva, D., and Arlant, P., 1980, Anterior callosotomy in epileptics with multiform seizures and bilateral synchronous spike and wave pattern, *Acta Neurochir. (Wien)* 30:127-135.

Luessenhop, A.J., 1970, Interhemispheric commissurotomy: (the split brain operation) as an alternate to hemispherectomy for control of intractable seizures, *Amer. Surg.* 36:265-268.

Luessenhop, A.J., dela Cruz, T.C., and Fenichel, F.M., 1970, Surgical disconnection of the cerebral hemispheres for intractable seizures: results in infancy and childhood, *JAMA* 213(1):1630-1636.

Marino, R., Jr., 1985, Surgery for epilepsy. Selective partial microsurgical callosotomy for intractable multiform seizures: criteria for clinical selection and results, *Appl. Neurophysiol.* 48:404-407.

Murro, A.M., Flanigin, J.R., Gallagher, B.G., King, D.W., and Smith, J.R., 1988, Corpus callosotomy for the treatment of intractable epilepsy, *Epilepsy Res.* 2:44-50.

Oguni, H., Olivier, A., Andermann, F., and Comair, J., 1991, Anterior callosotomy in the treatment of medically intractable epilepsies: a study of 43 patients with a mean follow-up of 39 months, *Ann. Neurol.* 30(3):357-364.

Papo, I., Quattrini, A., Provinciali, L., Rychlicki, F., Paggi, A., Del Pesce, M., and Ortenzi, A., 1989, Callosotomy for the management of intractable non-focal epilepsy: a preliminary personal assessment, *Acta. Neurochir.* 96:46-53.

Purves, S., Wada, J.A., Woodhurst, W.B., Moyes, P.D., Strauss, E., Kosaka, B., and Li, D., 1988, Results of anterior corpus callosum section in 24 patients with medically intractable seizures, *Neurology* 38:1194-1201.

Rappaport, Z.H. and Lerman, P., 1988, Corpus callosotomy in the treatment of secondarily generalizing intractable epilepsy, *Acta. Neurochir.* 94:10-14.

Spencer, S.S., Spencer, D.D., Glaser, G.H., Williamson, P.D., and Mattson, R.H., 1984, More intense focal seizure types after callosal section: the role of inhibition, *Ann. Neurol.* 16:686-93.

Spencer, S.S., Spencer, D.D., Williamson, P.D., and Mattson, R.H., 1985, The effects of corpus callosum section on secondary bilateral synchronous interictal EEG discharges, *Neurology* 35:1689-1702.

Spencer, S.S., Spencer, D.D., Williamson, P.D., Sass, K.J., Novelly, R.A., and Mattson, R.H., 1987, Partial vs. total corpus callosotomy for epilepsy, *in:* "Advances in Epileptology, Vol. 16. The 16th Epilepsy International Symposium," Raven, New York, pp. 323-325.

Spencer, S.S., Spencer, D.D., Williamson, P.D., Sass, K., Novelly, R.A., and Mattson, R.H., 1988, Corpus callosotomy for epilepsy I: seizure effects, *Neurology* 38:19-24.

Spencer, S.S., Elquera, E.D., Williamson, P.D., Spencer, D.D., and Mattson, R.H., 1992, Evolution of seizure characteristics after callosotomy, *J. Epilepsy* 4:149-156.

Spencer, S.S., Spencer, D.D., Sass K., Westerveld, M., Katz, A., and Mattson, R.H., 1993, Anterior, total and two-stage corpus callosum section: differential and incremental effects on seizures,*Epilepsia* 34(3):561-567.

Spencer, S.S., Katz, A., Ebersole, J., Novotny, E., and Mattson, R.H., 1993, Ictal EEG changes after anterior and total callosotomy, *Epilepsia* 34(3):568-573.

Van Wagenen, W.P. and Herren, R.Y., 1940, Surgical division of commissural pathways in the corpus callosum: relation to spread of an epileptic attack, *Arch. Neurol. Psychiatr.* 44:740-759.

Wilson, D.H., Reeves, A.G., Gazzaniga, M., and Culver, C., 1977, Cerebral commissurotomy for control of intractable seizures, *Neurology* 27:708-715.

Wilson, D.H., Reeves, A.G., and Gazzaniga, M., 1978, Division of the corpus callosum for uncontrollable epilepsy, *Neurology* 28:649-653.
Wilson, D.H., Reeves, A.G., and Gazzaniga, M.S., 1982, "Central" commissurotomy for intractable generalized epilepsy: series two, *Neurology (NY)* 32:687-697.

FUNCTIONAL PATHWAYS UNDERLYING IPSILATERAL AND CONTRALATERAL SPREAD OF TEMPORAL LOBE SEIZURES

Charles L. Wilson, Ph.D.

Department of Neurology and Brain Research Institute
2155 Reed Neurological Research Center
UCLA School of Medicine
Los Angeles, CA 90024

INTRODUCTION

Although there are a finite number of interhemispheric pathways by which limbic seizures may propagate from a temporal lobe focus, the exact pathway followed in a given patient is impossible to determine electrophysiologically because of the limited number of depth or subdural electrodes that are available during diagnostic telemetry monitoring. It is not surprising then, that the pathways of seizure spread observed in different laboratories depend largely upon which areas of the brain into which monitoring electrodes have been implanted. As a result, previous studies of temporal lobe seizure spread have emphasized different candidate pathways which vary in the significance attributed to their role in interhemispheric seizure propagation. In addition, different techniques may be used for measuring seizure spread which may or may not correspond with one another, and may also provide differing perspectives.

The goal of this chapter is first, to briefly review the anatomy of candidate inter-hemispheric pathways which epileptic activity may traverse, based on studies carried out in primates. Second, I will describe some of the techniques that are available for measuring interhemispheric seizure spread in epileptic patients and animal models of epilepsy, as well as some of the earlier conclusions which have been drawn using these techniques. Third, I wish to update the results of studies providing information from evoked potential comparisons of putative seizure pathways in epileptic patients. Finally, I will attempt to correlate the evoked potential results with recent studies of interhemispheric seizure propagation patterns obtained from patients during seizure monitoring.

Epilepsy and the Corpus Callosum II, Edited by A.G. Reeves
and D.W. Roberts, Plenum Press, New York, 1995

THE MAJOR COMMISSURAL SYSTEMS: CANDIDATE PATHWAYS OF INTERHEMISPHERIC SEIZURE SPREAD

Given the assumption of homology between human and lower primate neuroanatomy, recent experimental anatomical studies of the old world monkey allow a finer-grained analysis of pathways within the mesial temporal lobe (Pandya and Rosene 1985; Pandya and Yeterian 1987). In terms of temporal lobe seizure spread, the sources of many of the cells of origin and the location of the axons of temporal lobe neurons traversing commissural pathways have been described in the genu, splenium and inferior forceps of the corpus callosum, in the anterior commissure, in the hippocampal decussation and in the dorsal and ventral hippocampal commissures. A summary of the temporal lobe structures projecting both homotopically and heterotopically through these interhemispheric pathways in the macaque is provided in Table 1, as described by Demeter, Rosene and Van Hoesen (1990). Such information on interhemispheric connections, along with increased knowledge of ipsilateral limbic and frontal lobe connections is valuable for tracking seizure spread, particularly when used in combination with electrophysiological or metabolic studies of both normal and pathological activity in the human brain.

Two structures of central importance in regard to seizure propagation are the hippocampal formation and the amygdala. This is because the great majority of seizures of temporal lobe origin arise in these structures (Crandall 1987; Engel, Jr. 1993; Lüders 1991). Therefore, their interconnections to one another and with the rest of the brain are essential to discussion of the interhemispheric spread of seizures. However, anatomical data from primates is best used for assessing hippocampal

TABLE 1. Candidate pathways for interhemispheric seizure spread from temporal lobe.[1]

COMMISSURAL PATHWAY	PROJECTION AREA(S)
Corpus Callosum	
Rostrum of Corpus Callosum	Medial orbital frontal cortex.
Body of Corpus Callosum	Caudal two thirds of temporal lobe.
Splenium of Corpus Callosum	Caudal third of temporal lobe, Posterior parahippocampal gyrus, Occipital-temporal corticies.
Anterior Commissure	
(No topographical Organization noted except olfactory fibers located on anterior surface of bundle)	Olfactory areas of orbital frontal cortex. Rostral third of the temporal lobe. Light projection from entire temporal lobe. Pyriform cortex, orbital frontal cortex. Amygdala (heterotopic), light projection.
Hippocampal Commissure	
Ventral hippocampal commissure	Rostral hippocampus ("uncal" area)
Hippocampal decussation	Hippocampus to septum
Dorsal hippocampal commissure	Posterior parahippocampal gyrus, Subicular corticies to entorhinal cortex Medial entorhinal cortex

[1](After Demeter et al. 1990)

formation and amygdala commissural connections in the human, since lower mammals exhibit some significant differences from primates. In rodents and carnivores, the hippocampal commissures form one of the most massive inputs to the hippocampus, joining virtually every homotopic hippocampal field and cell population on one side to the other (Rosene and Van Hoesen 1987). In the macaque, this commissural pathway is reduced to a small component of fibers which join the anterior hippocampi through the ventral hippocampal commissure (Amaral et al. 1984a; Demeter et al. 1985).

In the book "The Human Nervous System" (Paxinos 1990), Amaral and Insausti indicate in their review of hippocampal anatomy that:

> *In the monkey, only the rostral (or uncal) part of the hippocampus and associated dentate gyrus are connected by commissural fibers. This contrasts markedly with the rodent brain, where both CA3 and CA1 of the entire hippocampus receive a strong, topographically organized projection from the opposite CA3 and the dentate gyrus receives a major input from cells of the contralateral polymorphic layer (of the dentate hilus). As in the rat, the monkey presubiculum gives rise to a substantial commissural projection which terminates in layer III of the contralateral entorhinal cortex. It would appear that in primates, and presumably in humans, this projection constitutes the major link between the hippocampal formation of the two sides.* (Amaral and Insausti 1990, p. 735)

The absence of a prominent hippocampal commissural system does not obviate the presence of limbic interhemispheric projections however. As stated above, the dorsal hippocampal commissure is still prominent, carrying fibers from the presubiculum to the contralateral entorhinal cortex, while the entorhinal cortex and parahippocampal gyrus project homotopically. In addition, the hippocampal decussation provides a disynaptic contralateral connection via the septum (Demeter et al. 1985).

The connections of the primate amygdala have been described by several investigators who have documented the exceptionally diverse linkage joining amygdala, widespread areas of ipsilateral neocortex, and the ipsilateral limbic system (Amaral 1986; Amaral et al. 1984b; De Olmos 1990; Saunders and Rosene 1988a; Saunders et al. 1988b). However, in the macaque, Demeter and colleagues (1990) report:

> *Our observations strongly suggest that there are no major interamygdaloid connections. Although sparse labeling in the anterior commissure was obtained following amygdaloid injections, there was no evidence for homotopic connections.*
> *. . . The apparent absence of interamygdaloid connections suggests that whatever functions the amygdala serves, it is not directly influenced by its counterpart in the opposite hemisphere. The lack of interamygdaloid connections is reminiscent of a similar paucity of interconnections between the hippocampal formations of the two hemispheres (Demeter et al. 1985) and suggests that the amygdala too, may be subject to or involved in hemispheric specialization.* (p. 50)

A recent study by Gloor and colleagues directly addressed the question of the phylogenetic fate of the hippocampal commissure in the human brain by examination of normal human brain sections from the extensive Yakovlev Collection in Washington, D.C. They "found no convincing evidence for a significant remnant of a ventral hippocampal commissure in the human brain" (Gloor et al., 1993, p. 1253), but could not entirely rule it out. They found the presence of a *dorsal* hippocampal commissure was unmistakable, however, which was clearly visible in the same position as found in the monkey brain. This suggests that in the human, the hippocampus has no interhemispheric connections, because in lower primates the dorsal hippocampal commissure carries no fibers from the hippocampus proper but instead interconnects only the left and right entorhinal, presubicular and parahippocampal corticies.

These anatomical findings provide a basis for the not uncommon observation during recording in the mesial temporal lobe, of prolonged seizure discharge limited to the limbic system of one hemisphere which may not always propagate to the contralateral hemisphere. Thus, if the observations of Gloor et al. (1993) are correct, for a seizure originating in one hippocampus, its spread to these adjacent mesial temporal corticies must precede its spread to the other hippocampus so that contralateral propagation is mediated multisynaptically rather than monosynaptically. Therefore, although a pathway for spread via the dorsal hippocampal commissure may be present, it is not as readily traversed by seizure discharge, and thus a tendency exists to retard the expansion of paroxysmal activity outside of the amygdala-hippocampal-entorhinal cortex circuits into adjacent temporal neocortical areas or to the contralateral side.

RESULTS OF SEIZURE PROPAGATION MAPPING

Although epileptology has been concerned with pathways of seizure propagation at least since Hughlings Jackson, it was not until the advent of chronic intracranial recording with depth electrodes that such pathways could be studied in epileptic patients by following propagation from structure to structure during seizure monitoring (Brazier 1972). Since that time, a number of centers have attempted to evaluate seizure spread on the basis of visual inspection of intracranial and scalp EEG recordings. Brazier's seminal work was followed by Lieb et al. (1976), Spencer et al. (1982), and Wieser (1983), who described the seizures recorded from the basal medial temporal lobe and the pattern of spread to the contralateral hemisphere.

Lieb et al. (1986 a, b) reported the time of propagation from the epileptic hippocampal formation to the contralateral hippocampal formation was not only correlated with the outcome of surgical treatment, but also with the extent of damage found in the resected temporal lobe. Thus, rapid propagation (less than 5 sec.) was associated with poor outcome and greater evidence of pathological damage to the hippocampal formation, while propagation times from 50 to 100 sec. were accompanied by an excellent prognosis for seizure relief. Lieb and colleagues attributed the extended time required for spread to the paucity of commissural connections of the hippocampi in humans as described in the previous section.

While there is general agreement between Lieb's results and that of Spencer (1982) and of Wieser (1983) as to the significance of rapid interhemispheric spread, there is disagreement as to the pathways involved in seizure propagation. Lieb concluded that seizures spread to neocortical structures such as the lateral temporal gyri and then via the corpus callosum to the contralateral temporal lobe, rather than through hippocampal or anterior commissures, Spencer (1987, 1988), employing both depth and subdural electrodes, reported that 30% of temporal lobe seizures propagated to the contralateral hippocampus without involvement of the neocortex, a figure in agreement with that reported by Wieser (1983, 1988, 1991; Wieser and Kausel 1987). Gloor et al. (1993) reported approximately 10% of temporal lobe seizures propagated directly to the contralateral hippocampus.

EXPERIMENTAL METHODS OF SEIZURE MAPPING

Although the chief method for mapping seizure spread is visual inspection of EEG, discrimination of the exact onset of the seizure and the electrode sites at which it occurs earliest can be quite difficult (Gotman et al., 1993). As a result, numerous mathematical treatments have been developed for analyzing seizure spread from one brain area to another. Brazier (1973) introduced linear coherence measures in order to quantify the dynamic phase relations of epileptic tissue during the process of spread. Gotman (1983) later developed a linear coherence measure which involved statistical evaluation of interactions between interhemispheric temporal lobe recording sites. It allowed measurement of phase lead or lag in msec, thus providing a predictive basis for causality. If EEG activity from two separated electrodes are found to vary in phase with one another and one leads the other reliably, then it is likely that one is driving the other (Gotman 1991). These quantitative techniques, when used in combination with visual inspection, help improve reliability of qualitative judgments regarding the temporal and spatial sequence of seizure spread.

Another analysis approach is AAMI (average amount of mutual information), a non-linear estimation of the degree of association between two sites (Lopes da Silva et al. 1987; Mars et al. 1985). It has the advantage of sensitivity to both linear and non-linear EEG interaction, but the disadvantage of giving values which are not normalized and not comparable across varying conditions. A third technique is the non-linear correlation ratio h^2, which is similar to the linear coherence measure used by Gotman, but requires only a brief sample of EEG (0.5 sec.) to calculate, and is sensitive to non-linear as well as linear EEG relationships (Pijn et al. 1990).

The technique which has been used most fruitfully in seizure analysis of patients with implanted depth electrodes has been linear coherence, which was initially used by Gotman (1983,1987) and later by Lieb et al. (1987a, b) to evaluate interhemispheric propagation of temporal lobe seizures. During seizures, both investigators found high coherence between ipsilateral recording sites and low coherence between interhemispheric temporal lobe sites. The intracranial recording sites targeted in the study by Lieb are illustrated in Figure 1 on sagittal and axial sections derived from MRI scans used to locate the tips of recording electrodes in the brain (Levesque et al. 1990; Zhang et al. 1990). For each patient, up to five of the eight sites illustrated were placed bilaterally in each temporal lobe.

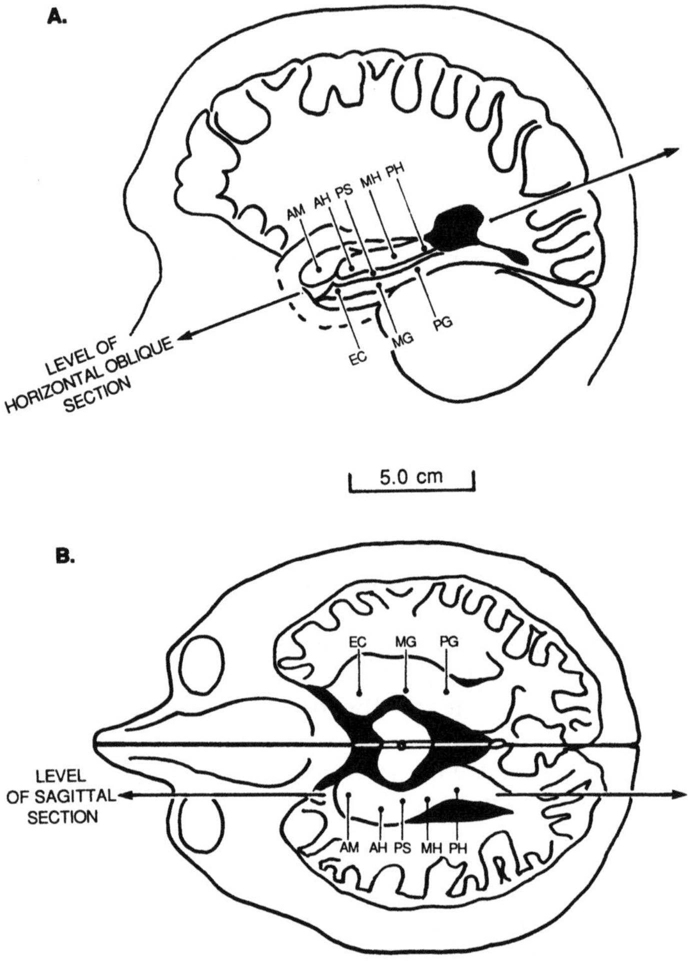

Figure 1. Standard array of depth electrode placements for temporal lobe seizure monitoring. 1A, a sagittal plane 2 cm from midline and 1B, an oblique axial plane parallel to the longitudinal extent of the hippocampus. Electrodes were introduced from the temporal aspect to sites in the amygdala (AM), anterior hippocampus (AH), subicular cortex (PS), middle hippocampus (MH), or posterior hippocampus (PH). Contralaterally, at a slice level 5 mm inferior to 1A, the position of electrodes in the entorhinal cortex (EC) and posterior parahippocampal gyrus (PG) are shown. Electrodes within both hippocampal and parahippocampal areas were placed bilaterally and targeted homotopically on preoperative MRI scans. (Reprinted from Wilson, Isokawa, et al., 1990, Functional Connections..., 82:279-292, with permission from Springer-Verlag.)

To calculate intra-ictal coherence between all of the individual recording sites participating in the seizure, Lieb selected three 9.6 sec. epochs of EEG from the seizure record. Epoch 1 was early in the seizure and Epoch 3 began about 20 secs. after seizure onset. In Figure 2, a map of recording sites in both left and right temporal lobes is shown containing those with significant phase spectra during the two epochs chosen. The lines connecting them are labelled with the phase lead in msec. The results show numerous significant correlations and their phase lags in each temporal lobe, but no significant phase relationships were observed interhemispherically during either the onset of the seizure, or during Epoch 3 when it had spread bilaterally from the left temporal lobe to right temporal lobe sites.

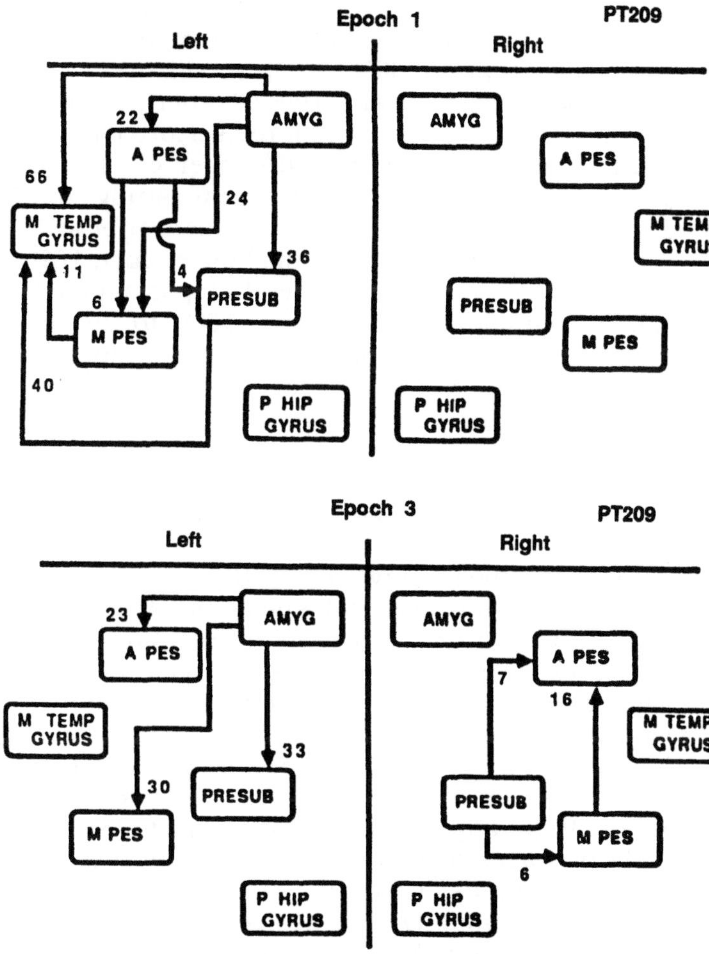

Figure 2. Flow charts summarizing the coherence/phase relationships derived from Epochs 1 and 3 for patient 209. The left amygdala appears to act as a pacemaker in both epochs. No clear linear relationships are observed to emerge between the left and right temporal lobes during the period when discharges clearly spread to the right temporal lobe by epoch 3. (Delay times between sites are in msec.) (Reprinted with permission from Lieb, et al. 1987a, Figure 6, p.111.)

Although this approach can detect relationships in activity on the order of msec. and provides dynamic measurements of the evolution ofseizure spread that are impossible to determine visually, it was not capable of detecting any relation between temporal lobes in seven out of ten patients evaluated in this study (Lieb et al. 1987a). For the three patients whose recordings did show significant coherence across the midline, the question remains whether the seizures were maintained in phase by connections of the dorsal hippocampal commissure, the anterior commissure or by connections through the corpus callosum. Because the anatomical data from monkeys indicates the ventral hippocampal commissure is reduced in the primate brain, Lieb and colleagues hypothesized that the corpus callosum must provide the pathway for spread of these seizures.

A report by Spencer et al. (1987) provided evidence to the contrary, however. Spencer et al. used visual inspection of combined depth and subdural electrode recordings to monitor neocortical activity in 11 patients during interhemispheric propagation of hippocampal formation seizures, and reported that in 30% of these seizures, there was no evidence of spread to lateral temporal cortex or other neocortical regions. These areas would necessarily be involved if the seizure followed a callosal route to the contralateral hippocampus (Spencer et al. 1987). This finding was further supported in a more recent study of 33 paticnts with depth electrodes and subdural strips (Spencer et al. 1990). In 23 patients with 157 seizures having hippocampal onset only, 10% remained unilateral, 23% spread first to the contralateral hippocampus, 56% spread first to the ipsilateral temporal cortex, and 10% to the contralateral hippocampus and the ipsilateral temporal cortex simultaneously. None of the seizures spread to contralateral temporal cortex prior to contralateral hippocampus as would be predicted if callosal propagation had occurred.

Animal Studies of Hippocampal Seizure Propagation

Newer animal studies have used the non-linear regression coefficient h^2 during inter-hippocampal spread of afterdischarges (ADs) evoked by electrical stimulation, employing more quantitative means to determine how seizures propagate in hippocampal circuits. In one study (Fernandes de Lima et al. 1990) recording electrodes placed in both dorsal hippocampi of rats under halothane anesthesia were used to measure responses to paired stimuli across the hippocampal commissure. Robust evoked responses of 3 to 7 msec. latency were recorded in the contralateral dorsal hippocampus. High frequency stimulation elicited seizure-like ADs, which immediately propagated bilaterally. During non-linear regression analysis, the AD activity occasionally showed a 3 to 7 second phase lag between hippocampi, but usually had a phase lag of 0. If the ventral hippocampal commissure was cut, the ADs became unilateral on the side of stimulation, phase relations between the two sides were lost, and evoked potentials between hemispheres were also blocked. If ventral hippocampal commissure remained intact, but the septum was lesioned, or the dorsal hippocampal commissure or corpus callosum was cut, the ADs remained bilateral and the two sides remained coherent.

In a second study (Nunes Filipe et al. 1989), electrodes were placed in the dorsal, posterior, and ventral hippocampus bilaterally, and stimulation was delivered to the dorsal or ventral hippocampus. Evoked potentials were easily obtained between the two dorsal hippocampi, but could not be recorded in the ventral hippocampus. Similarly, ventral hippocampal stimulation evoked clear local responses, but the responses did not reach the dorsal hippocampus or contralateral hippocampus.

Coherence analysis of the AD data matched the evoked potential data perfectly, including the absence of any propagation of ventral hippocampal seizure activity contralaterally, and showed phase lags of the same duration as the evoked potential latencies.

In the first study, it was concluded that cutting the corpus callosum had no effect on propagation of seizure activity from one hippocampus to the other, and that section of the ventral hippocampal commissure completely blocked ADs from reaching the contralateral hippocampus. In addition, it was concluded by the authors that such a finding would probably be true in humans as well. In the second study, it was concluded that while the two dorsal hippocampi of the rat are exceedingly well connected commissurally, the dorsal and ventral hippocampi are functionally unconnected, despite their proximity to one another, and the ventral hippocampi of each hemisphere are similarly unconnected to one another commissurally. Similar results regarding the absence of commissural transmission of evoked potentials or afterdischarge propagation from the ventral hippocampus of the cat was reported in two early studies by Elul (1964a,b).

These studies provide key information on the functional roles of the ventral and dorsal hippocampal commissures, as measured both by seizure discharge and evoked potential, but they are difficult to reconcile with the anatomical rearrangements that separate the rodent from the primate brain. First, the dorsal hippocampus in rat is homologous to the posterior

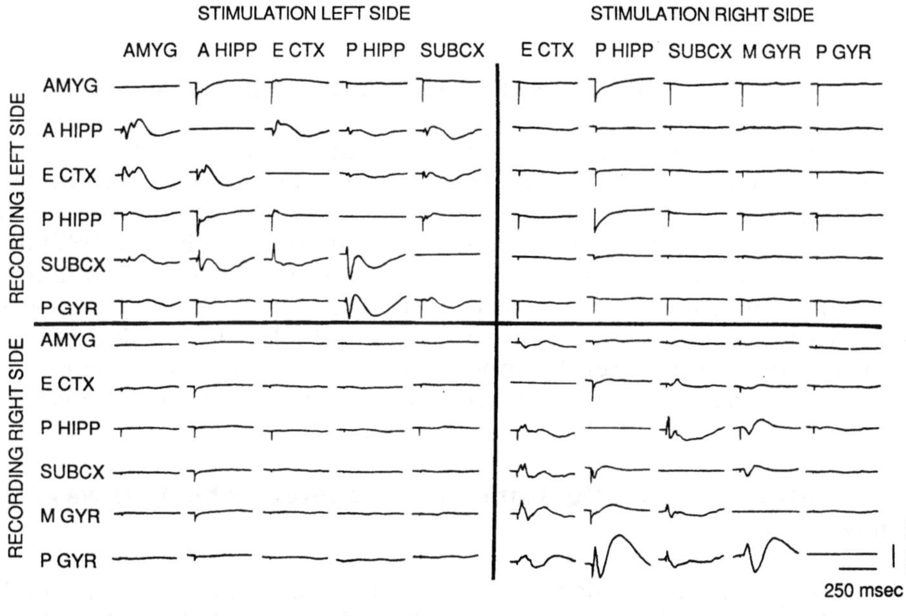

Figure 3. Field potential matrix for Patient 220 showing distribution of responses evoked in adjacent depth electrodes. Structures labeled at the top of each column are stimulation sites indicated to the left of each row. Straight lines indicate local responses were not recorded in this patient. Note that all clear responses are recorded ipsilateral to the site of stimulation; contralateral deflections are stimulus artifact. Each record is the average of 20 stimulations using a current strength of 5.0 mA, which occurs 50 msec. after the start of the sweep. All derivations are bipolar to insure origin of response is local. (Reprinted from Wilson, Isokawa, et al.,1990, Functional Connections..., 82:279-292, with permission from Springer-Verlag.)

hippocampus of primates, while the ventral hippocampus of the rodent is homologous to the anterior hippocampus in primates. This means that in lower primates, the only ventral hippocampal commissural connections which remain are those which arise from the anterior (ventral) hippocampus. However, the two electrophysiology studies cited above demonstrated that the rat ventral hippocampus is functionally disconnected from the contralateral ventral hippocampus. Conversely, the connections which show the greatest anatomical and functional strength in the rat are those commissural fibers which join the dorsal (posterior) hippocampi, and it has been anatomically demonstrated that fibers joining the posterior hippocampi are completely absent in primates (Amaral et al. 1984b; Demeter et al. 1985). These phylogenetic disparities illustrate the major reorganization of the commissural system that has taken place in primates, and suggest that conclusions about one species on the basis of another be drawn with care.

Another important point is the extremely high correlation between the evoked potential and afterdischarge data in the Elul studies of cats and the two newer studies of interhippocampal functional linkage in rats. Essentially identical estimates of functional connectivity were obtained in every case, whether high frequency AD stimulation or single pulse evoked potential stimulation were employed. This observation is germane to the choice of techniques available for mapping possible seizure pathways, and because a good deal of information has been obtained from evoked potential mapping of functional pathways in seizure pathways, this approach is discussed next.

EVOKED POTENTIAL STUDIES OF TEMPORAL AND FRONTAL LOBE

A primary source of data on functional anatomical connections are responses evoked by single pulse electrical stimulation of the brain. This approach was used for depth electrode sites in the temporal lobe by Brazier (1964), and later by Buser and Bancaud (1983), Wieser (1987) and Rutecki et al. (1989). This early work, and that of other investigators using this technique in the human brain has shown that response to homotopically placed electrodes during temporal lobe stimulation is rare, but nevertheless contralateral temporal responses have been reported in some patients, particularly those with bilateral seizure onsets (Wieser 1983; Wieser et al. 1987; Wilson et al. 1991, 1992).

Evoked Potential Investigations of Temporal Lobe Pathways in Epilepsy

Evoked potential studies at UCLA during stimulation of the eight temporal lobe areas shown in Figure 1 have been carried out in a large population of seizure patients. These studies have demonstrated strong ipsilateral functional association between hippocampal formation, parahippocampal gyrus, and amygdala, but virtually no functional connections to contralateral homotopic electrode sites (Wilson et al. 1990). An example of both of these results is shown in Figure 3, where the stimulation sites of a number of limbic areas are listed across the top of the figure for the left and right temporal lobes. Stimulation of these ten electrode sites was carried out one at a time. Response sites for the same areas are listed down the left side of the figure, so that ipsilateral responses appear in the upper left and lower right quadrants, while contralateral responses, if found, appear in the upper right and lower left quadrants. The numerous ipsilateral responses are more widely distributed than one might expect given the separation of 1 to 4 cm between each electrode, again emphasizing the degree of ipsilateral connectivity in the human limbic system. Data of the form shown in Figure 3 was collected from 74 temporal lobe depth electrode patients, and 394 electrodes were stimulated while recording in 581 sites ipsilaterally and 514 contralaterally. Of the ipsilateral sites, 438 (75%) responded while only 5 (1%) of the contralateral sites responded (Wilson et al. 1990).

Measurement of latencies and conduction velocities of the ipsilateral and contralateral responses are useful to demonstrate the contrast between the two kinds of responses. Ipsilateral responses ranged in onset latency and conduction velocity from 4.4 msec. and 3.6 meters/sec. in the perforant pathway to 24.8 msec. and 0.9 meters/sec. in the amygdalo-hippocampal pathway. Although long onset latencies were common for amygdala connections, latencies within hippocampal formation were almost always less than ten msec. and usually around five msec. (Wilson et al. 1990). The high degree of connectivity found in these studies raises the question of why isolated focal seizure onsets can be recorded from closely adjacent electrodes. The short latencies and rich connections of these areas are difficult to reconcile with the fairly common occurrence of focal seizure discharge at one electrode without spreading for several seconds or tens of seconds before propagating regionally or contralaterally. Such sites usually respond well to adjacent stimulating sites, but do not spread seizure activity with the same conduction velocity with which they spread evoked potentials.

Interhemispheric responses were evoked during stimulation of only two sites, presubiculum and amygdala, and these responses share two main attributes which contrast with ipsilateral responses: first, they have extremely long latencies, and second, they display unusual variability in response probability, suggestive of some type of "gating" mechanism. The contralateral responses to presubiculum stimulation were not homotopic, but were consistent with the anatomical description of Amaral et al. (1984) and Demeter et al. (1985), who described the primary interhemispheric connection between hippocampal areas as presubiculum to entorhinal cortex via the dorsal hippocampal commissure. The long onset latencies (90 and 105 msec.) the variability of the latencies, and the lability of response are all suggestive of a polysynaptic, not monosynaptic, pathway (Wilson et al. 1991).

A second, more recently observed example of contralateral response was evoked by amygdala stimulation (Wilson et al. 1992). Responsive sites included the contralateral amygdala, entorhinal cortex, and anterior hippocampus, with onset latencies that ranged from 65 to 100 msec. During both the presubiculum and amygdala stimulation, the reliability of the response was low, appearing only about 20 times during 177 stimuli in the first case (presubiculum), and in about one third of 80 stimulations in the second case (amygdala). Finally, the shape of each response waveform had an epileptiform "spike-like" appearance which was comparable to spontaneous interictal sharp waves, an indication that these responses might represent a pathophysiological process. In our laboratory, recent cross-correlation studies of a large population of single neurons in hippocampus, amygdala and entorhinal cortex have shown that 5% of the cells have significant firing correlations with the firing of cells in the contralateral homotopic structures. These interhemispheric cross-correlations are largely inhibitory (as shown in animals by Buzsaki, 1984) and have broad peaks indicative of multisynaptic interactions (Colder et al. 1994).

Orbitofrontal Pathways as a Route of Temporal Lobe Interhemispheric Seizure Propagation

Since there is little evidence in primates of a homotopic amygdaloid projection (Demeter et al. 1990), the functional pathway for the contralateral amygdala response is unknown. However, it requires only two to three synapses to reach the contralateral mesial temporal lobe via the orbitofrontal cortex. First, there are substantial projections from the amygdala to the medial and lateral orbitofrontal-frontal cortex (Amaral et al. 1984b), and other fibers project from amygdala to mesial frontal cortex via the ansa peduncularis (De Olmos 1990). Once the orbitofrontal cortex is reached, callosal fibers to the contralateral orbital cortex are present, and there are direct projections from both the ipsilateral and contralateral orbitofrontal cortex to the contralateral entorhinal cortex (Amaral et al. 1990).

Recently we studied frontal lobe pathways in ten patients who had orbitofrontal and temporal lobe depth electrodes implanted to evaluate frontal vs. temporal contributions to their seizure disorders. All patients were tested for callosal responses between orbitofrontal lobes, six of ten were tested for temporal lobe (amygdala) to frontal lobe pathways, and two with anterior or posterior cingulate electrodes were tested for connections across the corpus callosum (Wilson et al. 1992).

Figure 4 shows averaged recordings made during bipolar stimulation of the amygdala with single pulses, illustrating the large amplitude, short latency response of the left anterior pes hippocampi, in contrast to the longer latency but still prominent responses of the mesial and lateral orbitofrontal cortex. The electrodes were introduced through small twist drill holes through the temporal surface of the cortex such that the deepest contact, located at the midline was contact #1 and the most lateral, contact #8, was near the cortical surface. Electrode contacts were separated by 5.0 mm. As suggested by the latency data, the tip of the amygdala electrode was located about 1.0 cm. from the anterior pes hippocampus electrode and 10.0 cm. from the orbitofrontal electrode. It is clear from this response that a strong functional connection exists between amygdala and the frontal lobe. Amygdala stimulation was successful in evoking orbitofrontal responses in four of the six patients tested.

Having demonstrated a functional limbic to frontal lobe pathway from amygdala to orbitofrontal cortex, stimulation was next delivered to the frontal cortex electrodes to evaluate the strength of the orbitofrontal interhemispheric pathways. In the sagittal MRI slice shown in Figure 5, the positions of stimulating and recording electrodes used to

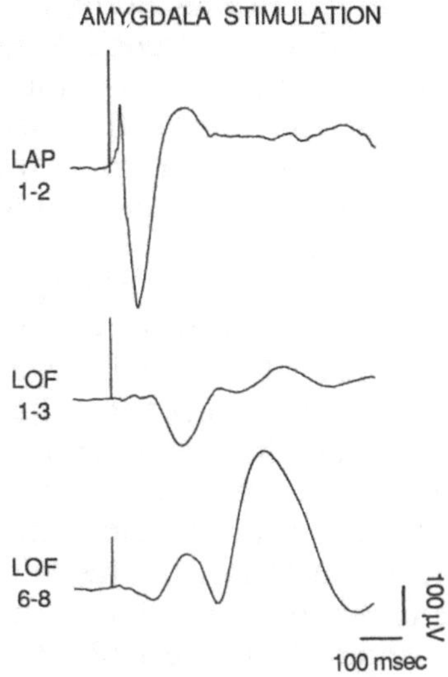

Figure 4. Hippocampal, medial and lateral orbitofrontal response to amygdala stimulation. In this and all subsequent figures experimental parameters were: averages of 80 trials, 1 pulse every 5 seconds., 5.0 mA current strength, positivity up.

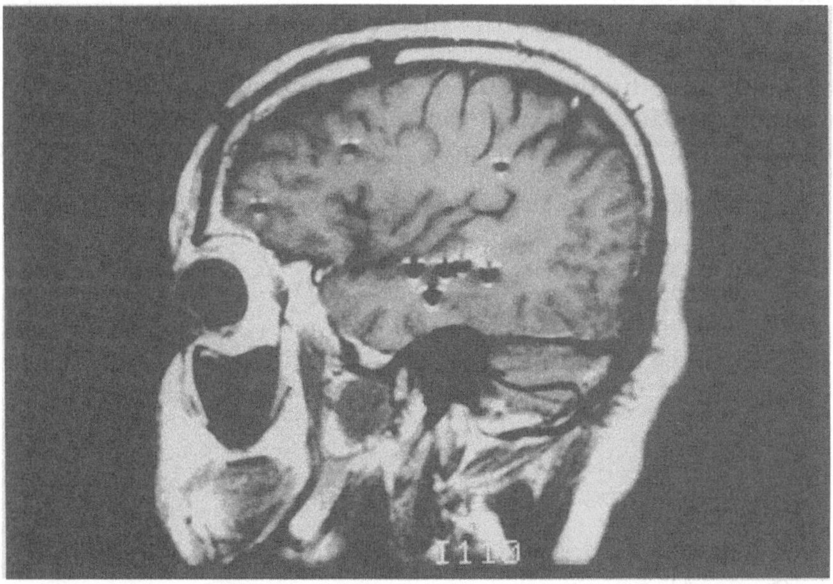

Figure 5. Sagittal MRI showing positions of depth electrodes in orbitofrontal, cingulate and temporal cortex from which the averaged responses in Figs. 6 and 7 were recorded.

produce the recordings shown in Figures 6 and 7 are illustrated. The siteof each electrode is marked by a dark area of signal loss which is much larger (3 to 5 times) than the size of the electrodes, which were 1.0 mm. in diameter. Also visible are the bright susceptibility artifacts on three sides of each area of signal loss. Of the extratemporal lobe electrodes, the most rostral is in the orbitofrontal cortex, the middle site is the anterior cingulate gyrus, and the posterior site is the posterior cingulate gyrus. Located within the temporal lobe, from front to back are the amygdala, entorhinal cortex, anterior hippocampus, middle hippocampus, and posterior parahippocampal gyrus.

All the recordings in Figure 6 are bipolar derivations, with the contact numbers marked to the left of the tracing. During stimulation of the right orbitofrontal cortex at contacts 2-3, a prominent evoked potential is recorded ipsilaterally from the adjacent contacts 4-5, and smaller responses are recorded in the contralateral orbitofrontal cortex at three sets of electrodes. There is a clear phase reversal between contacts 2-3 and 6-8, with 4-5 near the null, indicating the response is locally generated. During stimulation of the left orbitofrontal cortex at contacts 4-5, a prominent response is evoked at the adjacent contacts 2-3. Little response is noted from recording sites on the right side, an outcome that may be affected by several variables: 1) lack of homotopic electrode placements; 2) electrode positioning in gray versus white matter; and 3) differences in excitability associated with epileptic patho-physiology. It is possible that the probability of obtaining a frontal lobe commissural evoked potential is dependent on the direction of seizure propagation in patients with unilateral seizure onsets. It is clear that this is not always true, because in at least one of the six patients showing a frontal response, the direction was opposite that of seizure propagation.

An alternative limbic circuit which might support interhemispheric temporal lobe seizure spread is the cingulate gyrus. Wieser (1983, 1987) has reported that both anterior and posterior cingulate gyri may provide

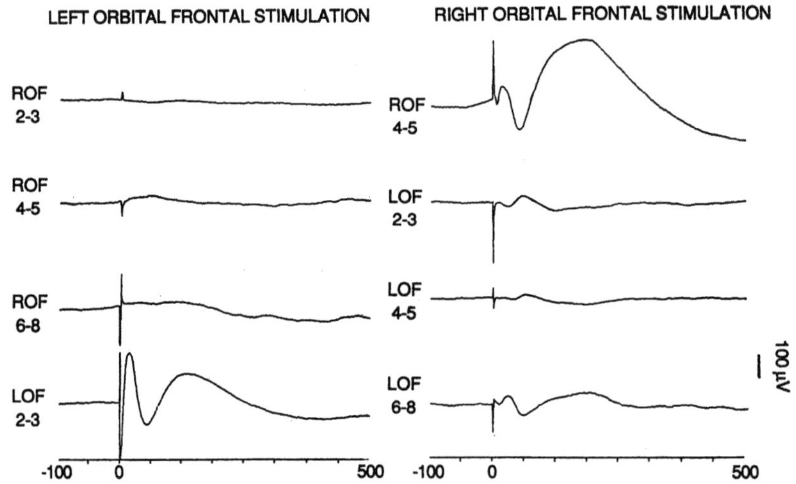

Figure 6. Responses evoked by left and right orbitofrontal stimulation. Depth electrodes crossed the anterior frontal lobe from lateral to medial, with the smallest contact number closest to the electrode tip. Electrode positions were verified with MR imaging. The tip of the right electrode was located 10.0 mm. from midline. The tip of the left electrode was located at midline. On the left, contact 4 was 15.0 mm. from midline and stimulation was delivered to electrode contacts 4-5 On the right, contact 2 was located 15.0 mm from midline and stimulation was delivered to contacts 2-3. See Figure 4 for parameters of stimulation.

avenues of seizure propagation. In order to examine the functional properties of cingulate cortex, two patients have been tested with single pulse stimulation, and both not only showed callosal responses but orbitofrontal responses as well. In Figure 7, the two cingulate electrodes shown in Figure 5 were stimulated, and produced clear responses in the ipsilateral orbitofrontal cortex and cross-callosal responses in the contralateral homotopic anterior cingulate gyrus.

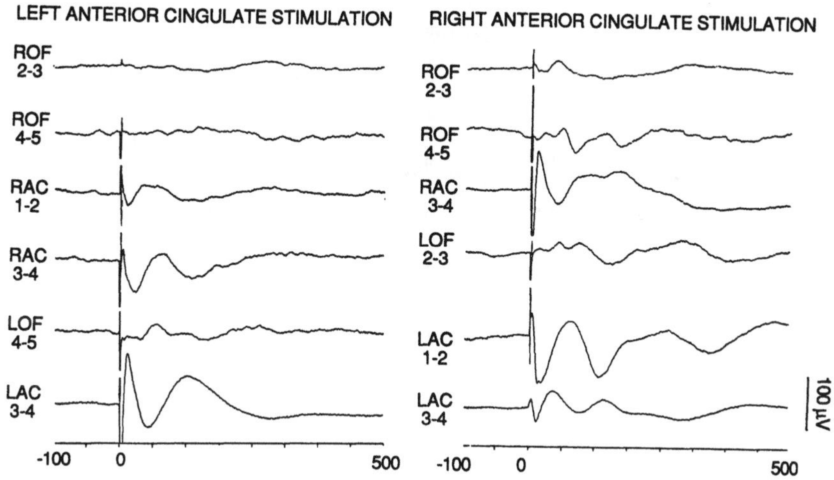

Figure 7. Responses evoked by left and right anterior cingulate stimulation. Recording as described in Figure 6, except anterior cingulate contacts 1-2 were stimulated on each side.

Figure 8 is a summary diagram illustrating the responsive pathways found for the patient of Figure 5, including responses to stimulation of the posterior cingulate gyrus as well. Although this patient showed prominent motor activity during his seizures, they arose from a right posterior temporal lesion which propagated rapidly to the right and left cingulate and orbitofrontal corticies, in the same direction as the most prominent evoked potentials summarized in Figure 8.

RECENT SEIZURE MAPPING STUDIES AND FRONTAL LOBE PROPAGATION

The most recent investigation to address these issues is that of Gloor et al. (1993) who, as mentioned earlier, examined the normal human hippocampal commissure anatomically, and concluded that the ventral hippocampal commissure "either does not exist, or is very rudimentary" (p. 1269). In regard to the study of Spencer et al. (1992), Gloor et al's. finding suggests that the decreased observation of direct propagation of seizures between left and right hippocampus does not stem from hippocampal cell loss in the hilar area which no longer possesses a ventral commissural connection. Those seizures which spread homotopically may instead originate from CA1 output fibers which synapse in presubiculum. From presubiculum seizure activity could enter the dorsal hippocampal commissure, and thence reach the contralateral entorhinal cortex.

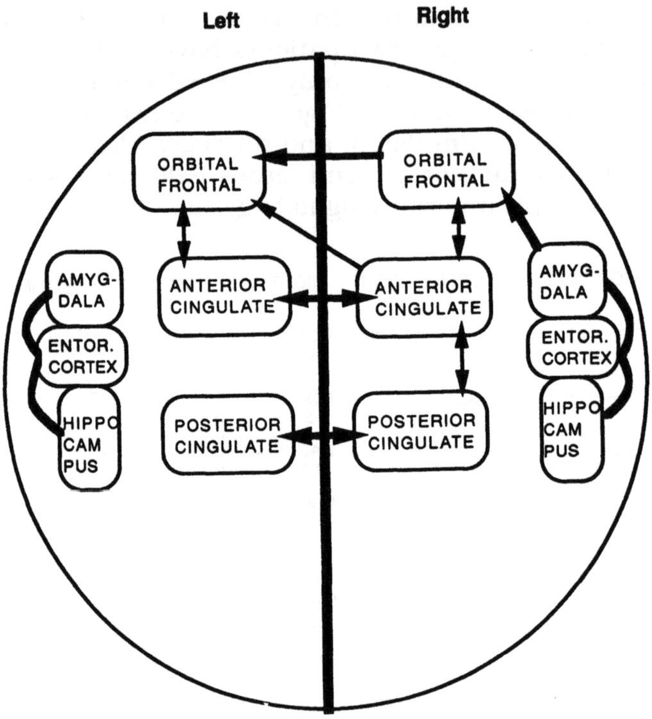

Figure 8. Evoked potential map for patient shown in Figure 5. The thickness of the arrows indicates the strength and reliability with which the pathway could be activated. (Reprinted from Wilson, Isokawa, et al., 1991, Functional Connections in the Human..., 85:174-187, with permission from Springer-Verlag.)

In addition to their anatomical studies, Gloor and colleagues (1993) evaluated the propagation pattern of 220 seizures in 53 patients with unilateral onsets in mesial temporal structures, and described four categories: Type I seizures were those in which spread was detected first in the contralateral hippocampus; Type II spread first to ipsilateral temporal cortex and then contralateral hippocampus; and, Type III spread diffusely to both mesial and lateral structures of the contralateral temporal lobe. Of the 220 seizures studied, 86% were designated Type IV: propagation in which it was unlikely that seizure activity traversed the hippocampal commissure. Interestingly enough, only 19% of Type IV seizures involved the frontal lobes, most showing ipsilateral temporal lobe spread, and then diffuse contralateral propagation. The dissimilar results obtained by Lieb et al. (1991) may be due to differences in frontal recording sites used by the two studies. This again relates to the problem described at the outset of the present paper: seizure spread patterns are dependent upon the choice of recording sites sampled.

A few other points from the paper by Gloor and colleagues require discussion. The first is that evoked potential and phase coherence methods may be less useful for mapping seizure spread than visual evaluation of seizure EEG records. As mentioned earlier, it is sometimes exceedingly difficult to determine when a seizure moves from one point to another. This is because changes signalling spread may consist of increases in amplitude, decreases in amplitude, increases in frequency, increases in synchrony or rhythmic activity, or a combination of these variables. This difficulty is the motivation for computer-aided techniques such as that by Gotman (1993) described earlier. Evoked potentials are useful in this regard because they are more objectively measured and can resolve the speed of neural conduction down to a few milliseconds. Also mentioned earlier is the use of H^2, the nonlinear coherence measure described by Pijn et al. (1990). It has the advantage of computing the phase delay with an EEG duration of one second or less, and in the rat shows phase delays in milliseconds which are comparable to those obtained with evoked potentials. This method also shows propagation of ADs over pathways which carry evoked potentials, but not between areas which show no evoked potential responses (Nunes Filipe et al. 1989). The upshot is that these three techniques, visual evaluation, evoked potentials and coherence, provide complementary data which together may help in visual evaluation of seizure propagation.

Second, Gloor and colleagues (1993) suggest that repetitive activity such as that produced by paired pulse stimulation or during frequency facilitation might be necessary to activate the functional potential of a pathway such as the dorsal hippocampal commissure (e.g., Wilson and Engel 1993). Although repetitive activity clearly has the capacity to facilitate the amplitude and sometimes the probability of responses to stimulation, the frequency of the repetitive signal is crucial, or suppression of a response will result instead of facilitation. The two occasions on which interhemispheric-evoked potentials were obtained during the studies described earlier (Wilson et al. 1990; 1992). did not appear to depend on repetition. Instead they seemed to require "gating" in which multisynaptic pathways which were not initially transmitting responses briefly opened and began to conduct responses. Whether this might involve disinhibition of inter-temporal inhibitory pathways (described in rodents by Buzsaki, 1984) or facilitation from the same factors that lead to seizure spread is unknown. A powerful argument for the inability of repetitive limbic activity to cross the dorsal hippocampal commissure or any other interhemispheric pathway is the extraordinary view of a 20 to 30 second-long paroxysmal afterdischarge evoked by stimulation that spreads instantly to adjacent ipsilateral limbic structures but leaves the contralateral homotopic areas serenely unperturbed. Clearly, some synaptic gate is closed during this period and opens during seizure spread, or on those occasions when a stimulation-evoked afterdischarge spreads interhemispherically and becomes a clinical seizure. Whether this process requires brainstem or thalamic involvement or is intrinsic to cortical pathways is a critical question which remains to be determined.

Finally, Gloor et al. (1993) conclude that the dorsal hippocampal commissure is "functional" on the basis of its clear anatomical presence and the highly probably likelihood that it participates in seizure spread. This conclusion is supported by evoked potential data obtained during stimulation of the presubiculum, as described by Wilson et al. 1990 and illustrated in Figure 9 (see the connection extending from subicular cortex to entorhinal cortex labeled "dhc"). However, on the basis of all available data, this pathway is activated to carry evoked potentials only rarely and a minority of seizures (10-30%) traverse it. Figure 9 has been modified to include orbitofrontal and cingulate corticies in this summary diagram of commissural pathways, as indicated by the evoked potential results reported in this paper. The ventral hippocampal commissure (vhc) is labeled as a pathway found in lower primates, where it connects only the anterior hippocampal fields. Based on the anatomical studies of Gloor and colleagues, as well as the functional measures described above, this pathway is not present in the human brain. This raises the question of what the normal function of the remaining dorsal hippocampal commissure might be given that it is the primary connection joining the limbic corticies in the human. One way of addressing this question is described in the chapter by Tramo in this volume in which differences in memory performance are compared in patients with anterior versus posterior callosotomy, since the dorsal hippocampal commissure is interrupted only with section of the splenial callosum.

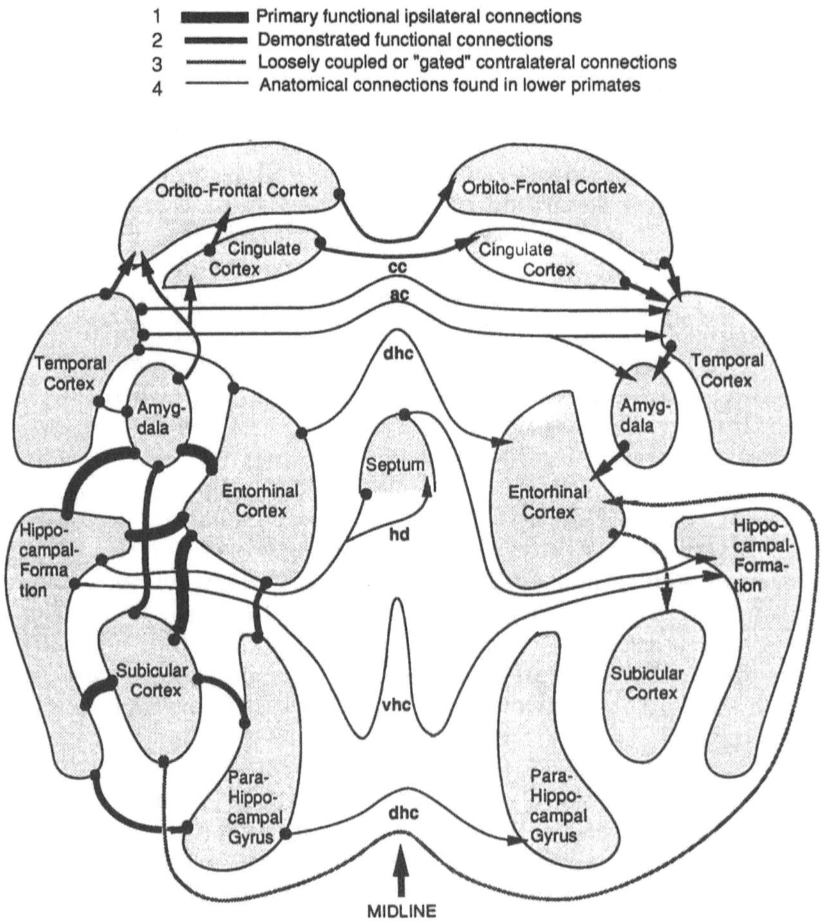

Figure 9. Summary diagram of limbic pathways related to interhemispheric seizure propagation. The first three sets of pathways were derived from electrical stimulation evoked potential studies carried out in seizure patients. The fourth set of pathways summarizes known anatomical connections in animals which were not found during evoked potential studies. (Adapted from Wilson et al. 1990, p. 183.)

Taken together, these findings indicate the complexity of the anatomical and functional issues that must be addressed in understanding normal temporal lobe commissural function and its role in interhemispheric spread of seizures. The main contribution of evoked potential studies for presurgical evaluation in epilepsy is in providing basic information about functional connections of the temporal and frontal lobes. It is hoped that further investigation will continue to improve our understanding of normal anatomical pathways as well as those participating in the pathophysiology of seizure spread.

ACKNOWLEDGMENTS

I thank all the faculty and staff of the UCLA Clinical Neurophysiology Program for their contributions to this research, Irene Wainwright for essential editorial assistance, and the patients of the surgical epilepsy program for their cooperation and participation.
Supported by NIH grant NS 02808.

REFERENCES

Amaral, D.G., Insausti, R., and Cowan, W.M., 1984a, The commissural connections of the monkey hippocampal formation, *J. Comp. Neurol.* 224:851.

Amaral, D.G. and Price, J.L., 1984b, Amygdalo-cortical projections in the monkey (*Macaca fascicularis*), *J. Comp. Neurol.* 230:463.

Amaral, D.G., 1986, Amygdalohippocampal and amygdalocortical projections in the primate brain, *in:* "Excitatory Amino Acids and Epilepsy," R. Schwarcz and Y. Ben-Ari, eds., Plenum, New York.

Amaral, D.G. and Insausti, R., 1990, The hippocampal formation, *in:* "The Human Nervous System," G. Paxinos, ed., Academic Press, New York.

Bertashius, K.M., 1991, Propagation of human complex-partial seizures: a correlation analysis, *Electroencephalogr. Clin. Neurophysiol.* 78:333.

Brazier, M.A.B., 1964, Evoked responses from the depths of the human brain, *Ann. NY Acad. Sci.* 112:33.

Brazier, M.A.B., 1972, Spread of seizure discharge in epilepsy: anatomical and electrophysiological considerations, *Exp. Neurol.* 36:263.

Brazier, M.A.B., 1973, Electrical seizure discharges within the human brain: the problem of spread, *in:* "Epilepsy, Its Phenomenon in Man," M.A.B. Brazier, ed., Academic Press, New York.

Buser, P. and Bancaud, J., 1983, Unilateral connections between amygdala and hippocampus in man: study of epileptic patients with depth electrodes, *Electroenceph. Clin. Neurophysiol.* 55:1.

Buzsaki, G., 1984, Feed-forward inhibition in the hippocampal formation, *Progr. Neurobiol.* 22:131-153.

Colder, B.W., Wilson, C.L., Frysinger, R.C., Chao, L.C., Harper, R.M., and Engel Jr., J., 1994, Neuronal synchrony and relation to burst discharge in epileptic temporal lobes, *J. Neurophysiol.* (In press).

Crandall, P.H., 1987, Cortical resection, *in:* "Surgical Treatment of the Epilepsies," J. Engel Jr., ed., Raven, New York.

De Olmos, J., 1990, Amygdala, *in:* "The Human Nervous System," G. Paxinos, ed., Academic, San Diego.

Demeter, S., Rosene, D.L., and Van Hoesen, G.W., 1985, Interhemispheric pathways of the hippocampal formation, presubiculum, entorhinal and posterior hippocampal cortices in the rhesus monkey: the structure and function of the hippocampal commissures, *J. Comp. Neurol.* 233:30.

Demeter, S., Rosene, D.L., and Van Hoesen, G.W., 1990, Fields of origins and pathways of the interhemispheric commissures in the temporal lobe of macaques, *J.Comp. Neurol.* 302:29.
Elul, R., 1964a, Regional differences in the hippocampus of the cat. I. Specific discharge patterns of the dorsal and ventral hippocampus and their role in generalized seizures, *Electroenceph. Clin. Neurophysiol.* 16:470.
Elul, R., 1964b, Regional differences in the hippocampus of the cat. II. Projections of the dorsal and ventral hippocampus, *Electroenceph. Clin. Neurophysiol.* 16:489.
Engel Jr, J., ed., 1993, "Surgical Treatment of the Epilepsies 2," Raven, New York.
Fernandes de Lima, V.M., Pijn, J.P., Nunes Filipe, C., and Lopes da Silva, F., 1990, The role of hippocampal commissures in the interhemispheric transfer of epileptiform afterdischarges in the rat: a study using linear and non-linear regression analysis, *Electroencephalogr. Clin. Neurophysiol.* 76:520.
Gloor, P., Salanova, V., Olivier, A., and Quesney, L.F., 1993, The human dorsal hippocampal commissure: An anatomically identifiable and functional pathway, *Brain* 116:1249-1273.
Gotman, J., 1983, Measurement of small time differences between EEG channels: method and application to epileptic seizure propagation, *Electroencephalogr. Clin. Neurophysiol.* 56:501.
Gotman, J., 1987, Interhemispheric interactions in seizures of focal onset: data from human intracranial recordings, *Electroencephalogr. Clin. Neurophysiol.* 67:120.
Gotman, J., 1991, Seizure spread, *in:* "Epilepsy Surgery," H. Luders, ed., Raven Press, New York.
Gotman, J., Levtova, V., and Farine, B., 1993, Graphic representation of the EEG during epileptic seizures, *Electroencephalogr. Clin. Neurophysiol.* 87:206-214.
Levesque, M.F., Zhang, J.X., Wilson, C.L., Behnke, E.J., Harper, R.M., Lufkin, R.B., Engel Jr, J., and Crandall, P.H., 1990, Stereotactic investigation of limbic epilepsy using a multimodal image analysis system, *J. Neurosurg.* 75:364.
Lieb, J.P., Walsh, G.O., Babb, T.L., Walter, R.D., and Crandall, P.H., 1976, A comparison of EEG seizure patterns recorded with surface and depth electrodes in patients with temporal lobe epilepsy, *Epilepsia* 17:137.
Lieb, J.P., Engel Jr., J., and Babb, T.L., 1986a, Interhemispheric propagation time of human hippocampal seizures. I. Relationship to surgical outcome, *Epilepsia* 27:286.
Lieb, J.P. and Babb, T.L., 1986b, Interhemispheric propagation time of human hippocampal seizures. II. Relationship to pathology and cell density, *Epilepsia* 27:294.
Lieb, J.P., Hoque, K., Skomer, C.E., and Song, X.W., 1987a, Inter-hemispheric propagation of human mesial temporal lobe seizures: a coherence/phase analysis, *Electroenceph. Clin. Neurophysiol.* 67:101.
Lieb, J.P., Babb, T.L., Engel Jr, J., and Darcey, T.M., 1987b, Propagation pathways of interhemispheric seizure discharges compared in human and animal hippocampal epilepsy, *in:* "Fundamental Mechanisms of Human Brain Function," J. Engel Jr., ed., Raven, New York.
Lieb, J.P., Dasheiff, R.M., and Engel Jr, J., 1991, Role of the frontal lobes in the propagation of mesial temporal lobe seizures, *Epilepsia* 32:822.
Lopes da Silva, F.H. and Mars, N.J.I., 1987, Spread of epileptic seizure activity in experimental and clinical epilepsy: the use of mutual information analysis, *in:* "Presurgical Evaluation of Epileptics. Basics, Techniques, Implications," H.G. Wieser and C.E. Elger, eds., Springer-Verlag, Berlin.
Lüders, H., ed., 1991, "Epilepsy Surgery," Raven, New York.
Mars, N.J.I., Thompson, P.M., and Wilkus, R.J., 1985, The spread of epileptic seizure activity in humans, *Epilepsia* 26:85.
Morrell, M.J., Sperling, M.R.,and O'Conner, M.J., 1990a, Propagation pathways in partial seizures of mesial temporal lobe origin, *Epilepsia* 31:620.
Morrell, M.J., Sperling, M.R., and O'Conner, M.J., 1990b, The frontal lobe in the clinical expression and surgical evaluation of partial epilepsy, *Epilepsia* 31:637.
Nunes Filipe, C., Pijn, J.P., Fernandes de Lima, V.M., and Lopes da Silva, F.H., 1989, Propagation of afterdischarges along the septo-temporal axis of the rat hippocampus: a quantitative analysis, *Electroencephalogr. Clin. Neurophysiol.* 73:172.
Pandya, D.N. and Rosene, D.L., 1985, Some observations on trajectories and topography of commissural fibers, *in:* "Epilepsy and the Corpus Callosum," A.G. Reeves, ed., Plenum, New York. Pandya, D.N. and Yeterian, E.H., 1987, Hodology of limbic and

related structures: cortical and commissural connections, *in:* "Presurgical evaluation of epileptics," H.G. Wieser and C.E. Elger, eds., Springer, Berlin, Heidelberg.
Paxinos, G., ed., 1990, "The Human Nervous System," Academic, New York.
Pijn, J.P.M., Vijn, P.C.M., Lopes da Silva, F.H., van Emde Boas, W., and Blanes, W., 1990, Localization of epileptogenic foci using a new signal analytical approach, *Neurophysiol. Clin.* 20:1.
Rosene, D.L. and Van Hoesen, G.W., 1987, The hippocampal formation of the primate brain: a review of some comparative aspects of cytoarchitecture and connections, *in:* "Cerebral Cortex, Volume 6," E.G. Jones and A. Peters, eds., Plenum, New York.
Rutecki, P.A., Grossman, R.G., Armstrong, D., and Irish-Loewen, S., 1989, Electrophysiological connections between the hippocampus and entorhinal cortex in patients with complex partial seizures, *J. Neurosurg.* 70:667.
Saunders, R.C. and Rosene, D.L., 1988a, A comparison of the efferents of the amygdala and the hippocampal formation in the rhesus monkey. I. Convergence in the entorhinal, prorhinal, and perirhinal cortices, *J. Comp. Neurol.* 271:153.
Saunders, R.C., Rosene, D.L., and Van Hoesen, G.W., 1988b, Comparison of the efferents of the amygdala and the hippocampal formation in the rhesus monkey. II. Reciprocal and non-reciprocal connnections, *J. Comp. Neurol.* 271:185.
Spencer, S.S., Spencer, D.D., Williamson, P.D., and Mattson, R.H., 1982, The localizing value of depth electroencephalography in 32 refractory epileptic patients, *Ann. Neurol.* 12:248.
Spencer, S.S., Williamson, P.D., Spencer, D.D., and Mattson, R.H., 1987, Human hippocampal seizure spread studied by depth and subdural recording: the hippocampal commissure, *Epilepsia* 28:479.
Spencer, S., 1988, Cortical and intercortical seizure spread, *in:* "The Anatomy of Epileptogenesis," B.S. Meldrum and J. Ferrendelli, eds., John Libbey-Eurotext, London.
Spencer, S.S., Spencer, D.D., Williamson, P.D., and Mattson, R., 1990, Combined depth and subdural electrode investigation in uncontrolled epilepsy, *Neurol.* 40:74.
Spencer, S.S., Marks, D., Katz, A., Kim, J., and Spencer, D.D., 1992, Anatomic correlates of interhippocampal seizure propagation time, *Epilepsia* 33:862.
Wieser, H-G., 1983, "Electroclinical features of the psychomotor seizure," G. Fischer-Butterworths, Stuttgart London.
Wieser, H-G. and Kausel, W., 1987, Limbic seizures, *in:* "Presurgical Evaluation of Epileptics.," H.-G. Wieser and E.C. Elger, eds., Springer, Berlin Heidelberg.
Wieser, H., 1988, Human limbic seizures: EEG studies, origin, and patterns of spread, *in:* "The Anatomy of Epileptogenesis," B. Meldrum, J. Ferrendelli, and H. Wieser, eds., John Libbey-Eurotext, London.
Wieser, H.-G., 1991, Ictal manifestations of temporal lobe seizures, *in:* "Advances in Neurology," D. Smith, D. Treiman, and M. Trimble, eds., Raven, New York.
Wilson, C.L. and Engel, J. Jr., Electrical stimulation of the human epileptic limbic cortex, *in:* "Electrical and Magnetic Stimulation of the Brain and Spinal Cord," O. Devinski, A. Beric, and M. Dogali, eds., Raven Press, New York, pp. 103-113.
Wilson, C.L., Isokawa, M., Babb, T.L., and Crandall, P.H., 1990, Functional connections in the human temporal lobe. I. Analysis of limbic system pathways using neuronal responses evoked by electrical stimulation, *Exp. Brain Res.* 82:279.
Wilson, C.L., Isokawa, T.L., Babb, P.H., Crandall, P.H., Lévesque, M.F., and Engel Jr, J., 1991, Functional connections in the human temporal lobe. II. Evidence for a loss of functional linkage between contralateral limbic structures, *Exp. Brain Res.* 85:174.
Wilson, C.L., Colder, B., Branch, G., Khan, S., Behnke, E., and Levesque, M., 1992, Evoked potential analysis of interhemispheric connections mediating limbic seizure propagation, *Epilepsia* 33:59.
Zhang, J., Lévesque, M.F., Wilson, C.L., Harper, R.M., Engel Jr, J., Lufkin, R., and Behnke, E., 1990, Multimodality imaging of brain structures for stereotactic surgery, *Radiology* 175:435.

CORPUS CALLOSUM SECTION FOR COMPLEX PARTIAL SEIZURES

S. J. Purves, M.D., FRCP(C),[1] J. A. Wada, M.D., ScD, FRCP,[2] and W. B. Woodhurst, M.D.[3]

[1]Sioux City Neurology
500 Jackson Street
Sioux City, IA 51101

[2]Department of Neurology
Health Sciences Centre Hospital
University of British Columbia
Vancouver, British Columbia V6T 2A1

[3]Department of Neurosurgery
Vancouver General Hospital
700 W. 10th Ave., Suite 300
Vancouver, British Columbia V5Z 4E5

INTRODUCTION

Corpus Callosum Section (CC Section) has attracted renewed interest in recent years as an option for surgical treatment for patients with medically intractable seizures who are not eligible for resective surgery. A variety of seizure types have been considered for treatment and this topic reviewed extensively elsewhere (Purves 1991, Spencer 1988, Gates 1992). There is some agreement emerging that specific types of generalized seizure patterns such as atonic or tonic drop attacks respond well to CC Section. The application of this procedure for complex partial seizures is still poorly understood. Much of the interest in CC Section for CPS has centered on those of presumed frontal origin but the extent of documentation of this assumption is often unclear.

Many patients selected for this procedure have multiple seizure types and the extent of pre-operative ictal recording required to precisely define seizures patterns has varied widely from one center to another and has evolved considerably in the last two decades. The "drop attack" which is

Epilepsy and the Corpus Callosum II, Edited by A.G. Reeves
and D.W. Roberts, Plenum Press, New York, 1995

one of the most common patterns targeted for treatment is extraordinarily difficult to capture and interpret with video and ictal EEG because of contamination with muscle and movement artifact obscuring the EEG, and concerns regarding patient safety during the recording period, with or without intracranial electrodes in place. The movement of the patient also requires individualized video camera control work to capture clearly and the patient needs to be examined at the time of the drop and immediately after by a skilled observer to determine the degree, if any, of loss of consciousness if these spells are to be clearly classified. Hence, many reported series have failed to precisely characterize all of the preoperative ictal seizure patterns with such direct recordings and have had to rely more on patient history and non-recorded observed seizures. It is only within the last decade that most patients have had ictal events recorded preoperatively to confirm some of the seizure patterns and exclude seizures which would be amenable to resective surgery. The extent of such recordings and the criteria for what is amenable to resective surgery is still inconsistent from one centre to another making comparison in the literature very difficult. Decisions to progress to invasive recording for more precise definition of seizure patterns or foci of origin also vary from one center to another depending on the individual patient's suitability for such recordings, availability of such techniques within the health care system, patient motivation and inclination to travel to referral centers and the availability and results of other neurological, neuroimaging, or neuropsychological evaluations.

Additionally, the remarkable progress in the past decade in the neuroimaging field (MRI, PET, SPECT) must also be recognized as contributing to the current selection of patients for treatment by CC Section. Much more subtle cerebral abnormalities are now identified which may serve to better focus the invasive recording and hence lead to or exclude resective surgery. The anatomy of the corpus callosum along with the post-operative extent of section can be clearly visualized with MRI. Functional neuroimaging (PET, SPECT) is playing a significant role in the identification of temporal CPS (Berkovic et al. 1992, Henry et al. 1992) but it's role for evaluation of extratemporal CPS evaluation is not yet clear.

Within the population of patients who are candidates for corpus callosum section because of medically intractable generalized seizures, a clear division between patients with seizures which are generalized from onset and those which are rapidly secondarily generalized from a CPS originating in a focus remote from the surface recording electrodes is not always possible preoperatively. The distinctions sometimes become clearer retrospectively post CC Section when the patient is left with a partial seizure pattern which appears to be a fragment of the onset of the preoperative generalized one (Gates et al. 1987). Whether the wider availability of invasive recording will identify these complex partial seizures with rapid generalization well enough to advise treatment with resection instead of CC Section remains to be determined. Often the surface recordings do not provide a clear indication of where to place the depth or subdural electrodes. Such intracranial electrodes are well known to have a very limited recording sphere and thus may not provide the overview of seizure propagation that is necessary in these complex patients. For these reasons it may be reasonable in selected cases with negative neuroimaging and documented rapid bifrontal spread of ictal activity to consider an anterior CC Section as an alternative to invasive studies of the frontal lobes.

RATIONALE FOR CC SECTION FOR CPS

Complex partial seizures can originate from frontal lobe but are much less common than those of temporal origin (Morris 1992, Williamson 1987). The anterior 2/3 of the corpus callosum is largely made up of crossing fibres between the frontal lobes, so that complex partial seizures of frontal origin have been the CPS type targeted for treatment by this surgery (Wada 1980, St. Hilaire 1985, Williamson 1985, and Turmel 1992). Frontal lobe seizures have presented a considerable challenge to identify and classify (Williamson 1992, Weiser et al. 1992). Data derived from the study of patients with intracerebral electrodes has lead to the International Committee of the ILAE's most recent classification (1989) based on anatomical divisions but this is still considered only as work in progress (Bancaud & Talairach 1992, Swartz et al. 1987). This classification system has been derived from selected patients studied with depth electrodes whose seizures began regionally and who in some cases could be treated successfully with a partial frontal lobectomy. These patients may represent only the subset of those with frontal lobe epilepsy whose seizures do not spread too rapidly from the focus of origin, thus allowing the identification of the focus for study. The patients who have partial seizures post CC Section that are the same as the onset of the preoperative pattern suggest that some patients have partial frontal seizures that propagate to the opposite hemisphere so rapidly that they cannot be clearly characterized preoperatively with surface electrodes (Gates 1987). In theory preoperative study of such patients with depth electrodes might allow treatment by resective surgery. But the identification of frontal lobe ictal onset with intracerebral electrodes is considerably more difficult than for temporal lobe seizures because of the need for greater numbers of electrodes to demonstrate the variable patterns of spread in these areas.

The results of frontal lobe resections have been much less encouraging than resective surgery in temporal lobe (Engel 1987) and particularly discouraging without an anatomical lesion identified (VanNess et al. 1989). There is also some neuropsychological deficit after frontal lobe resection whereas there is no identified deficit after anterior CC Section. Thus, if it can be demonstrated that the clinical expression of a patient's frontal lobe complex partial seizures are dependent on interhemisphere transmission which could be presumed to be via the anterior corpus callosum, then interrupting these pathways may prove to be an effective way of treating such seizures. It may be found that such surgery would modify the spread so that the clinical expression of the seizure is "milder" and this has been verified in review of clinical experience. The fact that the seizures may not stop completely seems a reasonable compromise in order to avoid the neuropsychological consequences of a frontal lobe resection. However, the uncertainty of the outcome; and the low rate of "cure" with CC Section has made this approach less desirable in the opinion of some. There may also be some inhibitory connections severed, along with pathways of excitatory transmission which is the explanation of the increased intensity of focal seizures post-operatively reported in a few patients (Spencer 1984, 1988).

RESULTS

Some of the more detailed series reporting the results of CC Section have identified the pre-operative seizure patterns in sufficient detail that the results regarding its use for treatment of CPS can be evaluated. Sometimes the CPS appear to be the primary type of seizures targeted for treatment and in others they are listed as an additional seizure type in a patient who has generalized seizures targeted for treatment with CC Section. In her 1988a review, Spencer reports 28 patients treated with anterior CC Sections and 90 patients treated with complete CC Section. Of the 28 anterior CC Sections, 11 had complex partial seizures, seven were improved. Of 90 patient with complete CC Sections, 42 had CPS of whom again approximately half (19) were improved. There were ten with increased seizures. In her own series of 22 patients (Spencer 1988b), nine of 19 patients had elimination of preoperative CPS but three experienced increased intensity or new kinds of partial seizures. Roberts and Reeves (1987) reported 25 patients with CPS among their seizure types preoperatively, ten of 25 had no further CPS post partial or complete CC Section and three more had greater than 80% reduction in this seizure type. Subsequent reports do not have such a predominance of complete CC Sections vs. anterior CC Sections. In fact, most are restricted to results anterior CC Section. Oguni's report (1991) documents multiple seizure types in 85 percent of the patients submitted to a partial CC Section. Eight of these 43 patients had CPS amongst the pre-operative seizure type and 50% showed improvement. None were worse but three developed some "new" mild type of post-op CPS thought to be similar to those reported by Gates in 1987 as fragments of their previous generalized seizures.

Fuiks (1991) has a series with over half of the patients studied with depth electrodes. Only 11 of 78 patients had CPS in their pre-op seizure types and these were all part of mixed patterns. Eighty-eight percent of the patients who had CPS were improved from the point of view of this particular seizure pattern. Nordgren et al. (1991) report a series of 18 patients in the pediatric age group with six showing pre-op CPS without generalization and five achieving excellent or good results with greater than 80 percent seizure reduction in this subgroup. CPS with secondary generalization also had a good outcome with three out of three improved.

Analysis of our own series at the University of British Columbia with regards to the results of complex partial seizures is summarized in Table 1. This series includes 21 patients with CPS in our previously reported series of 24 anterior callosotomies (Purves et al. 1988) and 25 patients with CPS of 30 patients subsequently operated on with greater than one year follow-up. The patients are classified into four groups based on the electroclinical assessment of CCTV-EEG recordings done preoperatively. Only surface and sphenoidal electrodes were used in these recordings with the exception of one recent patient. We found 12 of 14 patients with a single pattern of CPS believed to be frontal in origin were improved but none were completely seizure free. In these patients, the severity and/or the frequency was reduced by more than 80%, for example, by modifying a CPS pattern so that the patient no longer fell during seizures. Others went from several seizures per month to two or three per year. None of these 14 had lesions on CT or MRI and hence were not eligible for resective surgery at our centre. The one patient studied with subdural recordings had partial seizures suspected to involve supplementary motor area; however, these invasive recordings showed such rapid bilateral spread that we decided to offer an anterior callosotomy. This did stop the patient falling because it reduced the violent trunk muscle involvement but did not reduce the frequency. Post-operatively the ictal clinical activity also changed from bilateral to unilateral contractions of right face, arm, and leg. The one patient in this group who had a poor outcome with no change in her seizure pattern has subsequently been reinvestigated with improved MRI and repeat ictal EEG. On the basis of these studies, it was decided that her CPS pattern was more likely of temporal origin. She had a temporal lobectomy and became seizure free. Such a patient would now have been submitted for invasive recordings because the improved MRI imaging demonstrating a smaller temporal lobe was not congruent with the ictal clinical pattern suggesting frontal lobe origin and we feel that this would avoid such incorrect localization in future. Interestingly, there was no evidence for any neurological or functional deficit in this patient after an anterior CC section apart from the disappointment in the continued complex partial seizures characterized as an abrupt onset of motor automatisms frequently leading to a fall.

Table 1. CPS Results after anterior CC Section at the University of British Columbia.

Pre-Op Seizure Patterns	**Good**	**Poor**
Frontal CPS Only	12	1
Frontal CPS + 2° generalized seizures	6	9
CPS of multiple patterns +/- 2° generalized seizures	1	7
SUBTOTAL	19	17
Generalized with suspected focal onset	8	2
TOTAL	27	19

Good: >80% reduction in seizure frequency or severity
Poor: no change or minimal reduction

In the 15 patients with both CPS suspected to be of frontal origin and secondary generalized convulsive seizures, just six of 15 experienced reduced CPS after the anterior CC section. Some of the nine with lack of reduction of CPS patterns did experience benefit in reduction of the generalized convulsions. One of these patients had a subsequent resection of lateral frontal lobe because the residual clinical pattern could be more clearly localized to lateral frontal cortex and excision guided with acute corticography. Most of these whose partial seizures were not improved were quite low functioning suggesting diffuse cerebral pathology and one had hemiatrophy. There was one failed temporal lobectomy and another patient with a temporal lesion in this group whose CPS seizures did not improve.

Of the eight patients with more than one pattern of CPS, with or without secondary generalized seizures, seven did not experience any reduction in their partial seizure pattern although some had benefit in less frequent generalized convulsive attacks.

The fourth group of patients includes ten patients who had clinically generalized seizures but EEG features or imaging findings suggestive of a focal onset. These patients did not have clinical complex partial seizures preoperatively although it was suspected they had complex partial seizure onset before their secondarily generalized seizures. This group did very well with 8/10 much improved. One of the failures was a patient with generalized seizures after a brain tumor resection and the other one a failed temporal lobectomy.

If the last group is excluded since they did not have clinical CPS preoperatively, the overall results are quite similar to those of other series for anterior CC Section treatment of complex partial seizures with 20/36 or 55% of the patients improved. Inclusion of the group of patients with partial onset with rapid secondary generalization improves the series outcome slightly to 61% who are better.

CONCLUSIONS

Based on the preceding discussions, it can be concluded that the role of treatment of medically intractable complex partial seizures of frontal origin by corpus callosum section is still unclear and no simple recommendations should be made. The determinants of what is medically intractable and what is not amenable to resective surgery may vary from one institution to another and have been in a rapid evolution through the past 15 years while the series in the literature accumulated. There are three frontiers where the boundaries of selection are still changing. There have been new anticonvulsants released during the last two decades and more expected in the next decade making fewer (and perhaps more selected) patients medically intractable. The availability and approaches to non-invasive and invasive ictal EEG recordings have widened allowing more precise characterization of pre-operative seizure patterns. The dramatic developments in MRI neuroimaging technique has already changed our ability to identify focal lesions and the application of the functional imaging techniques of PET and SPECT may extend further the definitions of origin in CPS in the future. With all of these selection factors under such rapid evolution, many reports in the literature must be viewed with a critical eye and the results might not be the same in a series of patients with medically intractable CPS of suspected frontal lobe origin selected with current methodologies.

However, despite these evolving criteria for selection, there appears to be some agreement from both a theoretical and empirical viewpoint that anterior CC Section is a useful treatment for patients with complex partial seizures suspected to be of frontal lobe origin based on non-invasive EEG-CCTV ictal studies, particularly when these also demonstrate rapid propagation to contralateral frontal regions with or without the clinical manifestation of widespread generalization (tonic clonic convulsion). A review of the literature and our experience suggests that at least half of these patients will achieve considerable improvement in the CPS pattern after anterior CC Section without suffering any neurological deficit. Whether it is appropriate to proceed to anterior CC Section without invasive recordings in these patients remains controversial and will depend on the clinical experience and expertise of the investigator, MRI findings and suitability of the individual for such studies. Careful exclusion of temporal lobe CPS seems important, as these seizures are not reduced after anterior CC Section in our experience.

Most of the reports of CC Section for complex partial seizures are confined to the results of an anterior CC Section. There are insufficient cases in the literature who have extrafrontal CPS treated with posterior or complete CC Section to know if that will be effective. With the current approach of two-stage callosotomies, some experience may become available through the next decade. However, it will be necessary for all centers performing this procedure to continue the challenging and demanding task of careful documentation and classification of patients pre-operative and post-operative seizure patterns for any meaningful progress to occur in this area.

REFERENCES

Bancaud, J., Talairach, J., 1992, Clinical semiology of frontal lobe seizures, *in:* "Adv. Neurol.," P. Chauvel, and A.V. Delgado-Escueta, eds., Raven Press, New York, 57:3-58.

Berkovic, S.F., Newton, M.R., and Rowe, C.R., 1992, Localization of epileptic foci using SPECT, *in:* "Epilepsy Surgery," H. Luders, ed., Raven Press, New York, pp. 251-258.

Commission on Classification and Terminology of the International League Against Epilepsy, 1989, Proposal for revised classification of epilepsies and epileptic syndromes, *Epilepsia* 30(4):389-399.

Engel, J., 1987, Outcome with respect to epileptic seizures, *in:* "Surgical Treatment of the Epilepsies," J. Engel, ed., Raven Press, New York, pp. 553-571.

Fuiks, K.S., Wyler, A.R., Hermann, B.P., and Somes, G., 1991, Seizure outcome from anterior and complete corpus callosotomy, *J. Neurosurg.* 74:573-578.

Gates, J.R., Rosenfeld, W.E., Maxwell, R.E., and Lyons, R.E., 1987, Response of multiple seizure types to corpus callosum section, *Epilepsia* 28:28-34.

Gates, J.R., 1992, Corpus callosum candidacy, *in:* "Epilepsy Surgery," H. Luders, ed., Raven Press, New York, pp. 119-125.

Henry, T.R., Sutherling, W.W., Engel, Jr., J., Risinger, M.W., and Levesque, M.F., 1992, The role of positron emission tomography in presurgical evaluation of partial epilepsies of neocortical origin, *in:* "Epilepsy Surgery," H. Luders, ed., Raven Press, New York, pp. 243-249.

Morris, H., 1992, Frontal lobe epilepsies, *in:* "Epilepsy Surgery," H. Luders, ed., Raven Press, New York, pp. 157-165.

Nordgren, R.E., Reeves, A.G., Viguera, A.C., and Roberts, D.W., 1991, Corpus callosotomy for intractable seizures in pediatric age group, *Arch. Neurol.* 48:364-372.

Oguni, H., Olivier, A., Andermann, F., and Comair, J., 1991, Anterior callosotomy in the treatment of medically intractable epilepsies: a study of 43 patients with a mean follow-up of 39 months, *Ann. Neurol.* 30:357-364.

Purves, S.J., 1991, Selection of patients for corpus callosum section, *in:* "Surgery for Epilepsy," S.S. Spencer and D.D. Spencer, eds., Blackwell, Cambridge, MA, pp. 69-86.

Purves, S.J., Wada, J.A., Woodhurst, W.B., Moyes, P.D., Strauss, E., Kosaka, B., and Li, D., 1988, Results of anterior corpus callosum section in 24 patients with medically intractable seizures, *Neurology* 38:1194-1201.

Roberts, D.W. and Reeves, A., 1987, Effect of commissurotomy on complex partial epilepsy in patients without a resectable seizure focus, *Appl. Neurophysiol.* 50:398-400.

Saint-Hilaire, J.M., Giard, N., Bouvier, G., and Labrecque, R., 1985, Anterior callosotomy in the frontal lobe epilepsies, *in:* "Epilepsy and the Corpus Callosum, " A.G. Reeves, ed., Plenum Press, New York, pp. 303-314.

Spencer, S.S., Spencer, D.D., Glaser, G.H., Williamson, P.D., and Mattson, R.H., 1984, More intense focal seizure types after callosal section: the role of inhibition, *Ann. Neurol.* 16:686-693.

Spencer, S.S., 1988a, Corpus callosum section and other disconnection procedures for medically intractable epilepsy, *Epilepsia* 29(Suppl 2):585-599.

Spencer, S.S., Spencer, D.D., Williamson, P.D., Sass, K., Novelly, R.A., and Mattson, R.H., 1988b, Corpus callosotomy for epilepsy: I. seizure effects, *Neurology* 38:19-24.

Turmel, A., Giard, N., Bouvier, G., Labrecque, R., Veilleux, F., Roulear, I., and Saint-Hilaire, J.M., 1992, Indications for cortectomies or callosotomies, *in:* "Adv. Neurol.," P. Chauvel and A.V. Delgado-Escueta, eds., Raven Press, New York 57:689-706.

VanNess, P.C., Estes, M.L., Salanova, V., Morris, H.H., and Luders, H.O., 1989, Seizure reductions and pathologic findings in 37 operated patients with suspected neocortical epilepsy, *Epilepsia* 30:728.

Wada, J.A., 1980, New surgical treatment through experimental models, *in:* "Advances in Epileptology: The Xth Epilepsy International Symposium," J.A. Wada and K. Penry, eds., Raven Press, New York, pp. 195-213.

Weiser, H.G., Swartz, B.E., Delgado-Escueta, A.V., Bancaud, J., Walsh, G.O., Maldonado, H., and Saint-Hilaire, J.M., 1992, Differentiating frontal lobe seizures from temporal lobe seizures, *in:* "Adv. Neurol.," P. Chauvel and A.V. Delgado-Escueta, eds., Raven Press, New York 57:267-286.

Williamson, P.D., 1985, Corpus callosum section for intractable epilepsy, *in:* "Epilepsy and the Corpus Callosum," A.G. Reeves, ed., Plenum Press, New York, pp. 243-279.

Williamson, P.D., 1992, Frontal lobe seizures: problems of diagnosis and classification, *in:* "Adv. Neurol.," P. Chauvel and A.V. Delgado-Escueta, eds., Raven Press, New York 57:289-310.

Williamson, P.D., Weiser, H.G., and Delgado-Escueta, A.V., 1987, Clinical characteristics of partial seizures, *in:* "Surgical Treatment of the Epilepsies," S. Engel, ed., Raven Press, New York, pp. 101-120.

THE ROLE OF POSTERIOR CALLOSOTOMY IN PATIENTS WITH SUBOPTIMAL RESPONSE TO ANTERIOR CALLOSOTOMY

David W. Roberts, M.D.,[1] Alexander G. Reeves, M.D.,[2] and Richard E. Nordgren, M.D.[2]

[1]Section of Neurosurgery
[2]Section of Neurology
Dartmouth-Hitchcock Medical Center
One Medical Center Drive
Lebanon, NH 03756

The current practice of sectioning the anterior one-half to two-thirds of the corpus callosum in selected patients with intractable seizure disorders has been predicated on the findings that partial section alone is often effective in controlling certain seizure types and that anterior section is better tolerated than complete callosotomy. In those patients in whom an unsatisfactory seizure response to anterior section has been obtained, completion of the section may then be performed. The benefit of such second stage surgery, in terms of seizure frequency reduction by seizure type, is the subject of this study.

MATERIALS AND METHODS

Of 98 patients who have undergone surgery of the corpus callosum for treatment of medically intractable seizure disorders at our institution, 74 patients have had initial division of the anterior portion (the remainder had initial posterior section or complete section at one operation). Of these, follow-up of at least one year is available in 64 patients. A second operation for division of the remaining posterior corpus callosum was performed in 27 of these 64 patients, the major criterion for completion of the commissurotomy being unacceptable persistence of generalized seizures. These patients were operated upon for both their first and second procedures between 1982 and 1991. The mean interval between surgeries was 8 months; the mean follow-up after completion of the callosotomy was 28 months.

Frequency of major motor, drop attack, complex partial and focal motor seizures was determined for all patients for the time periods prior to either surgery, following anterior callosal section, and following second stage surgery. This was obtained by interview with the patient and his or her family by the patient's neurologist in the outpatient department. Response of each seizure type to surgery was graded as 100%, ≥80 but <100%, ≥50 but <80%, and <50% reduction in frequency. Seizure response was also correlated with extent of corpus callosum sectioned at first surgery, as determined by operative report and/or MRI.

RESULTS

In those patients subsequently undergoing completion of their callosal section, reduction in seizure frequency by at least 80% following anterior section had been noted in 6 of 21 patients with major motor seizures, 6 of 16 patients with drop attacks, 6 of 8 with complex seizures, and 1 of 7 with focal motor seizures. Following their second procedure, comparable improvement in seizure frequency was noted in 13 of 21 patients with major motor seizures, 13 of 16 patients with drop attacks, 8 of 8 patients with complex partial seizures, and 1 of 7 with focal motor seizures. Seizure responses for each of these seizure types, broken down by percent reduction in frequency, are given in Tables 1 - 4.

New onset of focal motor seizures, consistent with the attenuated expression of previously generalized seizure activity, occurred in 12 of these 27 patients. Considering all seizure types together, the percentage of these patients obtaining 80-100% reduction of generalized seizures approximately doubled following the second surgery (62% versus 29%).

Seizure response with respect to length of corpus callosum sectioned at first surgery is provided in Tables 5 - 7. This data demonstrates a tendency towards improved response with greater length of initial section, but the number of patients in each group was small.

DISCUSSION

The earliest clinical experiences documenting the effectiveness of corpus callosum section in the treatment of intractable seizure disorders varied in the extent of section, comprising both partial and complete callosal divisions, with inconsistent inclusion of the anterior commissure or one of the fornices. Through the late 1970s and early 1980's, division of the complete corpus callosum in either one or two stages, but with preservation of the anterior commissure and fornices, became more common. These reports led to wider recognition that commissurotomy was effective in selected patients.[1-26]

Over the same period, recognition of the neuropsychological consequences of commissurotomy also became widespread.[27-29] Although the majority of patients appeared to function satisfactorily in their usual activities of daily living or in standardized examinations,[30] the possibility of deleterious effects independent of beneficial seizure control was appreciated.[31] These effects were clearly more common in patients undergoing complete rather than partial callosal section.[27-29]

Since many patients undergoing such surgery were noted to do well after anterior section alone and since anterior section usually was free of associated disconnection sequelae, many centers proceeded with only partial sections.[32-42] Some of these series report reasonably high success. Huck et al. described excellent responses in five of five patients.[32] Purves et al. noted improvement in 75% of patients,[39] and Fuiks et al. described efficacy in 78% of anteriorly sectioned patients.[42] Other reports of partial section have been less encouraging. Blume reported eight anterior callosal sections with three patients improved and two patients with equivocal response.[35] Makari et al. noted worthwhile improvement in 40%,[40] and Oguni noted success in 56%.[41]

Other centers, and particularly those who had had previous experience with complete sections, adopted the strategy of performing partial section and subsequently proceeding to completion of the section only if the initial operation produced less than the desired benefit.[18,19,21-26,43,44] Dartmouth has employed this strategy since 1982, and many other series now similarly include both partial and total callosal sections. In the series presented here, the percent of patients achieving ≥80% reduction in major motor seizures improved from 29% to 62% following completion of section. A similar degree of reduction in drop attacks improved from 38% to 81%, and for complex partial seizures improvement was seen from 75% to 100%. There was no expectation of significant improvement in focal motor seizure activity and none was seen.

These findings are consistent with the experience at Yale, where total section was much more effective than partial section in abolishing secondarily generalized seizures (77% versus 35%).[24] Although a significant difference in overall response was not noted by Gates and colleagues in those patients found to have inadvertent sparing of a portion of the corpus callosum, two patients who were re-operated for completion of section did improve.[20] The series of Fuiks et al. cited earlier does include ten patients who went on to completion and five patients were further improved.[42] With the possible exception of that series whose authors expressed reservations about proceeding to completion based on increased complications and only partial response, the collected experience would suggest that completion of callosal section in those patients obtaining suboptimal benefit after partial sectioning has a very reasonable chance of producing worthwhile further improvement.

The performance of partial section, with or without intent to consider later completion, would -- for determining the length of initial section -- benefit by understanding the possible relationship between length of section and seizure response. It is of interest that in some series of partial callosotomy with less encouraging effect, such as that of Blume,[35] only anterior one-half callosal section was performed, while in those series of partial section with greater efficacy, such as Fuiks et al. (anterior 80% section)[42] and Huck et al. (the anterior 6-7 cm),[32] more extensive division was performed. Oguni et al, who had reported 56% overall worthwhile improvement in their series, analyzed their seizure responses in terms of extent of callosotomy and reported 89% success in those who had undergone anterior two-thirds section and only 47% in those with anterior one-half section.[41] The experience reported in the present series is consistent with the hypothesis that greater extent of section is associated with better seizure response but the number of patients is small. As the more recognized neuropsychological changes associated with commissurotomy are often spared even with extensive but incomplete section, it may well be most appropriate in those patients who are undergoing partial callosal section to perform a more extensive partial division for optimal seizure improvement.

CONCLUSION

These findings are consistent with those of a number of other institutions. In general, results from partial section have not matched those achieved by complete section. Our experience favors an approach utilizing an aggressive although still partial anterior callosal section at initial surgery, followed by completion of the section in selected patients. Although posterior section may carry an increased neuropsychological cost relative to anterior section, for many patients the benefit of a second procedure for completion of callosal section exceeds any negative effect.

Table 1. Response of major motor seizures to initial anterior and subsequent posterior callosal section in 21 patients with that seizure type who underwent staged section.

100%	reduction	5
≥80%	reduction	1
≥50%	reduction	5
<50%	reduction	10
After completion of section:		
100%	reduction	9
≥80%	reduction	4
≥50%	reduction	1
<50%	reduction	7

Table 2. Response of drop attacks to initial anterior and subsequent posterior callosal section in 16 patients with that seizure type who underwent staged section.

After anterior section:		
100%	reduction	5
≥80%	reduction	1
≥50%	reduction	4
<50%	reduction	6
After completion of section:		
100%	reduction	7
≥80%	reduction	6
≥50%	reduction	2
<50%	reduction	1

Table 3. Response of complex partial seizures to initial anterior and subsequent posterior callosal section in eight patients with that seizure type who underwent staged section.

After anterior section:		
100%	reduction	3
≥80%	reduction	3
≥50%	reduction	2
<50%	reduction	0
After completion of section:		
100%	reduction	4
≥80%	reduction	4
≥50%	reduction	0
<50%	reduction	0

Table 4. Response of focal motor seizures to initial anterior and subsequent posterior callosal section in seven patients with that seizure type who underwent staged section.

After anterior section:		
100%	reduction	1
≥80%	reduction	0
≥50%	reduction	0
<50%	reduction	6
After completion of section:		
100%	reduction	0
≥80%	reduction	1
≥50%	reduction	1
<50%	reduction	5

Table 5. Response of major motor seizures to length of initial anterior section and to subsequent posterior callosal section in 21 patients with that seizure type who underwent staged section.

After anterior one-third section (n = 1):			After completion of section:		
100%	reduction	0	100%	reduction	0
≥80%	reduction	0	≥80%	reduction	0
≥50%	reduction	0	≥50%	reduction	0
<50%	reduction	1	<50	reduction	1
After anterior one-half section (n = 8):			After completion of section:		
100%	reduction	1	100%	reduction	1
≥80%	reduction	0	≥80%	reduction	2
≥50%	reduction	3	≥50%	reduction	0
<50%	reduction	4	<50	reduction	5
After anterior two-thirds section (n = 6):			After completion of section:		
100%	reduction	2	100%	reduction	3
≥80%	reduction	1	≥80%	reduction	2
≥50%	reduction	1	≥50%	reduction	1
<50%	reduction	2	<50	reduction	0
After anterior three-quarters section (n = 6):			After completion of section:		
100%	reduction	2	100%	reduction	5
≥80%	reduction	0	≥80%	reduction	0
≥50%	reduction	1	≥50%	reduction	0
<50%	reduction	3	<50	reduction	1

Table 6. Response of drop attacks to length of initial anterior section and to subsequent posterior callosal section in 16 patients with that seizure type who underwent staged section.

After anterior one-third section (n = 0):			After completion of section:		
100%	reduction	0	100%	reduction	0
≥80%	reduction	0	≥80%	reduction	0
≥50%	reduction	0	≥50%	reduction	0
<50%	reduction	0	<50	reduction	0
After anterior one-half section (n = 6):			After completion of section:		
100%	reduction	1	100%	reduction	3
≥80%	reduction	0	≥80%	reduction	0
≥50%	reduction	3	≥50%	reduction	2
<50%	reduction	2	<50	reduction	1
After anterior two-thirds section (n = 5):			After completion of section:		
100%	reduction	1	100%	reduction	1
≥80%	reduction	1	≥80%	reduction	4
≥50%	reduction	1	≥50%	reduction	0
<50%	reduction	2	<50	reduction	0
After anterior three-quarters section (n = 5):			After completion of section:		
100%	reduction	3	100%	reduction	3
≥80%	reduction	0	≥80%	reduction	2
≥50%	reduction	0	≥50%	reduction	0
<50%	reduction	2	<50	reduction	0

Table 7. Response of complex partial seizures to length of initial anterior section and to subsequent posterior callosal section in eight patients with that seizure type who underwent staged section.

After anterior one-third section (n = 0):			After completion of section:		
100%	reduction	0	100%	reduction	0
≥80%	reduction	0	≥80%	reduction	0
≥50%	reduction	0	≥50%	reduction	0
<50%	reduction	0	<50	reduction	0
After anterior one-half section (n = 3):			After completion of section:		
100%	reduction	1	100%	reduction	1
≥80%	reduction	2	≥80%	reduction	2
≥50%	reduction	0	≥50%	reduction	0
<50%	reduction	0	<50	reduction	0
After anterior two-thirds section (n = 3):			After completion of section:		
100%	reduction	2	100%	reduction	3
≥80%	reduction	1	≥80%	reduction	0
≥50%	reduction	0	≥50%	reduction	0
<50%	reduction	0	<50	reduction	0
After anterior three-quarters section (n = 2):			After completion of section:		
100%	reduction	0	100%	reduction	0
≥80%	reduction	0	≥80%	reduction	2
≥50%	reduction	2	≥50%	reduction	0
<50%	reduction	0	<50	reduction	0

REFERENCES

1. W.P. Van Wagenen and R.Y. Herren, Surgical division of the commissural pathways in the corpus callosum: relation to spread of an epileptic attack, *Arch. Neurol. Psych.* 44:740-759 (1940).
2. J.E. Bogen and P.J. Vogel, Cerebral commissurotomy in man: preliminary case report, *Bull. Los Angeles Neurol. Soc.* 27:169-172 (1962).
3. J.E.Bogen, E.D. Fisher, and P.J. Vogel, Cerebral commissurotomy: a second case report, *JAMA* 194:1328-1329 (1965).
4. J.E. Bogen, R.W. Sperry, and P.J. Vogel, Addendum: commissural section and propagation of seizures, *in:* "Basic Mechanisms of the Epilepsies," H.H. Jasper, A.A. Ward, Jr., and A. Pope, eds., Little, Brown, and Company, Boston (1969) p. 439.
5. A.J. Luessenhop, Interhemispheric commissurotomy: (the split brain operation) as an alternative to hemispherectomy for control of intractable seizures, *Am. Surg.* 36:265-268 (1970).
6. A.J. Luessenhop, T.C. dela Cruz, and G.M. Fenichel, Surgical disconnection of the cerebral hemispheres for intractable seizures, *JAMA* 213:1630-1636 (1970).
7. A.L. Amacher, Midline commissurotomy for the treatment of some cases of intractable epilepsy, *Childs Brain* 2:54-58 (1970).
8. D.H. Wilson, C. Culver, M. Waddington, and M. Gazzaniga, Disconnection of the cerebral hemispheres: an alternative to hemispherectomy for the control of intractable seizures, *Neurology* 25:1149-1153 (1975).
9. D.H. Wilson, A.G. Reeves, M.S. Gazzaniga, and C. Culver, Cerebral commissurotomy for control of intractable seizures, *Neurology* 27:708-715 (1977).
10. D.H. Wilson, A. Reeves, and M. Gazzaniga, Division of the corpus callosum for uncontrollable epilepsy, *Neurology* 28:649-653 (1978).
11. D.H. Wilson, A. Reeves, and M. Gazzaniga, Corpus callosotomy for control of intractable seizures, *in:* "Advances in Epileptology: The Xth Epilepsy International Symposium," J.A. Wada and J.K. Penry, eds., Raven Press, New York (1980) pp. 205-213.
12. D.H. Wilson, A.G. Reeves, and M.S. Gazzaniga, "Central" commissurotomy for intractable generalized epilepsy: series two, Neurology 32:687-697 (1982).
13. R.E. Harbaugh, D.H. Wilson, A.G. Reeves, and M.S. Gazzaniga, Forebrain commissurotomy for epilepsy: review of 20 consecutive cases, *Acta Neurochir.* 68:263-275 (1983).
14. G. Geoffroy , M. Lassonde, F. Delisle, and M. Decarie, Corpus callosotomy for control of intractable epilepsy in children, *Neurology* 33:891-897 (1983).
15. M. Rayport, S.M. Ferguson, and W.S. Corrie, Outcomes and indications of corpus callosum section for intractable seizure control, *Appl. Neurophysiol.* 46:47-51 (1983).
16. M. Rayport, W.S. Corrie, and S.M. Ferguson, Corpus callosum section for control of clinically and electroencephalographically classified seizures, *in:* "Epilepsy and the Corpus Callosum," A.G. Reeves, ed., Plenum Press, New York, (1985) pp. 329-337.
17. A.G. Reeves and P.M. O'Leary, Total corpus callosotomy for control of medically intractable epilepsy, *in:* "Epilepsy and the Corpus Callosum," A.G, Reeves, ed. Plenum Press, New York (1985) pp. 269-280.
18. S.S. Spencer, D.D. Spencer, P.D. Williamson, and R.H. Mattson, Effect of corpus callosum section on secondary bilaterally synchronous interictal EEG discharges, *Neurology* 35:1689-1694 (1985).
19. S.S. Spencer, J.R. Gates, A.G. Reeves, and D.D. Spencer, R.E. Maxwell, and D.W. Roberts, Corpus callosum section for uncontrolled epilepsy, *in:* "Surgical Treatment of Epilepsy," J. Engel, ed., Raven Press, New York (1987) pp.425-444.
20. J.R. Gates, W.E. Rosenfeld, R.E. Maxwell, and R.E. Lyons, Response of multiple seizure types to corpus callosum section, *Epilepsia* 28:28-34 (1987).
21. D.W. Roberts and A.G. Reeves, Effect of commissurotomy on complex partial epilepsy in patients without a resectable seizure focus, *Appl. Neurophysiol.* 50:398-400 (1987).

22. D.D. Spencer and S.S. Spencer SS, Corpus callosotomy in the treatment of medically intractable secondarily generalized seizures of children, *Cleve. Clin. J. Med.* 56 (Suppl 1):S69-S77 (1988).
23. S.S. Spencer, Corpus callosum section and other disconnection procedures for medically intractable epilepsy, *Epilepsia* 29(Suppl. 2):S85-S99 (1988).
24. S.S. Spencer, D.D. Spencer, P.D. Williamson, et al., Corpus callosotomy for epilepsy. I. Seizure effects, *Neurology* 38:19-24 (1988).
25. D.W. Roberts, Section of the corpus callosum for epilepsy, *in:* "Operative Neurosurgical Techniques: Indications, Methods, and Results, ed 2," H. Schmidek and W.H. Sweet, eds., Grune and Stratton, Inc., Orlando (1988) pp. 1243-1250.
26. R.E. Nordgren, A.G. Reeves, A.C. Viguera, and D.W. Roberts, Corpus callosotomy for intractable seizures in the pediatric age group, *Arch. Neurol.* 48:364-372 (1991).
27. M.S. Gazzaniga, "The Bisected Brain," Appleton-Century-Crofts, New York (1970).
28. M.S. Gazzaniga, J.E.Bogen, and R.W. Sperry, Observations on visual perception after disconnection of the cerebral hemispheres in man, *Brain* 88:221-236 (1975).
29. B.T. Volpe, J.J. Sidtis, and J.D. Holzman, Cortical mechanisms involved in praxis: observations following partial and complete section of the corpus callosum in man, *Neurology* 32:645-650 (1982).
30. R.B. Ferrell, C.M. Culver, and G.J. Tucker, Psychosocial and cognitive function after commissurotomy for intractable seizures, *J. Neurosurg.* 58:374-380 (1984).
31. S.M. Ferguson, M. Rayport, and W.S. Corrie, Neuropsychiatric observations on behavioral consequences of corpus callosum section for seizure control, *in:* "Epilepsy and the Corpus Callosum," A.G. Reeves, ed., Plenum Press, New York, pp. 501-514 (1985).
32. F.R. Huck, J. Radvany, J.O. Avila, C.H. Pires de Camargo, R. Marino, Jr., P.C. Ragazzo, D. Riva, and P. Arlant, Anterior callosotomy in epileptics with multiform seizures and bilateral synchronous spike and wave EEG pattern, *Acta Neurochirurgica* Suppl 30:127-135 (1980).
33. J.O. Avila, J. Radvany, F.R. Huck, C.H. Pires de Camargo, R. Marino, Jr., P.C. Ragazzo, and D. Riva, Anterior callosotomy as a substitute for hemispherectomy, *Acta Neurochirurgica* Suppl. 30:137-143 (1980).
34. G. Bouvier, C. Mercier, J.M. St Hilaire, N. Giard, R. Labrecque, R.A. Beique, Anterior callosotomy and chronic depth electrode recording in the surgical management of some intractable seizures, *Appl. Neurophysiol.* 46:52-56 (1983).
35. W.T. Blume, Corpus callosum section for seizure control: rationale and review of experimental and clinical data, *Cleve. Clin. Q.* 51:319-332 (1984).
36. R. Marino Jr. and P.C. Ragazzo, Selective criteria and results of selective partial callosotomy, *in:* "Epilepsy and the Corpus Callosum," A.G. Reeves, ed., Plenum Press, New York, (1985) pp. 281-301.
37. J.M. Sainte-Hilaire, N. Giard, G. Bouvier, and R. Labrecque, Anterior callosotomy in frontal lobe epilepsies, *in:* "Epilepsy and the Corpus Callosum," A.G. Reeves, ed., Plenum Press, New York (1985) pp. 303-314.
38. A.M. Murro, H.F. Flanigan, B. Gallagher, D.W. King, and J.R. Smith, Corpus callosotomy for the treatment of intractable epilepsy, *Epilepsy Res.* 2:44-50 (1988).
39. S.J. Purves, J.A. Wada, W.B. Woodhurst, P.D. Moyes, E.Strauss, B. Kosaka, D. Li, Results of anterior corpus callosum section in 24 patients with medically intractable seizures, *Neurology* 38:1194-1201 (1988).
40. G.S. Makari, G.L. Homes, A.M. Murro, et al., Corpus callosotomy for the treatment of intractable epilepsy in children, *J. Epilepsy* 2:1-7 (1989).
41. H. Oguni, A. Oliviér, F. Andermann, and J. Comair, Anterior callosotomy in the treatment of medically intractable epilepsies: a study of 43 patients with a mean follow-up of 39 months, *Ann. Neurol.* 30:357-364 (1991).
42. K.S. Fuiks, A.R. Wyler, B.P. Hermann, and G. Somes, Seizure outcome from anterior and complete corpus callosotomy, *J. Neurosurg.* 74:573-578 (1991).
43. E. Garcia-Flores E, Corpus callosum section for patients with intractable epilepsy, *Appl. Neurophysiol.* 50:390-397 (1987).
44. Z.H. Rappaport and P. Lerman, Corpus callosotomy in the treatment of secondary generalizing intractable epilepsy, *Acta Neurochir.* 94:10-14 (1988).

ANTERIOR CALLOSOTOMY ADDED TO FRONTAL LOBECTOMY IN FRONTAL LOBE EPILEPSY

N. Giard M.D., F.R.C.P.C.,[1] I. Rouleau, Ph.D.,[2]
A. Bouthillier, M.D., F.R.C.P.C.,[3] G. Bouvier, M.D., F.R.C.P.C.,[3] and J.M. Saint-Hilaire, M.D., F.R.C.P.C.[1]

[1]Service de Neurologie
Hôpital Notre-Dame, Université de Montréal

[2]Service de Neurologie
Hôpital Notre-Dame, Université du Québec à Montréal

[3]Service de Neurochirurgie
Hôpital Notre-Dame, Université de Montréal

INTRODUCTION

Frontal lobe epilepsies are much more complex to investigate than those epilepsies of temporal lobe origin. Rapid diffusion of ictal discharge, typical of frontal lobe epilepsy, often renders a clear definition of the epileptogenic zone impossible unless invasive recordings are performed.[1] Even with the use of such invasive techniques, the results of frontal lobe cortectomy are not always satisfactory[2] and this is certainly so when compared with the results of temporal lobe cortectomies.

During the past decades, sectioning of the corpus callosum has been used for the treatment of many types of epilepsies, including frontal lobe epilepsy.[3] Experimental data[4,5,6] have suggested a major participation of the corpus callosum in the production of bilaterally synchronous EEG manifestations. Sectioning of the corpus callosum has been shown to decrease these bilaterally synchronous discharges in humans.[7,8]

Our experience in the investigation of frontal lobe epilepsies with depth electrodes has shown that for the majority of patients, the epileptogenic zone is large and not clearly defined, involving simultaneously both internal and external surfaces of the frontal lobe. Despite extensive cortectomies, a high percentage of patients continues to have seizures. The most probable hypothesis for this persistence of

Epilepsy and the Corpus Callosum II, Edited by A.G. Reeves
and D.W. Roberts, Plenum Press, New York, 1995

seizures is that a portion of the epileptogenic zone remains after the cortectomy. Rationally speaking, the addition of an anterior callosotomy (a.c.) to the frontal cortectomy should interrupt a sufficient number of callosal fibers to reduce the number of residual epileptic discharges diffusing to the contralateral frontal lobe and, consequently, prevent the occurrence of clinically significant seizures.

This study has been undertaken to demonstrate whether there are any possible benefits to the addition of an a.c. to a frontal lobe cortectomy for the treatment of frontal lobe epilepsy.

MATERIALS AND METHODS

This study includes 28 patients who underwent surgery for frontal lobe epilepsy between the years 1980 and 1990 at Notre-Dame Hospital, Montreal. Patients who were treated with callosotomy alone were excluded from this population. Fourteen patients were treated with frontal lobectomy alone and 14 patients were treated with a combined frontal lobectomy with a.c. (Table 1) Patients with an a.c. added to the frontal lobectomy were not selected in any way. This group consisted of 14 consecutive patients investigated for frontal lobe epilepsy surgery. The second group of 14 patients without callosotomy were also not selected in any particular way. It consisted of 14 consecutive patients investigated both before and after those patients treated with an additional a.c. The two groups of patients studied were comparable with regard to age of seizure onset, duration of epilepsy, and age at time of surgery. The mean post-operative follow-up period was seven years, with a minimal follow-up period of two years.

Table 1. Patient population (N = 28)

	Frontal lobectomy	**Combined anterior callosotomy with frontal lobectomy**
Number of patients	14	14
Mean age of onset of epilepsy	10 (6-33)	8 (0-24)
Mean duration of epilepsy (years)	8 (3-16)	15 (5-27)
Mean age at time of surgery	18 (10-36)	23 (10-41)
Mean post-operative follow-up period (years)	8 (3-14)	7 (4-12)

Presurgical evaluations[9] included clinical assessment, neuroradiological investigation (contrast C.T. scan and M.R.I.), intensive EEG monitoring with telemetry and video recording, with the use of both surface and depth electrodes, depth electrode stimulation, and neuropsychological testing.

Although these 28 patients had seizures with different clinical manifestations, the motor manifestations were found to be the most frequent:[10] tonic flexion or extension of the body; elevation of one or both arms; head deviation; and, secondarily generalized tonic-clonic seizures.

Each patient was studied with scalp EEG monitoring with telemetry and video recording until at least three clinical seizures were recorded. In the majority of patients, interictal and ictal EEG discharges were bilaterally synchronous involving both frontal lobes. The EEG investigation with depth electrode recording was complemented by detailed electrical stimulation. The typical implantation method for frontal lobe evaluation consisted of at least six depth electrodes for both frontal lobes. Temporal lobes were implanted bilaterally, internally at the amygdala and at the hippocampus, and correspondingly, externally, at the second temporal gyrus (T2).

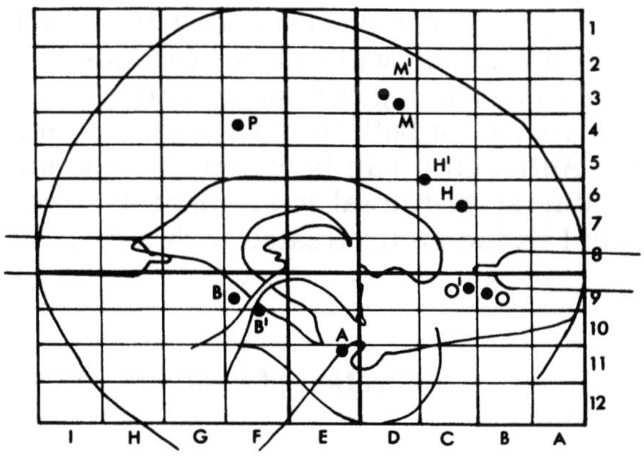

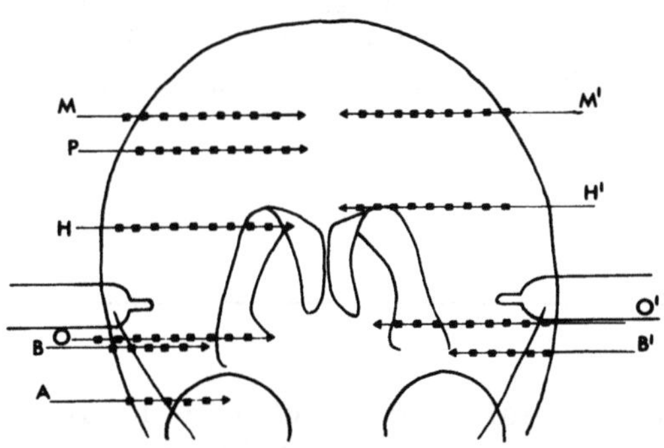

Figure 1. Implantation sites for frontal lobe evaluation

P:	right paracentral lobule (internal) + right post central gyrus (external)
A:	right amygdala (internal)+ right anterior T2 (external)
M:	right supplementary motor area (internal) + right F2 (external)
B:	right hippocampus (internal) + right T2 (external)
O:	right orbital gyrus (medial surface) + right orbital gyrus (lateral)
H:	right cingulate gyrus (internal) + right F2 (external)
B':	left hippocampus (internal) + left T2 (external)
O':	left orbital gyrus (medial surface) + left orbital gyrus (lateral)
H':	left cingulate gyrus (internal) + left F2 (external)
M':	left supplementary motor area (internal) + left F2 (external)

In most patients, a complete neuropsychological test battery was administered prior to and following surgery. This battery included standardized tests of intellectual functioning, memory, language, perceptual and constructional abilities, as well as specific tests to detect frontal lobe dysfunction. Preoperative psychological assessment failed to yield any evidence of focal cerebral dysfunction, differing in this aspect from patients with temporal lobe epilepsy. Many patients presented with a neuropsychological profile compatible with a diffuse cerebral dysfunction and had a history of long-term learning disabilities. Intracarotid Sodium Amobarbital testing was performed when an atypical pattern of cerebral dominance for language was suspected.

Subsequently, patients were classified into three categories defined according to the localization of ictal onset as recorded with depth electrode recordings, and defined by cortical stimulation: (1) Focal frontal lobe onset; (2) Regional frontal lobe onsets; (3) Diffuse bifrontal onset that was lateralized to one frontal lobe on the basis of stimulation results (Table 2). The focal ictal onset corresponded to a well-localized onset without an early diffusion, i.e., within three seconds, to other structures in the same frontal lobe. The regional onset corresponded to ictal onset recorded simultaneously from both mesial structures and external cortex of one frontal lobe. Although it can be noted that the only eight patients associated with a diffuse bi-frontal ictal onset as recorded with depth electrodes, are found in the group treated with a.c., a selection bias has not been made.

Table 2. Localization and lateralization of ictal onset by depth electrode recordings and stimulation.

Patients (N=28)	Focal frontal lobe onset	Regional frontal lobe onset	Diffuse bi-frontal onset; lateralization by stimulation
Frontal lobectomy N = 14	6	8	0
Combined frontal lobectomy and ant. callosotomy N = 14	1	5	8

When the focal frontal lobe onset was localized to the mesial structures, only a mesial frontal lobectomy including the supplementary motor area and the cingulate gyrus was performed. Conversely, in patients with a regional frontal lobe onset, a large frontal lobectomy was done in all patients with the exception of two who had only a mesial resection.

The length of the a.c. ranged from 4 to 8 cms. A 5 cm length was used when the a.c. was performed at the same time of the frontal cortectomy, which included all but two of the patients. One patient had an 8 cm a.c. in a second operation after a prior frontal lobectomy; the second patient had a 7 cm a.c. in a first operation, followed by a second operation with a frontal lobectomy.

RESULTS

Our results show that whether or not an a.c. is combined with frontal lobectomy, seizure control is similar in both groups. (Table 3) When the total numbers of patients are studied: excellent results were found in thirteen of twenty-eight (46%) (Engel, Class I and II)[19]; four of twenty-eight had a significant seizure reduction (14%); eleven of twenty-eight patients received no benefit (40%). The two groups differed by the fact that the eight cases with diffuse ictal onset as recorded with depth electrodes are found in the group treated with a.c. Even with the exclusion of these eight cases, the results still remain the same, that is, a.c. adds no beneficial effect. Three of the six remaining patients in the group with the additional callosotomy showed good results; six of fourteen patients in the group without callosotomy also had good results.

Table 3. Post-operative seizure outcome (Engel's classification)

Frontal lobectomy Patients (N = 14)		Combined frontal lobectomy with anterior callosotomy Patients (N = 14)
I.	3	6
II.	3	1
III.	2	2
IV A.	3	2
IV B.	3	3

Table 4. Results of post-operative seizure outcome with focal or regional frontal onset, or diffuse bi-frontal onset in patients treated by frontal lobectomy or by combined frontal lobectomy and anterior callosotomy (Engel's classification)

	Focal or regional frontal onset		Diffuse bi-frontal onset (localized only by stimulation)
Class	**Lobectomy**	**Combined lobectomy with ant. callosotomy**	**Combined lobectomy with anterior callosotomy**
I	3	3	3
II	3	0	1
III	2	0	2
IV A	3	2	0
IV B	3	1	2

Our data show no difference in the post-operative seizure outcome when comparing the results of frontal cortectomy performed alone, with frontal cortectomy combined with a.c. There is also no difference for right or left frontal lobe resections. Post-operative outcome was the same for both the groups of eight patients with diffuse bi-frontal ictal onset, as with the group with focal or regional ictal onset.

Ten patients with a structural lesion restricted to a single frontal lobe, as demonstrated with CT scan or MRI and confirmed at pathology (e.g., cavernous angioma, oligodendroglioma, astrocytoma), were shown to have a much better post-surgical outcome. In this group, excellent results (Engel's class I and II) were obtained in seven of ten patients (70%). In contrast, eighteen patients without a localized structural lesion in one frontal lobe, had excellent results in only six of eighteen patients (33%). In both groups, the addition of an a.c. did not affect the outcome. As is generally accepted,[7] and has also been shown in our study,[2] the presence of a structural lesion localized in one frontal lobe is the best predictive factor for excellent post-operative results in frontal lobe epilepsies.

Functional outcome

Functional outcome was assessed clinically by the investigators. Judgment was based on behavioral changes, work and/or school performances and social adaptation. These results are presented in Table 5. Only two of fourteen patients with frontal lobectomy alone showed a deterioration in post-operative functional status, whereas similar results were seen in five of fourteen patients with an additional (a.c.). However, one cannot conclude that the a.c. was responsible for this deterioration since many patients with combined a.c. and frontal lobectomy did show an *improvement* in post-operative functional status. A strong correlation is apparent between improvement in functional status and reduction in seizure frequency (Table 6). In fact, the four patients who did present with an improvement in functional status following a combined frontal lobectomy and a.c. all belong to Class I of Engel.

Table 5. Relationship between type of surgery and global functional outcome (N = 28)

	GLOBAL FUNCTIONAL OUTCOME		
Type of surgery	**improved**	**unchanged**	**worse**
Frontal lobe alone (N = 14)	1	11	2
Frontal lobe + a.c. (N = 14)	4	5	5

Table 6. Relationship between seizure outcome and global functional outcome (N = 28)

	GLOBAL FUNCTIONAL OUTCOME		
Seizure outcome (Engel classification)	**improved**	**unchanged**	**worse**
Class I and II	5	8	0
Class III	0	3	1
Class IV	0	5	6

Neuropsychological investigation

A comprehensive neuropsychological test battery was administered pre- and post-operatively to seventeen patients. The proportion of left handers was considerably higher than is generally found in the general population (6/17: 35%). In addition, 50% of patients with a left frontal focus had either a bilateral or right hemisphere dominance for language as determined by Intracarotid Sodium Amytal (ISA) test. None of the right frontal lobe patients in whom ISA test was performed had this profile, although the proportion of left handers was equivalent in Right and Left frontal lobe groups. This finding suggests a transfer of all or some linguistic functions to the right hemisphere in cases of left hemisphere dysfunction. Although the small number of patients in each surgical group and the intra-group heterogeneity precluded statistical analysis of the neuropsychological tests' results, a qualitative analysis of the individual profiles can be performed.

Intellectual functioning: The mean pre-operative IQ was 86.6 (range: 52-119). In our sample, ten patients (59%) had an IQ of 85 or less. In general, IQ remained unchanged post-operatively, and even improved in some patients. In one patient, however, a significant drop was observed (C.B.: 103 - 84). This patient underwent a LF lobectomy associated with a.c. Both verbal and performance scales significantly worsened. A comprehensive examination of the medical and surgical charts did not reveal any factors that might have predicted or explained such a deterioration.

Frontal lobe semiology: the battery we used included a number of tests sensitive to frontal lobe dysfunction, such as the Wisconsin Card Sorting Test (WCST), Trail A & B, Mazes, Luria's graphic series and motor sequences. Signs of frontal lobe dysfunction, such as intrusions in memory tests, perseverations and impulsivity were also taken into account in the evaluation of the presence and severity of frontal lobe pathology. Pre-surgical evaluation revealed a mild to moderate frontal lobe semiology in ten of seventeen patients. No changes were observed post-operatively in ten patients (6/12 with a.c., 4/5 without a.c.). Improvement was observed in only one patient (RF with a.c.) whereas deterioration was noted in six patients. This deterioration was minimal in four (all with a.c.) and marked in two patients. These two patients were not able to complete a single category on the WCST, showing numerous perseverative errors and intrusions on memory tests, and demonstrating severe problem-solving deficits. One patient had a LF lobectomy with a.c and the other, a RF lobectomy without a.c. Thus, the addition of a.c. cannot account for the emergence of a severe frontal lobe semiology in these two patients. The role of the a.c. in the causation of the mild deterioration observed in four patients (all with a.c.) cannot be assessed because of the small sample, and the unequal distribution of patients with (N = 12) and without (N = 5) a.c.

Language functions: In most patients, expressive and receptive language abilities remained unchanged post-operatively. Subtle improvement was observed in four patients, all following non-dominant frontal lobe resection. This could be attributed to an improvement in general functioning, i.e., increase in speed, better attentional abilities, etc. In three patients, certain deficits not observed before surgery were noted. One patient developed a transcortical sensory aphasia following LF lobectomy (without a.c.). Two other patients showed errors in language tasks that could be attributed to frontal lobe dysfunction such as perseverative errors, difficulties in sentence production, word finding deficits, attentional difficulties in oral comprehension tasks, slowing of reading. One of these patients had a LF lobectomy with a.c., and the other, a RF lobectomy without callosotomy. In general, verbal fluency decreased after dominant F. lobe resection whereas no changes were observed following non-dominant F lobectomy. Results were similar in groups with or without a.c.

Memory functions: Average memory quotient (Wechsler Memory Scale: WMS) was 90.8 pre- and 85.2 post-operatively. There was a significant decrease (more than 10 points) in six patients (5 LF, 1 RF) and a significant improvement in only one patient (RF with a.c.). A mild to moderate decline in verbal memory (logical memory and associate learning subtests of the WMS, and Rey Auditory Verbal Learning Test) was observed in most patients following dominant left frontal lobectomy. Pre- and post-operative results were equivalent in the right frontal group, with the exception that in two patients a marked frontal lobe semiology could account for the deficits observed. Interestingly, a verbal memory deficit was also observed following non-dominant left frontal lobectomy. In fact, the profile presented by these patients was essentially similar to what was observed following dominant left frontal lobectomy. A slight decline in visuo-spatial memory, as assessed by the delayed visual reproduction of the WMS and delayed recall of the Rey Complex figure, was observed in some patients irrespective of the side (dominant vs. non-dominant) or the type (with or without a.c.) of surgery.

DISCUSSION

Depth electrode investigations in frontal lobe epilepsies have shown that the epileptogenic zone, in the majority of cases, involves either the entire frontal lobe or at least a large portion of the frontal lobe.[1] In our study a regional or diffuse onset was found in twenty-one of twenty-eight patients. Therefore it is not surprising that despite a large resection, a portion of the epileptogenic zone persists which may explain the non-successful results of surgery. From our experience and the experience of others[2] showing unsatisfactory results of surgery for frontal lobe epilepsies, we have suggested that the addition of a partial section of the anterior callosum at the time of the frontal cortectomy might be a possible solution. This operation was initiated with the objective of sectioning pathways for propagation of epileptic discharge to the contralateral hemisphere and, hence, improving the clinical results of surgery. However, we have limited the anterior section of the corpus callosum to its anterior half (5 cm) instead of two-thirds, as defined in the classical a.c. without cortectomy.[1,2,3] We believed that the restriction of the section in this manner might have prevented neuropsychological complications. A very extensive sectioning of the interhemispheric association pathways, added to the frontal cortectomy itself could exaggerate a pre-existing neuropsychological deficit. We know that these frontal lobe patients frequently demonstrate signs of diffuse brain damage as shown by a generally low IQ. These results are in contrast to those obtained with the temporal lobe epilepsies.

The hypothesis that the addition of a partial a.c. would reduce seizure occurrence post-operatively was not confirmed. Certain points can be discussed to explain these results. Although the corpus callosum is known as the main structure responsible for the synchronization of EEG,[4,5,6,8,13] it is accepted that there exist other subcortical pathways, independent of the corpus callosum, which are capable of transmitting epileptic discharge to the contralateral hemisphere.[14,15,16] One might also raise the possibility of homolateral diffusion, as a first step, i.e., via the intrahemispheric pathways. For example, epileptic discharge would propagate from the frontal to the temporal lobe and, subsequently, to the other hemisphere via the posterior part of the corpus callosum or the hippocampal commissure.[17,18] Spencer[7] has shown that complete sectioning of the corpus callosum in humans is two times more efficient than partial sectioning for the control of secondarily generalized seizures. These results suggest the probable participation of each portion of the corpus callosum during the contralateral propagation of the epileptic discharge. Another explanation might be the presence of additional independent extra-frontal epileptic foci that would continue their epileptic activity following the initial frontal resections. It is very probable that a more extensive anterior resection of the corpus callosum, such as a two-thirds section, instead of the one-half section, might have given better results when compared with the group treated with frontal cortectomies alone.[19] Wada[20] has shown that the section of the anterior two-thirds of the corpus callosum is sufficient to control the contralateral diffusion in an experimental kindling model. Although relevant, this hypothesis cannot be addressed in our study. A prospective study in which a more extensive resection of the anterior corpus callosum, i.e., in which a two-thirds section is added to a frontal cortectomy, might very well answer this question.

REFERENCES

1. A. Turmel, N. Giard, G. Bouvier, R. Labrecque, F. Veilleux, I. Rouleau, and J.M. Saint-Hilaire, Frontal lobe seizures and epilepsy: indications for cortectomies, *in:* "Advances in Neurology," P. Chauvel et al., eds., Raven Press, New York, Vol. 57, pp. 689-705 (1992).
2. A. Rougier, F. Richer, M. Martinez, G. Bouvier, A. Bouthillier, R. Labrecque, and J.M. Saint-Hilaire, Résultats, *in:* Investigation et traitement chirurgical de l'épilepsie, Rougier, A., Saint-Hilaire, J.M., Bouvier, G., et Loiseau, P., eds., *Neurochirurgie* 38:79-97 (1992).
3. J.M. Saint-Hilaire, N. Giard, G. Bouvier, and R. Labrecque, Anterior callosotomy in frontal lobe epilepsies, *in:* "Epilepsy and the Corpus Callosum," A.G. Reeves, ed., Plenum Press, New York, pp. 303-314 (1985).
4. E.M. Marcus and C.W. Watson, Bilateral synchronous spikes wave electrographic patterns in the cat, *Arch. Neurol.* 14:601-610 (1966).
5. E.M. Marcus and C.W. Watson, Bilateral symmetrical epileptogenic foci in central cortex of the monkey, *E.E.G. Clin. Neurophysiol.* 23:77-97 (1967).
6. J. Musgrave and P. Gloor, The role of the corpus callosum in bilateral interhemispheric synchrony of spikes and wave discharge in feline generalized penicillin epilepsy, *Epilepsia* 21:369-378 (1980).

7. S.S. Spencer, D.P. Spencer, P.D. Williamson, K. Sass, R.A. Novelly, and R.H. Mattson, Corpus callosotomy for epilepsy. I. Seizure effects, *Neurology* 38: 19-24 (1988).
8. M.E. Fiol, J.R. Gates, R. Mireles, R.E. Maxwell, and D.M. Erickson, Value of intraoperative EEG changes during corpus callosotomy in predicting surgical results, *Epilepsia* 34:74-78 (1993).
9. J.M. Saint-Hilaire and G. Bouvier, Presurgical evaluation protocols. Notre-Dame Hospital, Montreal, *in:* "Surgical Treatment of the Epilepsies, " J. Engel, Jr., ed. (second edition), Raven Press, New York, pp. 732-733 (1993).
10. F. Veilleux, J.M. Saint-Hilaire, N. Giard, A. Turmel, G.P. Bernier, I. Rouleau, M. Mercier, and G. Bouvier, Clinical manifestations of seizures of the human medial frontal lobe,*in:* "Advances in Neurology," P. Chauvel et al. (eds.), Vol. 57, Raven Press, New York, pp. 245-255 (1992).
11. J. Engel, Jr., Outcome with respect to epileptic seizures, *in:* "Surgical Treatment of the Epilepsies," J. Engel, Jr., ed. (second edition), Raven Press, New York, pp. 553-571 (1987).
12. J.J. Purves, J.A. Wada, W.B. Woodhurst, W.B., P.D. Moyes, E. Strauss, B. Kosaka, and D. Li, Results of anterior corpus callosum section in 24 patients with medically intractable seizures, *Neurology* 38:1194-1201 (1988).
13. J.R. Gates, J.E. Lippick, J. Yap, and R.J. Gummit, Corpus callosotomy: clinical and electroencephalographic effects, *Epilepsia* 25:308-316 (1984).
14. W.T. Blume, Corpus callosum section for seizure control: rationale and review of experimental and clinical data,*Cleve. Clin. Q.* 51:319-332 (1984).
15. J.A. Wada and M. Sato, Generalized convulsive seizures induced by daily electrical stimulation of the amygdala in cats. Correlation electrographic and behavioral features, *Neurology* 24:565-574 (1974).
16. R.C. Collins, C. Kennedy, L. Sokioloff, and F. Plum, Metabolic anatomy of focal motor seizures, *Arch. Neurol.* 33:536-542 (1976).
17. J.P. Lieb, K. Hoque, and C.E. Skommer, Interhemispheric propagation of human mesial temporal lobe seizures: a coherence/phase analysis, *EEG Clin. Neurophysiol.* 67:101-119 (1987).
18. C. Adam, F. Richer, and J.M. Saint-Hilaire, Seizure propagation and resistance to surgery for temporal lobe epilepsy, *Epilepsia* 33 (Suppl. 3):93 (1992).
19. H. Oguni, A. Olivier, F. Andermann, and J.Comair, Anterior callosotomy in the treatment of medically intractable epilepsies: a study of 43 patients with a mean follow-up of 39 months, *Ann. Neurol.* 30:357-364 (1991).
20. J. A. Wada, New surgical treatment through experimental models. *in:* "Advances in Epileptology: the Xth Epilepsy International Symposium," J.A. Wada and K. Penry, eds., Raven Press, New York, pp. 195-213 (1980).

THE ROLE OF FRONTAL LOBE RESECTIONS COMBINED TO CALLOSAL SECTIONS IN THE TREATMENT OF SECONDARY GENERALIZED EPILEPSIES

Arthur Cukiert, M.D.[2]
Andre Olivier, M.D., Ph.D., F.R.C.S.(C)[1]
Frederick Andermann, M.D., F.R.C.P.[1,]

[1]Montreal Neurological Institute, McGill University
[2]Department of Neurology and Neurosurgery,
University Sao Paulo

The medical treatment of refractory secondary generalized epilepsies (SGEs) has always been a challenge.

These patients usually present with moderate to severe mental retardation and multiform generalized seizures which may be extremely difficult to control with the presently available anti-epileptic drugs (AEDs). These patients are usually treated with a high dosage politherapy which includes the association of modern AEDs in various combinations (such as carbamazepine and valproate) and benzodiazepines. This regimen leads frequently to severe side effects. This aggravates significantly the already severe clinical picture characterized by a diffuse brain pathology causing seizures and frequent secondary injuries.

Interictal electroencephalographic (EEG) abnormalities are generally diffuse. Multifocal activity is often seen, as well as diffuse background slowing. Secondary bilateral synchrony (SBS) is often obvious interictally. Ictal onset is often bilateral and synchronous but partial seizures with a more localized onset can also be present. In a significant number of cases, EEG abnormalities predominate over one or both frontal lobes.

The epileptogenic threshold of the different cortical regions and their contribution for SBS has been extensively studied.[1-3] The important role of the mesial and parasagittal cortex in the genesis of SBS has already been emphasized.[4-7]

Restricted cortical resections are the ideal modality of treatment for medically refractory epilepsies. SGE is associated with diffuse cortical pathology and it is not often amenable to this kind of localized resection.

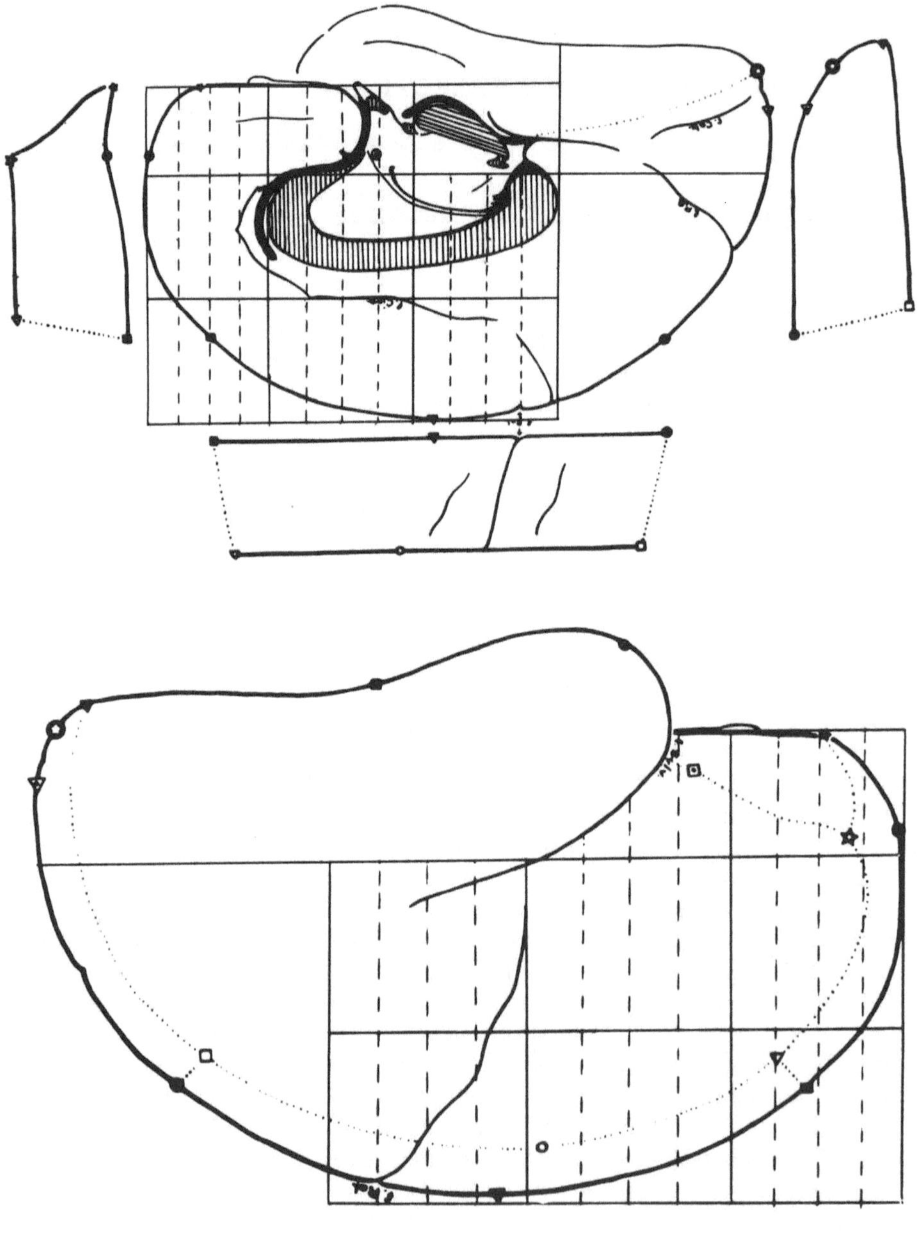

Figure 1. Frontal lobe planes and compartments based on the callosal proportional system.

Callosal sections have been proposed as a palliative procedure in these cases.[8-12] The rationale for this operation is mainly to avoid spread and interactions of the multiple foci present in these patients.

Patients with severe SGE have also been submitted to callosal sections associated with frontal resections.[13] The rationale for this approach would be to couple the desynchronizing effect of callosotomy to a decreased seizure tendency generated by a cortical removal. The clinical usefulness of this procedure has not yet been fully demonstrated.

We present the long-term clinical results of a series of patients with SGE submitted to callosal sections associated with frontal lobe resections. This series is then compared to a series of similar patients submitted only to a callosal section.

METHODS

Sixteen patients with SGE who were submitted to frontal cortical resections combined with a callosal section and had at least one year of follow-up were included in this study (7 males, 9 females). The age at the time of surgery ranged from 0.8 to 34 years (mean 17.9). The duration of the epileptic syndrome ranged from 0.6 to 27 years (mean 8.8). Nine patients were submitted to a left and seven to a right frontal resection.

The pre-operative seizure patterns, pre-operative EEG findings, preoperative electrocorticography (ECoG), surgical complications, resection size, early post-operative seizures (defined as seizures in the first two weeks post-operatively) and etiological data were studied. The relationship of the clinical outcome to early post-operative seizures and resection size was addressed. The clinical outcome in relation to seizures was rated according to the scale proposed by Engel et al.[14]

The resection size was determined by means of the planar intraoperative brain map. The frontal lobe was divided in approximately 45 compartments according to proportional volumes derived from landmarks in the corpus callosum.

A horizontal callosal plane (HCP) was defined by means of the most inferior part of the splenium and genu of the corpus callosum. Anterior and posterior callosal planes (ACP and PCP) were then created perpendicularly to the horizontal callosal line (HL) and passing through the anterior part of the genu and the posterior aspect of the splenium, respectively. A middle callosal plane (MCP was created in between the two planes described above. The volumes between the most anterior point of the frontal lobe and the ACP, between the ACP and MCP and between the MCP and PCP were further subdivided by means of three additional planes parallel to the previous ones into four compartments each. These procedures created eight different compartments within the corpus callosum. In a similar way, the space between the plane passing through the most cranial point of the brain parallel to the HCP itself was further subdivided in two by a plane between them. The central sulcus can be delineated by a line passing through the mid-callosal point (defined by the intersection of the HCP and MCP) and through a point 5 mm in front of the PCP at its most cranial portion, thus defining the posterior border of the frontal lobe (Fig. 1). The number of the cortical and callosal volumes created by this approach which have been resected in each case was used to approximately rate the extension of the resection. Statistical analysis using qui-square and T-test transformations was carried out.

RESULTS

The pre-operative seizure patterns comprised simple partial (n=4), complex partial (n=6), generalized tonic-clonic (n-7), atonic (n=4), tonic (n=3), and atypical absence (n=5) seizures.

All patients had diffuse, bilateral and multifocal EEG abnormalities, predominating over one frontal lobe. SBS was evident in all patients. Pre- and post-resection ECoG was available in five cases. In two, post-resection spikes were present.

Surgical complications included a subdural hematoma (n=1), meningitis (n=1), and brain swelling (n-1). Early post-operative seizures were present in 12 cases. The mean number of resected (Fig. 2) frontal lobe compartments was 31 (out of 45). The resection always included at least the mesial and parasagittal cortex. The mean extension of the callosal section was 3.7 compartments (47%).

Pathological analysis of the cortical specimens showed gliosis (n=5), chronic encephalitis (n=1), AVM (N=1), tuberous sclerosis (n-1), or absence of abnormality (n=7).

No patient became seizure free post-operatively; 37% (n=6) were in class II, 43% (n=7) in class III and 18% (n=3) in class IV. There was a significant relationship between the presence of early post-operative seizures and poorer outcome. Despite the fact that the values did not reach statisticald significance, there was a trend towards poorer results when no abnormality was seen in the specimen. There was no relationship between the extent of the cortical resection and the clinical outcome.

FRONTAL RESECTIONS COMBINED WITH CALLOSOTOMIES
Resection Size (number of compartments)

	1-5	6-10	11-15	16-20	21-25	26-30	31-35	36-40	41-45
N			1	3	2		4	2	4

Figure 2. Size of the frontal resections as determined by the compartments described in the text (N=16).

DISCUSSION

Frontal lobe resections combined with callosal sections performed in patients with SGE are not curative procedures. No patient became seizure free after the operation.

Interesting data are obtained by comparing these results with those obtained in a similar population submitted to a callosotomy alone (Oguni et al.).[10] In the latter, a subpopulation of 15 patients with bilaterally asymmetrical independent and generalized spike and wave discharges with SBS with obvious lateralization were studied. The mean extent of callosal section was 52% as compared to 47% in the present series. Despite the fact that the seizure types for this specific subpopulation was not specified by Oguni, his overall series included patients with simple partial (9.3%), complex partial (18%), generalized tonic-clonic (72%), atonic (60%), and atypical absence (23%) seizures. Clinical outcome in both series can be summarized as in Figure 3.

	This series (N)	This series (%)	Oguni's (N)	Oguni's (%)
I	0	0	2	13
II	6	37	3	20
III	7	43	4	26
IV	3	18	6	40

Figure 3. Summary of the clinical outcome in relation to seizures rated using Engel's scale in the present series and in Oguni's.

There is no statistically significant difference in outcome between these two series. Two patients in Oguni's series became seizure free after callosotomy, a result rarely achieved after callosal section. It is also a result difficult to explain since the cortical generators were still in place and theoretically might generate at least partial seizures. Probably a more prolonged follow-up of these two cases would show recurrence of seizures.

The analyses of these series show that even extensive frontal cortical removals including the mesial and parasagittal cortex are not effective in improving the outcome of patients with refractory SGE when compared to callosal sections alone. This is probably explained by the diffuse cortical pathology present in these patients. It should also be noted that if extensive frontal lobe resections are carried out, there is no need to perform an anterior callosotomy since the main origin of anterior callosal axons is, in fact, removed.

Restricted frontal resections can yield good results in other types of epileptic syndromes.[14-20] In six patients with refractory frontal lobe seizures related to direct-impact head trauma with frontal depressed skull fractures who were submitted to localized resections, five were seizure free and the last one was markedly improved after surgery (personal communication).

On the other hand, the identification of epileptogenic zones involving the frontal lobes has always been more difficult than those involving the temporal lobe. The results of epilepsy surgery vary proportionally to our ability to remove a specific epileptogenic area. The several series reporting the clinical outcome after frontal lobe resections underscore the poorer results when compared to temporal lobe removals.[21,22] These results are very likely related to the fact that well-localized frontal foci associated to refractory epilepsy are not documented so frequently.[23-26] Specific frontal lobe epileptic syndromes have only recently been extensively studied.[27-33] Clinical outcome after frontal lobe resections will only improve as far as we can succeed in further understanding the anatomical, physiological, and pathological aspects of these brain areas.

There appear to be no role for the association of frontal lobe resections and callosal sections in patients with SGE. These patients get maximal benefit from a callosal section alone but even the latter procedure remains palliative. On the other hand, the significant decrease in atonic, tonic-clonic, and atypical absence seizures after callosotomy clearly improves these patients' quality of life and this procedure represents an excellent therapeutic option until better and more effective AEDs are developed.

REFERENCES

1. P. Gloor, Generalized cortico-reticular epilepsies. Some considerations on the pathophysiology of generalized bilaterally synchronous spike and wave discharge, *Epilepsia* 9:249-263 (1969).
2. P. Gloor, Generalized epilepsy with spike and wave discharge: a reinterpretation of its electrographic and clinical manifestations, *Epilepsia* 20:571-588 (1979).
3. A. Cukiert, S.W. Baumel, M. Andrioli, and R. Marino, Jr., Effects of corpus callosum stimulation on the morphology, synchrony and frequency of the epileptic bursts in the feline topical penicillin generalized epilepsy model, *Stereotact. Funct. Neurosurg.* 52:18-25 (1989).
4. B.L. Ralston, Cingulate epilepsy and secondary bilateral synchrony, *Electroencephalogr. Clin. Neurophysiol.* 13:591-598 (1961).
5. R.L. Isaacson, H. Schwartz, N. Persoff, and L. Pinson, The role of the corpus callosum in the establishment of areas of secondary epileptiform activity, *Epilepsia* 12:133-146 (1971).
6. B.R. Tharp, Orbito-frontal seizures. A unique electroencephalographic and clinical syndrome, *Epilepsia* 13:627-642 (1972).
7. T.A. Pedley, B.R. Tharp, and K. Herman, Clinical and electroencephalographic characteristics of midline parasagittal foci, *Ann. Neurol.* 9:142-149 (1981).
8. F.R. Huck, J. Radvany, J.O. Avila, et al., Anterior callosotomy in epileptics with multiform seizures and bilateral and synchronous spike and wave EEG pattern, *Acta Neurochir.* 30(Suppl.):127-135 (1980).
9. I. Papo, A. Quattrini, L. Provinciali, et al., Callosotomy for the treatment of drug resistant generalized epilepsies, *Acta Neurochur.* 50(Suppl.):134-135 (1990).
10. H. Oguni, A. Olivier, F. Andermann, and J. Comair, Anterior callosotomy in the treatment of medically intractable epilepsies: a study of 43 patients with a mean follow-up of 39 months, *Ann. Neurol.* 30:357-364 (1991).
11. R.E. Nordgren, A.G. Reeves, A.C. Viguera, and D.W. Roberts, Corpus callosotomy for intractable seizures in the pediatric age group, *Arch. Neurol.* 48:364-372 (1991).
12. K.S. Fuiks, A.R. Wyler, B.P. Hermann, and G. Somes, Seizure outcome from anterior and complete corpus callosotomy, *J. Neurosurg.* 74:573-578 (1991).
13. A. Turmel, N. Giard, G. Bouvier, et al., Frontal lobe seizures and epilepsy. Indications for cortectomies or callosotomies, *in:* "Advances in Neurology, Vol. 57," P. Chauvel and A. Delgado-Escueta, eds., Raven Press, New York (1992) pp. 689-705.
14. J. Engel., Outcome with respect to epileptic seizures, *in:* "The Surgical Treatment of Epilepsy," J. Engel, ed., Raven Press, New York (1987) pp. 553-572.
15. T. Rasmussen, Surgical therapy of frontal lobe epilepsy, *Epilepsia* 4:181-198 (1963).
16. T. Rasmussen, Surgical treatment of frontal lobe epilepsy, *Epilepsia* 11:27-28 (1970.
17. J. Talairach, J. Bancaud, G. Szikla, et al., Approche nouvelle de la neurochirurgie de l'epilepsie (methodologie stereotaxique et resultats therapeutiques), *Neurochirurgie* 20 (Suppl. 1):240 (1974).
18. T. Rasmussen, Surgery of frontal lobe epilepsy, *Adv. Neurol.* 8:197-205 (1975).
19. T. Rasmussen, Characteristics of a pure culture of frontal lobe epilepsy, *Epilepsia* 24:482-493 (1983).
20. C.E. Polkey, Cortical resections outside the temporal lobe for intractable epilepsy -- excluding multilobar resections and hemispherectomy, *Acta Neurochir.* 50(Suppl.):128-130 (1990).
21. J. Talairach, J. Bancaud, A. Bonis, et al., Surgical therapy for frontal lobe epilepsies, *in:* "Advances in Neurology, Vol. 57," P. Chauvel and A. Delgado-Escueta, eds., Raven Press, New York (1992) pp. 707-732.

22. J. Adler, G. Erba, K.R. Winston, et al., Results of surgery for extratemporal partial epilepsy that began in childhood, *Arch. Neurol.* 48:133-140 (1991).
23. L. Fegersten and A. Roger, Frontal epileptogenic foci and their clinical correlations, *Electroencephalogr. Clin. Neurophysiol.* 13:905-913 (1961).
24. G. Mazars, Criteria for identifying cingulate epilepsies, *Epilepsia* 11:41-47 (1970).
25. B.L. Ludwig, C. Ajmone-Marsan, and J. van Buren, Cerebral seizures of probably orbitofrontal origin, *Epilepsia* 16:141-158 (1975).
26. S. Geier, J. Bancaud, J. Talairach, et al., The seizures of frontal lobe epilepsy. A study of the clinical manifestations, *Neurology* 27:951-958 (1977).
27. K. Tukel and H. Jasper, The electroencephalogram in parasagittal lesions, *Electroencephalogr. Clin. Neurophysiol.* 4:481-494 (1952).
28. P.D. Williamson, D.D. Spencer, S.S. Spencer, et al., Complex partial seizures of frontal lobe origin, Ann. Neurol. 18:497-504 (1985).
29. K. Waterman, S.J. Purves, B. Kosaka, et al., An epileptic syndrome caused by a mesial frontal lobe seizure foci, *Neurology* 37:577-582 (1987).
30. L.F. Quesney, A. Olivier, F. Andermann, and P. Gloor, Pre-operative EEG investigation in patients with frontal lobe epilepsy: trends, results and pathophysiologic considerations, *J. Clin. Neurophysiol.* 4:208-209 (1987).
31. L.F. Quesney and A. Olivier, Pre-operative EEG evaluation in frontal lobe epilepsy, *Acta Neurol. Scand.* 78 (Suppl.):61-72 (1988).
32. L.F. Quesney, M. Constain, D.R. Fish, and T. Rasmussen, The clinical differentiation of seizures arising in the parasagittal and anterolateral frontal convexities, *Arch. Neurol.* 47:677-679 (1990).
33. W.W. Sutherling, M.W. Risinger, P.H. Crandall, et al., Focal functional anatomy of dorsolateral fronto-central seizures, *Neurology* 40:87-98 (1990).

OPEN AND STEREOTACTIC SEGMENTAL CALLOSOTOMY: EFFECTS ON SEIZURE FREQUENCY

Raul Marino, Jr., M.D., Arthur Cukiert, M.D., and Gary Gronich, M.D.

Division of Functional Neurosurgery
University of Sao Paulo, School of Medicine
Sao Paulo SP, Caixa Postal 8091
Brazil

INTRODUCTION

Corpus callosum section has been used as a palliative procedure in the treatment of the secondary generalized epilepsies and secondarily generalized seizures.

The corpus callosum is the thickest brain commissure in humans and is related to more than 90% of the interhemispheric transfer of information. Its role in the bilateral spread and synchronization of epileptic discharges has been extensively studied in animal[1, 2] and clinical[3, 4] studies. Almost nothing was known about its physiology and functions in the beginning of this century. It was only after the study of patients submitted to total and partial callosal sections (mainly in the treatment of epilepsy) that the interhemispheric disconnection syndrome could be better defined.

Patients with unilateral or bilateral extensive epileptogenic areas frequently disclosed secondary bilateral synchrony (SBS) and might represent up to 20% of the outpatient epileptic population. Many of these patients show a poor response to standard antiepileptic drugs (AEDs) regimens in mono- and polytherapy. They are potential candidates for surgical procedures aiming to interrupt or reduce the spreading of the epileptic discharges through the corpus callosum.

In our service, callosotomy was performed both by open and stereotactic surgical procedures. This paper reports the effects on seizure frequency obtained comparatively from these surgical approaches.

Epilepsy and the Corpus Callosum II, Edited by A.G. Reeves
and D.W. Roberts, Plenum Press, New York, 1995

TABLE 1. Clinical summary. Open callosotomy. n = countless; SMA = supplementary motor seizure (Reprinted from Surg. Neurol. 34:219-228, 1990.)

PATIENT	(pre-post) ABSENCES	(pre-post) TONIC-CLONIC	(pre-post) TONIC	(pre-post) ATONIC	(pre-post) SMA	(pre-post) MYOCLONIC
1	20/day - 1/day	1/year - 3/year	0	1/week - 0	1/day - 2/week	0
2	n/day - 0	10/day - 2/month	0	2/day - 2/month	2/month - 0	10/day - 2/wek
3	20/day - 1/month	2/year - 0	0	3/week - 0	1/month - 2/month	0
4	n/day - 5/day	1/year - 0	0	5/day - 1/month	1/year - 3/year	0
5	n/day - 0	2/month - 0	0	5/day - 1/day	0	5/week - 0
6	n/day - 3/day	0	0	0	0	10/day - 3/month
7	0	0	0	0	6/day - 6/day	0
8	n/day - 0	1/year - 0	0	0	0	0
9	n/day - 5/day	2/year - 0	0	2/week - 0	6/day - 6/day	0
10	40/day - 2/day	1/year - 1/year	0	0	4/day - 0	0
11	n/day - 0	0	0	0	0	0
12	n/day - 2/week	0	0	2/day - 1/month	3/day - 0	0
13	n/day - 10/day	1/year - 0	0 - 2/year	1/year - 2/year	1/year - 2 year	0
14	n/day - 1/day	1/week - 1/month	2/month -0	2/month - 0	2/month -1/month	0

15	n/day - 1/week	3/year - 0	1/week - 0	1/week - 0	2/year - 0	1/week - 0
16	20/day - 2/month	1/week - 3/month	0	1/day - 0	0	0
17	0	0	5/day - 0	0	0	0
18	n/day - 0	1/year - 0	0	0	0	2/day - 1/day
19	n/day - 5/day	1/day - 1/month	1/day - 0	1/day - 0	1/day - 4/week	0
20	2/day - 2/day	2/month - 1/month	0	3/week - 0	3/week - 1/month	5/day - 1/month
21	n/day - 2/month	6/year - 0	0	1/week - 2/month	1/month - 0	1/week - 0
22	0	1/day - 1/year	0	0	0 - 1/year	15/day - 3/day
23	4/day - 3/week	1/year - 1/year	0	3/day - 6/year	4/week - 0	0
24	n/day - 0	0	3/week - 0	0	0	40/day - 0
25	n/day - 0	1/year - 0	2/day - 0	2/month - 0	1/month - 1/day	1/month - 0
26	15/day - 0	0	4/day - 4/day	4/day - 4/day	0	0
27	n/day - 0	6/year - 0	0	4/day - 1/day	0	3/day - 0
28	1/month - 0	10/month - 2/month	2/month - 0	0 0	0	0

METHODS

Patient selection criteria included:

1. Secondary generalized epilepsies with a duration of the seizure disorder longer than five years;
2. Generalized seizures (tonic-clonic, tonic, atonic, myoclonic and absences) refractory to current AEDs regimens in mono- and polytherapy associated with adequate AED plasma levels;
3. EEG recordings disclosing bilateral and synchronous epileptic activity. This activity could be symmetric and synmorphic or not and associated or not with focal epileptic activity. Its amplitude predominated in the more anterior regions. Ictal recordings showed a generalized onset of the habitual seizure types in each case. Many of these patients disclosed a multifocal EEG pattern and there was always the involvement of at least one frontal lobe;
4. No progressive neurologic disease;
5. Cognitive deficits were present in all cases in different degrees and were defined by extensive neuropsychological testing;
6. Negative neuroradiological findings or positive studies where the lesion(s) is (are) topographically related to the epileptogenic process; and
7. At least one-year post-operative follow-up.

Twenty-eight out of thirty-five patients submitted to open anterior partial callosotomy had close follow-up. A parasagittal frontoparietal craniotomy was performed over the non-dominant hemisphere. The dura was opened, the ipsilateral mesial surface was retracted and the corpus callosum was exposed by means of microsurgical techniques. Electrocorticographic (ECoG) monitoring of the secondary bilateral synchrony was obtained by means of platinum electrodes applied over both mesial surfaces. The callosal section was carried out under continuous ECoG recording and its extension defined by the disruption of the SBS. A marker was left over the lesioned corpus callosum for future radiological control in all cases. Post-operative MRI scans were available in 13 cases.

Twelve other patients were submitted to stereotactic anterior partial callosotomies. Under general anesthesia, the patient was submitted to a pneumoencephalogram and a non-dominant carotid artery angiogram. The limits and topography of the corpus callosum were then carefully defined. Twist-drill openings were made in the skull after stereotactic preparation using the Tallairach system. Radiofrequency lesions were generated during 60 seconds at 65-70 degrees Celsius, spaced 0.5 cm from each other. The callosal lesions were verified by MRI in all cases.

RESULTS

In 27 patients (94%) submitted to open anterior partial callosotomy there was a marked decrease in the frequency of the generalized seizures. New seizure types appeared in seven cases: in four cases the new seizure pattern represents a modification (usually in the duration and motor components) of the pre-operative patterns and in three cases there was no

TABLE 2. Clinical summary. Stereotactic callosotomy. n = countless; SMA = supplementary motor seizures; pre- and post-op seizure frequencies per month. (Reprinted from Surg. Neurol. 34:219-228, 1990.)

PATIENT	MYOCLONIC	TYPICAL ABSENCE	ATYPICAL ABSENCE	TONIC	ATONIC	PARTIAL MOTOR	SMA	TONIC-CLONIC
1	0	n / 0	15 / 1	0	0	0	0	0
2	0	0	90 / 0	0	0	0	90 / 0	0
3	0	n / 0	30 / 200	0	0	0	0	4 / 0
4	0	n / 0	n / 0	7 / 3	900 / 0	0	0	0
5	0	0	30 / 15	0	150 / 5	n / 0	0	0
6	0	n / 0	0	0	17 / 4	0	0	0
7	10 / 0	0	20 / 0	0	n / 0	0	0	45 / 100
8	0 / 30	n / 10	n / 0	0	0	0	0	0
9	0	n / 0	0 / 10	0	60/ 20	0	0	0
10	10 / 0	n / 0	0 / 6	0	0	0	0	0 / 1
11	0	15 / 0	0	0	15 / 0	0	30 / 60	15 / 7
12	0	n / n	n / n	30 / 0	0 / 100	0	0	30 / 0
13	0	60 / 0	0 / 5	150 / 0	120 / 0	0 / 10	0	2 / 1

clear relationship between the pre- and post-operative seizure patterns (all focal) (Table 1). Clinical improvement was related to a variable disruption of secondary bilateral synchrony in 27 cases. The improvement in the daily performance was related to a marked reduction in the frequency of generalized seizures (especially absences and atonic seizures). A disconnection syndrome was detected in one case. There was no mortality in this series.

Eleven patients (91%) submitted to stereotactic segmental callosotomy showed a marked decrease in the frequency of generalized seizures. Two patients suffered from new seizure patterns post-operatively (Cases VIII and XII). Clinical improvement was related to the disruption of SBS in different degrees. There was no disconnection syndrome or mortality in this series (Table 2).

DISCUSSION

Corpus callosum section effectively reduced the frequency of generalized seizures in both open and stereotactic series. Clinical results after these procedures seem to be similar. A lower morbidity was associated with the stereotactic procedure, which has been the sole procedure used in our center since 1986.[5]

The effect on the EEG of both procedures are also similar. It is characterized by a partial or complete disruption of the SBS. Both partial or complete SBS disruption may be associated with clinical improvement.

Total callosotomy has been performed in some centers.[6] Only segmental callosotomies have been performed in our center. The intervention may cause a disruption in SBS, especially between both frontoparietal regions and is rarely associated with disconnection syndromes.[7] Our previous experience using an ECoG-guided procedure to determine the extent of the callosal sections[8] was extremely helpful in defining a more or less standard anterior two-thirds callosal lesion to be used in stereotactic callosotomy.

The development of new seizure patterns might represent a modification in the epileptogenic process where focal activity which was not previously detected or clinically relevant could be seen.[9]

Both procedures led to a marked decrease in the generalized seizures frequency and improvement in the daily performances. The cognitive deficits in these series were related to the severity of the epileptic syndrome.[10] Even in the mentally retarded patients the decrease in the generalized seizure frequency led to worthwhile clinical improvement.

Callosotomy is a palliative procedure. It is effective in the control of generalized seizures in cases of severe secondary generalized epilepsies refractory to current AEDs regimens. The potential candidates should be carefully studied and selected according to the above criteria. The use of MRI-guided stereotactic surgery will soon turn stereotactic segmental callosotomy into an easier and safer procedure.

SUMMARY

Corpus callosum section is an effective palliative procedure in medically refractory generalized epilepsies. Twenty-eight patients were studied after open and 12 after stereotactic callosotomy. In 94% and 91% of the cases submitted to open or stereotactic callosotomy, respectively, there was a marked decrease in the frequency of the generalized seizures. Clinical improvement was related to the decrease in the generalized seizure frequency, especially absences and atonic seizures. Both procedures caused a disruption in the secondary bilateral synchrony in these cases. The lower morbidity of stereotactic procedures led to a preference of this type of procedure in our center.

REFERENCES

1. A. Cukiert and C. Timo-Iaria, Electrophysiological evidence for an L-shaped interhemispheric connection in the cat, *Arq. Neuropsiquiat.* 47:381-384 (1989).
2. E.M. Marcus and C.W. Watson, Symmetrical epileptogenic foci in monkey cerebral cortex. Mechanisms of interaction and regional variation in capacity for synchronous discharges, *Arch. Neurol.* 19:99-111 (1968).
3. W.P. Van Wagenen and N.Y. Harren, Surgical division of commissural pathways in the corpus callosum. Relation to the spread of an epileptic attack, *Arch. Neurol. Psychiatry* 7:40-59 (1940).
4. R. Marino, Jr., Surgery for epilepsy: selective partial microsurgical callosotomy for intractable multiform seizures: criteria for clinical selection and results, *Apl. Neurophysiol.* 48:404-449 (1985).
5. R. Marino, Jr. and G. Gronich, Corpus callosum stimulation and stereotactic callosotomy in the management of refractory generalized epilepsy, *Arq. Neuropsiquiat.* 47:320-325 (1989).
6. D.H. Wilson, A. Reeves, and M. Gazzaniga, Division of the corpus callosum for uncontrollable epilepsy, *Neurology* 28:649-653 (1978).
7. R. Marino, Jr. and P.C. Ragazzo, Selection criteria and results of selective anterior partial callosotomy, *in:* "Epilepsy and the Corpus Callosum," A.G. Reeves, ed., Plenum Press, New York, pp. 281-301 (1985).
8. R. Marino, Jr., J. Radvany, F. Huck, C.H.P. de Camargo, and G. Gronich, Selective EEG-guided microsurgical callosotomy for refractory generalized epilepsy, *Surg. Neurol.* 34:219-228 (1990).
9. S.S. Spencer, D.D. Spencer, G.H. Alaser, P.D. Williamson, and R.H. Mattson, More intense focal seizures after callosal section: the role of inhibition, *Arch. Neurol.* 50:686-693 (1984).
10. C.H.P. Camargo, R.M.A. Makray, J. Radvany, and R. Marino, Jr., Partial frontal callosotomy: a neuropsychological frontal lobe syndrome detected in the background of improvement of epilepsy, *in:* "Epilepsy and the Corpus Callosum," A.G. Reeves, ed., Plenum Press, New York, pp. 435-444 (1985).

PREOPERATIVE EVALUATION OF CHILDREN FOR CORPUS CALLOSOTOMY

Richard E. Nordgren, M.D.

Section of Neurology
Dartmouth-Hitchcock Medical Center
One Medical Center Drive
Lebanon, New Hampshire 03756

The majority of patients with epilepsy in childhood have benign seizure types and eventually outgrow their seizures. Even young patients with seizures that are considered refractory to medication have a good chance of eventually having a remission.[1] However, there is a group of children with severe and truly intractable epilepsy who have multiple seizure types that are devastating to their lives. These children with the most severe of epilepsy may in fact be candidates for corpus callosotomy since it is very unlikely that their seizures will improve as they mature. While the majority of patients who have had corpus callosotomy performed to date have been young adults or adolescents, it is appropriate to consider the surgery in patients of a younger age. Callosotomy in young patients may improve not only seizure control, but also behavior and alertness, which allows better functioning in the school and home. This improvement may have several reasons, including fewer clinical seizures, less subclinical interictal epileptiform activity, and the ability to manage the patient on less medication. Also, the younger patient may tolerate the procedure better than the older patient and have less deleterious effects from the procedure (the "disconnection syndrome," new speech or memory problems). In fact, none of our young patients have had any of these problems persist postoperatively.[2]

It can be conservatively estimated that there are several thousand children in the United States who might be candidates for corpus callosotomy and several hundred patients per year who will develop severe epilepsy who eventually might be considered candidates for this operation.[3] As more centers are performing callosotomies in young patients, it is important to consider the differences in selection criteria in children as compared to adults who are being evaluated for the

Epilepsy and the Corpus Callosum II, Edited by A.G. Reeves and D.W. Roberts, Plenum Press, New York, 1995

procedure.[2,4] Indications for callosotomy at our Center are:

- Medical intractability of seizures (at least four years of intractable seizures with the attempted use of all standard anticonvulsants).
- Seizure type potentially amenable to callosotomy (particularly generalized atonic, tonic, or tonic-clonic seizures).
- Potential functional benefit if improvement in seizure control is achieved.
- No single epileptiform focus is identified.

In addition, it is extremely important to ascertain whether the child's seizure disorder is of a type that is likely to improve as he or she matures. It usually is not difficult to diagnose children who have truly intractable epilepsy with little possibility of later spontaneous improvement in their condition. If at all possible, it is important to determine the etiology of the patients seizures, since patients with diffuse pathology (i.e., seizures secondary to diffuse hypoxia-ischemia) in our experience are less likely to respond to callosotomy than patients with unilateral lesions or "hemiatrophy." The procedure would rarely be considered if the child has a CNS degenerative disorder. When the child is examined, one especially looks for evidence of unilaterality (i.e., a hemiparesis), evidence of systemic illness and the presence of a neurocutaneous disorder.

We have not considered retardation and even severe retardation as a contraindication to the procedure. Several of our most retarded patients have had an excellent response to the procedure with a greater than 80% reduction in their seizure frequency (Ref. 2, cases 7,8,13,17 &18). Another patient (Ref. 2, case 12) had no improvement in seizure frequency, but his seizures postoperatively were less severe and he had a marked improvement in his functional capabilities, improving from not being testable by a psychologist prior to surgery to being able to attend school and learn to read. Similar results in severely retarded patients have been reported by others.[5] Our consideration for surgery in these patients strongly depends on the potential for functional improvement if their most disabling seizures are controlled (i.e., a severely retarded patient with generalized atonic seizures or generalized tonic seizures who is ambulatory but having multiple injuries from falls).

An important part of the evaluation of young patients undergoing callosotomy is MRI scanning. MRI may delineate an etiology for the patient's seizures, is essential to exclude focal lesions amenable to a direct surgical approach, and also is necessary to exclude agenesis of the corpus callosum. Patients may also undergo CT scanning to augment the MRI. Usually cerebral angiography is not essential other than as part of the Wada test.

Prolonged Video-EEG monitoring should be performed in young patients undergoing callosotomy, recording a reasonable sample of the patient's seizures. This usually is not a problem since most patients who are being evaluated for callosotomy have more frequent seizures than patients who are being considered for a focal resection. Intuitively, one would assume that preoperative EEG's and especially recording of actual seizures would have predictive value as to results of the procedure. Unfortunately, this has not been the case and preoperative EEG data has mostly been of use to exclude pseudo seizures and to demonstrate that the patient does not have a single seizure focus that would be amenable to surgical resection.[6,7] Even with Video-EEG monitoring, it is often difficult to classify the exact type of seizure that has been recorded. For example, it is sometimes impossible to distinguish a generalized atonic seizure from a generalized myoclonic seizure with falling, especially when the recording is carried out with the patient in bed. Also, it is often difficult with surface electrodes alone to separate generalized onset tonic or tonic-clonic from focal onset seizures with very rapid generalization resulting in a secondarily generalized tonic or tonic-clonic seizure. In fact, we have found that for the purposes of evaluating patients for possible callosotomy, a clinical description of the most obvious clinical manifestation of the seizure is more important than what appears on the EEG during the seizure.[2] Many of our patients have had clearly synchronous discharges during preoperative seizures convert to unilateral bursts postoperatively, while others have had these synchronous bursts remain bilateral after surgery. The latter group has not usually had as good a response to the callosotomy, and unfortunately, we have not been able to predict from the preoperative recordings which will convert. Also, patients who have clearly had seizures that begin as focal discharges and then generalize have not all had a clear response to callosotomy, suggesting that other than callosal pathways are important for seizure propagation in these patients. In our experience, patients with bilaterally synchronous discharges preoperatively who have had clear unilateral pathology are more likely to have theses synchronous discharges convert postoperatively to unilateral discharges than patients with bilaterally synchronous discharges who have not had demonstrated unilateral pathology. It is hoped that newer methods of computerized EEG analysis such as Gotman's "phase and coherence" methods[8] or parametric analysis of EEG data[9] will allow better prediction of a patient's response to the procedure.

SPECT and PET scanning have not been available long enough to assess whether they have any predictive value in evaluating patients undergoing corpus callosotomy. PET scanning is not available at most centers performing callosotomy but has shown some promising findings in young patients, differentiating sub groups in infantile spasms.[10] We also have demonstrated frequent unilaterality in patients undergoing SPECT scanning prior to callosotomy.[11] These tests may prove useful in predicting unilateral seizure foci which then rapidly generate synchronous seizures using the corpus callosum for transmission.

Neuropsychologic testing is attempted in all young patients undergoing callosotomy but many are either too retarded or too young to cooperate with the testing. This is discussed in detail by Dr. Lassonde in another section of this symposium.

It has been recommended that all patients undergoing callosotomy have Wada testing performed prior to surgery in order to determine if there is bilateral speech and memory function.[12] Some patients with bilateral speech function and mixed dominance have developed new speech problems postoperatively, and this is more likely to happen after a complete rather than a partial section. However, the test is often impossible to perform in children or in retarded patients. Our policy is to try a "dry run" to see if the patient is capable in carrying out the actual test. Because of their age and ability to cooperate, only a very few of our

young patients have had the procedure performed prior to surgery and none of our young patients have had obvious new speech or memory problems postoperatively, whether the Wada test was performed or not. It is hoped that there will be a consensus developed as to the indications and utility of Wada testing prior to corpus callosotomy, since there is great variation between centers as to the use of this procedure in the preoperative evaluation of callosotomy patients,
especially in children.

REFERENCES

1. P.R. Huttenlocher and R.J. Hapke, A followup of intractable seizures in childhood, *Ann. Neurol.* 128:699-705 (1990).
2. R.E. Nordgren, A. G. Reeves, A.C. Viguera, and D. W. Roberts, Corpus callosotomy for intractable seizures in the pediatric age group, *Arch. Neurol.* 48:364-372 (1991).
3. R.E. Nordgren, Corpus callosotomy is an underutilized procedure in children, *J. Epilepsy* 4:73-80 (1991).
4. S.S. Spencer, Corpus callosum section and other disconnection procedures for medically intractable epilepsy, *Epilepsia* 29 [Suppl 2]:S85-S99 (1988).
5. G. Geoffroy, M. Lassonde, F. Delisle, and M. Decarie, Corpus callosotomy for control of intractable epilepsy in children, *Neurology* 33:891-897 (1983).
6. S.S. Spencer, D. D. Spencer, P.D. Williamson, and R.H. Mattson, Effects of corpus callosum section on secondary bilaterally synchronous interictal EEG discharges, *Neurology* 35:1689-1694 (1985).
7. J.R. Gates, I.E. Leppik, J. Yap, and R.J. Gumnit, Corpus callosotomy: clinical and electroencephalographic effects, *Epilepsia* 25:308-316 (1984).
8. J. Gotman, Interhemispheric interactions in seizures of focal onset: data from human intracranial recordings, *Electoenceph. Clin. Neurophysiol.* 67:120-133 (1987).
9. J. Lu, E.W. Hansen, and R.E. Nordgren, Computerized analysis of epileptic foci and propagation from multichannel EEG data. Abstract, Second Dartmouth International Conference, Epilepsy and the Corpus Callosum.
10. H.T. Chugani, W.D. Shields, D.A. Shewmon, D.M. Olson, M.E. Phelps, and W.J. Peacock, Infantile spasms: I. PET identifies focal cortical dysgenesis in cryptogenic cases for surgical treatment, *Annals Neurol.* 4:406-13 (1990).
11. D.W. Roberts, D. Herzberg, A.G. Reeves, R. Nordgren, and P. Bohdiewicz, Interictal single-photon emmision computed tomography in patients undergoing corpus callosum section for intractable seizure disorders, *Epilepsia* 32(3):77-78, (1991).
12. K.J. Sass, D.D. Spencer, S.S. Spencer, et al., Corpus callosotomy for epilepsy. II. Neurologic and neuropsychologic outcome, *Neurology* 38:24-28 (1988).

CORPUS CALLOSOTOMY IN CHILDREN

Frank Ritter, M.D.

The Minnesota Epilepsy Group, PA
310 N. Smith Ave., Suite 300
St. Paul, Minnesota 55415

The term "cerebral disconnection" has been used to describe surgical procedures that sever selected interhemispheric white matter tracts. The connections most often divided are the corpus callosum and hippocampal commissure, although the anterior commissure, massa intermedia, and one fornix have been sectioned in some patients. Currently, centers performing this type of surgery limit divisions to a partial anterior or complete corpus callosotomy, usually including the hippocampal commissure.

This type of surgery was not routinely performed on children until 1970[9] almost 30 years after Van Wagenen and Herren's initial series. Experience with corpus callosotomy in children has been limited.[9,10,11,16]

The term children will be limited to those patients who had surgery at age 12 years or younger. There are clinical and theoretical reasons for this age limit outlined by Spencer.[15] Table I shows seizure outcome following corpus callosotomy on 19 children at Minnesota and on 26 children from the combined literature.[3,9-16] Other cases exist, but there is not sufficient detail in those reports for assessment.

In the Minnesota series, nine patients had recurrent convulsive status epilepticus prior to surgery. Only two patients have had episodes of status epilepticus postoperatively. Sixteen of 19 children had Lennox-Gastaut syndrome; 12 of the 16 had a positive outcome. Atonic and tonic (drop) seizures have been the most responsive. The remaining tonic seizures are often transformed into a partial, much less violent episode. These are usually not associated with falling. Thus, injuries are greatly reduced or eliminated. Secondarily, generalized tonic-clonic seizures frequently improve with corpus callosotomy. Complex partial seizures have been reported as decreased in frequency following corpus callosotomy. In our cases, one has become seizure free.

Epilepsy and the Corpus Callosum II, Edited by A.G. Reeves
and D.W. Roberts, Plenum Press, New York, 1995

SUMMARY

1. Corpus callosotomy in children is most effective for tonic, atonic, and secondary generalized tonic-clonic seizures. Other seizure types may also decrease in frequency.
2. Children with Lennox-Gastaut syndrome do well, with a good outcome in 70 to 80%.
3. A lateralizing neurological deficit may predict a better outcome, although the number of patients is too small to be certain.
4. Many children with severe mental retardation respond well. Experience in children with normal intelligence is limited.
5. Status epilepticus decreases in most children after surgery.
6. Other variables including age of onset, history of infantile spasms, interictal EEG abnormalities, abnormal P.E. examinations, history of status epilepticus, degree of mental retardation, family history of seizures and known etiology of seizures, do not appear to have a predictive value so far.

Outcome for surgery will be based on the following scale:

Positive (+)	Greater than 80% reduction in seizure frequency of the type for which the surgery was performed, decreased severity of seizures and improvement in life.
Unchanged (O)	Results of surgery did not improve seizure control 80% or more, whether or not there was a positive impact on life.
Negative (-)	Results of surgery did not improve seizure control and there was an adverse impact on life due to surgery, or seizures worsened.

Table 1. Review of Minnesota experience and the previously published results of children undergoing corpus callosotomy.

	LITERATURE	MINCEP	COMBINED
Number	n (26)	n (19)	n (45)
Age at Surgery	3 mos - 12 yrs (mean 7.5 yrs)	3 mos - 12 yrs (mean 6.9 yrs)	3 mos - 12 yrs (mean 7.2 yrs)
Mental Retardation	23 (88%)	18 (95%)	41 (91%)
Partial Callosotomy	9 (34%)	5 (36%)	14 (31%)
Positive Outcome	19 (73%)	13 (68%)	32 (71%)
Unchanged	6 (23%)	4 (21%)	10 (22%)
Negative Outcome	1 (4%)	2 (10%)	3 (7%)
Lateralized deficit on examination	known: 13 (11+)	known: 6 (3+)	known: 19 (14) 73%
	unknown: 6 (6+)	unknown: 0--	unknown: (6) 100%
	none: 7 (3+)	none: 13 (10+)	none: 20 (13) 65%

REFERENCES

1. D.H. Wilson, A.G. Reeves, and M. Gazzaniga, Division of the corpus callosum for uncontrollable epilepsy, *Neurology* 28:649-653 (1978).
2. G. Bouvier, C. Mercier, J.M. St. Hilaire, N. Giard, R. Labrecque, and R.A. Beique, Anterior callosotomy and chronic depth electrode recording in the surgical management of some intractable seizures, *Appl. Neurophysiol.* 46:52-56 (1983).
3. R.E. Harbaugh, D.H. Wilson, A.G. Reeves, and M.S. Gazzaniga, Forebrain commissurotomy for epilepsy, *Acta Neurochirurgica* 68:263-275 (1983).
4. D. Wilson, C. Culver, M. Waddington, and M. Gazzaniga, Disconnection of the cerebral hemispheres:an alternative to hemispherectomy for the control of intractable seizures, *Neurology* 25:1149-1153 (1975).
5. A.L. Amacher, Midline commissurotomy for the treatment of some cases of intractable epilepsy, *Child's Brain* 2:54-58 (1976).
6. A.M. Murro, H.F. Flanigin, B.B. Gallagher, D.W. King, and J.R. Smith, Corpus callosotomy for the treatment of intractable epilepsy, *Epilepsy Res.* 2:44-50 (1988).
7. S.J. Purves, J.A. Wada, W.B. Woodhurst, P.D. Moyes, E. Strauss, B. Kosaka, and D. Li, Results of anterior corpus callosum section in 24 patients with medically intractable seizures, *Neurology* 38:1194-1201 (1988).
8. S.S. Spencer, D.D. Spencer, P.D. Williamson, K. Sass, R.A. Novelly, and R.H. Mattson, Corpus callosotomy for epilepsy, *Neurology* 38:19-24 (1988).
9. A.J. Luessenhop, T.C. dela Cruz, and G.M. Fenichel, Surgical disconnection of the cerebral hemispheres for intractable seizures, *JAMA* 213(10):1630-1636 (1970).
10. G. Geoffroy, M. Lassonde, F. Delisle, and M. Decarie, Corpus callosotomy for control of intractable epilepsy in children, *Neurology* 33:891-897 (1983).
11. G.S. Makari, G.L. Holmes, A.M. Murro, J.R. Smith, H.F. Flannigin, M.J. Cohen, K. Huh, B.S. Gallagher, A.B. Ackell, R. Campbell, and D.W. King, Corpus callosotomy for the treatment of intractable epilepsy in children, *J. Epilepsy* 2(1):1-7 (1989).
12. J.R. Gates, W.E. Rosenfeld, R.E. Maxwell, and R.E. Lyons, Response of multiple seizure types to corpus callosum section, *Epilepsia* 28(1):28-34 (1987).
13. A.G. Reeves, ed., "Epilepsy and the Corpus Callosum," Plenum Press, New York (1985).
14. W.P. Van Wagenen, and R.U. Herren, Surgical division of commissural pathways in the corpus callosum, *Acta. Neurol. Psychiatr.* 44:740-759 (1940).
15. D.D. Spencer and S.S. Spencer, Corpus callosotomy in the treatment of medically intractable secondarily generalized seizures of children, *Cleve. Clinic J. Med.* 56:S69-78 (1989).
16. R.E. Nordgren, A. Reeves, A.C. Viguera, and D.W. Roberts, Corpus callosotomy for intractable seizures in the pediatric age group, *Arch. Neurol.* 48:364-372 (1991).

ABSENCE OF DISCONNECTION SYNDROME

AFTER EARLY CALLOSOTOMY

Maryse Lassonde,[1] Hannelore Sauerwein,[1] Guy Geoffroy,[2] and Claude Mercier[2]

[1]Groupe de Recherche en Neuropsychologie Expérimentale, Département de Psychologie, Université de Montréal

[2]Département de Neurologie et Neurochirurgie, Hôpital Sainte-Justine de Montréal
Montréal, Québec

INTRODUCTION

Ever since the seminal studies of Sperry, Gazzaniga and Bogen (e.g., Sperry et al. 1969) it is well established that epileptic adults with therapeutic section of the corpus callosum show interhemispheric disconnection deficits. Typically, these patients are unable to compare sensory information separately conveyed to different hemispheres. This applies particularly to the visual and tactile modalities that are based upon predominantly crossed pathways.

In the tactile modality, callosotomy has been found to interfere with the patients' capacity to cross-match objects, shapes or other tactile stimuli placed in each hand out of the subject's view (e.g., Sperry et al. 1969). In the same condition, the right-handed patients also show a unilateral anomia for stimuli placed in their left hand (e.g., Gazzaniga 1970; Gazzaniga et al. 1963; Sperry et al. 1969). Furthermore, they display difficulties when asked to indicate on the other side of the body the localization of touch applied to distal areas of one side (e.g., Gazzaniga 1970; Geffen et al. 1985). In the visual modality, adult callosotomized patients are generally found to be unable to discriminate between pairs of stimuli presented one in each hemifield. Akin to the tactile modality, they also show difficulties in verbally identifying visual stimuli tachistoscopically presented in their left hemifield (e.g., Gazzaniga 1970).

Although partial recovery has been reported in some of these functions, most of the deficits in interhemispheric communication appear to

Epilepsy and the Corpus Callosum II, Edited by A.G. Reeves and D.W. Roberts, Plenum Press, New York, 1995

persist in the long term following the surgery (Goldstein and Joynt 1969, Goldstein, Joynt, and Hartley 1975; Bogen and Vogel 1975; Bogen 1985). Taken together, these findings suggest that the mature brain relies principally on the corpus callosum for transfer and integration of tactile or visual information.

In contrast, subjects born without a corpus callosum do not show the typical disconnection syndrome. The absence of disconnection signs in these subjects has been attributed to the fact that the acallosal brain has the possibility to make use of neural plasticity in order to compensate for the congenital absence of the main interhemispheric pathway. This plasticity hypothesis has been tested in animals. In cats, section of the corpus callosum effected before completion of the myelination cycle does not abolish interhemispheric transfer of visual learning (Ptito and Lepore 1983). The adoption of corpus callosotomy as a therapeutic measure for intractable epilepsy in children has allowed us to confirm the plasticity hypothesis in man.

In this chapter, we will present an overview of the results gathered over the last 15 years in a pediatric population who underwent callosotomy before or after completion of the myelination cycle. Our results have shown that when section of this commissure is done before puberty, a period at which callosal maturation is presumably completed, the patients do not exhibit any signs or interhemispheric disconnection (Lassonde et al. 1986, 1988, 1991) In fact, the capacity of these young callosotomized patients to effect interhemispheric transfer is comparable to that observed in cases of callosal agenesis.

Two experiments will be reported. In the first experiment we wanted to test whether the age at callosotomy (before versus after puberty) would play an important role in the preservation and/or recuperation of interhemispheric communication. Simple tactile transfer tasks were chosen in order to obtain data from as many patients as possible, including those whose attention span and/or mental deficiency precluded more sophisticated procedures. In the second experiment we intended to further explore the extent and limits of cerebral plasticity by comparing the performance of one of the subjects who underwent callosotomy at the age of six with that of adult subjects who naturally lack the corpus callosum (callosal agenesis). The callosotomized patient was tested at age 12, that is six years following the surgery when maximum compensation was probably achieved. We also added two visual tasks in order to investigate whether the results from tactile experiments would apply to another sensory modality that, by virtue of its anatomical wiring, relies principally on callosal transfer for bilateral integration of sensory messages.

METHODS

Experiment I. Simple Tactile Tasks

Subjects. The sample consisted of ten callosotomized patients, four who underwent callosotomy before or at age 11 and six patients operated on between the ages of 13 and 21. For the purpose of this study, the former will be referred to as the young group and the latter as the older group. The older group is further subdivided into three left- and three right-handers since our studies have consistently suggested that the left-handers may have some bilateral speech representation.

All patients underwent callosotomy for relief of uncontrollable epilepsy at the Hôpital Sainte-Justine in Montreal. Clinical and neuropsychological data of the patients are reported elsewhere (Geoffroy et al. 1990; Lassonde et al. 1990). For detailed case histories, we refer to Geoffroy et al. (1983, 1986) and Lassonde et al. (1986, 1988, 1991). Demographic data of the patients are shown in Table 1. The case history of the young patient KMS who has not yet been described is presented below:

KMS, a six-year-old, left-handed boy is the only child of a young healthy couple. He was born in the 35th week of gestation, and his perinatal history was considered uneventful although the use of forceps was deemed necessary. His postnatal development was reportedly normal until the age of 1 1/2 years. At this time the child presented with myoclonic and generalized seizures and his EEG showed a very active left centro-temporo-parietal focus. The seizures responded poorly to anticonvulsant therapy from the start. When the decision to operate was taken, the child had a combination of akinetic, tonic-clonic and absence seizures. EEG traces showed frequent secondary generalization of seizure discharges originating in the left hemisphere. A CT scan also revealed an anomaly of white matter, which was most marked over the parietal lobes. The aggravation of the epilepsy was accompanied by a progressive decline of acquisitions. Before surgery, the child was no longer able to ambulate, to communicate or to feed himself. At the age of four years, nine months, he underwent complete callosotomy. The surgery resulted in an important improvement of his neurological status in that generalized seizures were abolished. Medication was substantially reduced. During the first nine months following surgery, he had a right hemiparesis which has subsided with the help of physiotherapy. According to his mother, the child has regained all of his previous acquisitions. On formal testing, he obtains an IQ of 87 on the Leiter International Performance Scale and of 78 on the WPPSI. He presently attends a private kindergarten.

Apparatus, stimuli and procedure. The tests administered in this experiment were: 1-Unimanual naming of familiar objects; 2-Intermanual matching of objects; and 3-Intra- and intermanual localization of touch.

1. Unimanual Naming Task. The stimulus material for this experiment consisted of 30 familiar objects (e.g., a teaspoon, a key), measuring approximately 6-8 cm, a size that is sufficiently small to fit into the hand of a younger child.

The stimuli were presented behind a concealing screen consisting of an open cardboard box with two apertures of a radius of 6.5 cm through which the subjects could extend their hands. The distance between the two apertures was 30 cm to avoid touching of the hands during bilateral presentation of stimuli. The experimental table on which the box rested was lined with a styrofoam mat in order to eliminate auditory cues from the stimulus material.

Subjects were asked to extend their right or left hand through one of the openings of the box with palms facing upwards. A stimulus was then placed in the subjects' hand and they were asked to verbally identify the stimulus. Twenty trials with 20 different objects were administered to each hand in such a manner that all stimuli were palpated as often with each hand. The order of the hands receiving the stimulus was varied randomly. For children who were too immature to be tested on this apparatus, the stimuli were placed in a bag made of a silky material that could be closed by pulling a cord. After the hand of the child was guided to the stimulus the cord was pulled quickly before the child had a chance to retrieve the object for visual inspection. For intermanual matching two bags, one for each hand, were employed. The subjects were tested individually in one or two sessions not exceeding 30 minutes.

2. Intermanual Matching Task. Stimuli and procedure were the same as in task one with the exception that a stimulus pair was presented simultaneously to the hands (one stimulus to each hand). The subjects' task consisted in judging whether or not the two stimuli were identical or different. Twenty trials with ten identical and ten different pairs were administered in a random order.

3. Intra- and Intermanual Localization of Touch. The experimental set-up was the same as described for the matching and naming tasks. Subjects' hands were inserted in the apertures, with palms exposed and fingers slightly extended. A light, brief touch was applied to the upper segment of one of the subjects' fingers with the point of a sharpened pencil. Subjects had to designate the locus of stimulation by touching the stimulated finger with the thumb of the same hand (intramanual or uncrossed condition) or by touching the corresponding finger of the non-stimulated hand using the thumb of this latter hand (intermanual or crossed condition). Both conditions were first practiced in full view before being administered in the experimental apparatus. Each hand received 12 stimulations in each condition during which the four fingers were touched equally often. The order of hand and finger stimulated varied randomly.

Experiment II. Tactile and Visual Discrimination

Subjects

1. Callosotomized Subjects. The callosotomized subject SD was the only patient of the young group described above who was able to concentrate sufficiently to perform the tasks of this experiment. MRI examination revealed a complete callosotomy in this patient (Lassonde et al. 1988). At the time of testing, SD was 10 1/2 years old; almost four years had elapsed since his surgery.

2. Callosal Agenesis Subjects. This group was composed of one male and five females, aged 13 to 24 years with complete agenesis of the corpus callosum confirmed by computerized tomography. Four of the patients originated from the Saguenay area of Quebec, a region that has a relatively high prevalence of a hereditary form of callosal agenesis associated with progressive myelopathy known as the Andermann syndrome (see Andermann et al.1975). The other two patients were siblings of a family unrelated to the Andermann cohort. The demographic data of the

Table 1. Description of callosotomized patients (ages are given in years and months).

Case	Sex	Handedness	Age[1]	Mental Age	Age at Callosotomy	Age at Seizure Onset	Seizure Type	Seizure Frequency	Site(s) of initial focus	Other anomalies of the CNS
Young group										
KMS	M	L	6;0	3;6	4;9	18 mos	akinetic, tonic-clinic, absences	12/mo	left centro-temporo-parietal	anomaly of white matter
SD	M	R	11;11	9;0	6;10	3yrs	tonic-clonic,myoclonic nocturnal	frequent status bilateral	right centro parietal	none known
YL	M	R	11;7	3;6	10;8	5yrs	tonic-clonic	4/day	right posterior	none known
PL	F	R	15;6	7;6	7;6	3 yrs	akinetic, tonic-clonic, absences	frequent status	right centro-parietal	hydrocephalus (shunt)
Older group										
SL	M	L	16;4	5;6	13;6	5 yrs	tonic-clonic, Lennox-Gastaut	12/day	right fronto-temporal	none known
MB	F	R	23;2	9;6	15;7	3 yrs	tonic-clonic, complex partial	6/day	right temporal	anterior temporal lobectomy at 9
JL	F	L	19;9	6;0	16;3	7 yrs	tonic,tonic-clonic	5/day	right fronto-temporal, left temporal	microcephaly, coloboma, cataracts
YE	M	A[2]	21;10	11;6	18;10	3 yrs	tonic-clonic, absences	2/day	temporal bilateral	spastic left hemiparesis, optic atrophy
FP	M	R	20;11	9;0	20;0	at birth	tonic-clonic, complex partial	1/day	right anterior, left temporal	amblyopia
MP	M	R	23;8	2;6	21;0	4 yrs	tonic-clonic	5/night 2/day	right parasagittal, left frontal	right hemiphere trauma at 4

[1]Age at last testing session; [2] Ambidextrous

acallosal subjects are presented in Table 2. For a detailed case history we refer to Lassonde et al. (1988).

Table 2. Demographic data of the acallosal subjects.

Subject	Sex	Handedness	Age at Testing	IQ	Other Pathology of the CNS
MJ	F	Right	13 yrs	65	myelopathy
HB	F	Right	16 yrs	64	myelopathy
AM	F	Left	16 yrs	55	myelopathy
NL	F	Left	17 yrs	74	myelopathy
MG	M	Left	17 yrs	77	none known
LG	F	Right	24 yrs	78	none known

Apparatus, Stimuli and Procedure. Both the tactile and the visual tasks employed in this experiment involved uni- and bilateral matching of pairs of stimuli. Furthermore, the experimental set-up was designed to allow for the additional measurement of response times in both modalities.

1. Tactile Tasks. The three tests employed in this experiment required a discrimination of pairs of stimuli on the basis of: a) size (large/small); b) shape (e.g., triangle, cone); and c) texture (sandpaper). The stimuli of the first two tests were three-dimensional geometric forms made from wood. The small shapes measured approximately 2 cm^3 whereas the larger shapes measured on average 5 x 5 x 2 cm. The stimuli of the third category consisted of small squares (5 cm^2) of various sandpapers that were glued on cardboard of the same dimension.

Each test comprised 60 trials. On half of the trials, the two stimuli of a pair were identical and on the other half they were different. The pairs were presented under three conditions: intramanually to the left hand (20 trials) and to the right hand (20 trials) and intermanually (20 trials). The order of identical and different pairs varied randomly within and between conditions.

The apparatus used for stimulus presentation was a modified Lafayette Inter-Sensory Transfer Stimulator (model 16016), equipped with a timer for the measurement of response times. The stimulus pairs were fastened with velcro on a 23-cm-wide cloth that was drawn tightly over two metal drums. The latter were rotated by a motor connected to a Lafayette Repeat Cycle Timer (model 16016) which, when activated, allowed for the advancement of one stimulus pair at a time. The subject's hands had access to a stimulus pair through a rectangular opening (31 x 18 cm) in the front panel of the apparatus. A flap, partially covering the opening, concealed the stimuli from the subject's view. The extension of the subject's hand though the opening activated a photocell (Jana, model J9100) which in turn triggered a timer (Hunter clock-counter, model 120A). The subject's verbal response emitted in a microphone (Shure, model 255) stopped the timer via an electronic voice delay (Stoelting, model 17222).

Subjects were seated in front of the stimulator with one or both arms resting in an armrest. Upon a signal from the experimenter, subjects extended one hand or both hands though the opening in order to explore the stimuli. In the intermanual condition each hand explored one stimulus. For unimanual exploration, the subject was required to rapidly alternate between the two stimuli of a pair with the same hand. Subjects were instructed to use their fingertips to explore the stimuli and to report as quickly as possible whether the two stimuli were identical or different.

2. Visual Tasks. These tests involved a discrimination of pairs of stimuli on the basis of a) color, and b) shape. The stimuli of the first test were small squares (1 cm^2) of ten different colors. The stimuli of the second test consisted of ten geometric shapes of the same color (black), approximately measuring 1 $cm.^2$ All stimuli were mounted on 10 x 15 cm white cardboard cards. For interhemispheric discrimination, the two stimuli of a pair were mounted 5 cm to the left and the right of a central fixation point, corresponding to 3° of eccentricity. For the intrahemispheric condition, they were mounted one on top of the other 5 cm to either the left (LVF) or the right (RVF) of the fixation point. Forty (40) stimulus cards were prepared for each condition, 20 cards with identical and 20 cards with different pairs. The order of identical and different pairs varied in a random manner within and between conditions.

The stimuli were presented via a four-field Gerbrand tachistoscope (model T-4A) for an exposure time of 150 msec. Stimulus presentation activated a Gerbrands digital clock-counter (model 127) which was arrested by the subject's oral response into a Shure microphone (model 5755) connected to a Gerbrands Electronic Voice Relay (model 1341).

Testing was carried out in a light-attenuated room. The presentation of a central fixation point, marked by a red dot, preceded each stimulus pair. Subjects were reminded on each trial to maintain their central fixation and to report as quickly as possible whether the two stimuli of a pair were identical or different.

RESULTS

Experiment I

1. Unilateral Naming Task. The percentages of correct responses obtained on the intermanual object naming tasks are presented in Table 3. As can be seen from this table, only the older subjects and the young patient SD were testable before surgery at which time most of them performed the task equally well with either hand (mean: dominant hand: 99%; non-dominant hand: 95%). The other children in the young group did not have the attention span required to produce consistent results or refused to cooperate. Nevertheless, their first postoperative performance showed that the section of the corpus callosum did not interfere to any extent with their ability to name objects in either hand at a high level of accuracy (mean: dominant hand: 87%; non-dominant hand: 76%). Moreover, their results at later testing sessions revealed further improvements to the effect that the two patients who were retested six to nine years after surgery obtained perfect scores with their non-dominant hand.

Table 3. Percentages of correct responses obtained on the unilateral object naming task.

Groups	Handedness	Age at Testing[1]	Pre-op DH	Pre-op NDH	Post-op DH	Post-op NDH
Young group						
KMS	Left	4;9	-	-		
		6;0			100	86
SD	Right	6;10	100	83		
		7;10			90	70
		11;11			90	100
PL	Right	7;6	-	-		
		7;8			70	70
		15;6			100	100
YL	Right	10;8	-	-		
		11;0			88	81
		11;7			88	86
Older group						
SL	Left	13;6	100	92		
		14;7			90	54
		16;4			90	90
MB	Right	15;7	100	100		
		16;7			100	40
		23;2			100	50
JL	Left	16;3	92	92		
		16;9			92	92
		19;9			100	100
YE	Ambidextrous	18;10	100	100		
		20;6			100	80
FP	Right	19;11	100	100		
		20;5			80	0
		20;11			100	20
MP	Right	21;2	100	100		
		22;7			100	70
		23;8			100	70

[1]Ages are given in years and months; DH: Dominant hand; NDH: Non-dominant hand

The postoperative performance of the young group contrasted with that of the six patients who underwent callosotomy in adulthood. As shown in Table 3, five out of the six showed a 20 to 100% reduction of correct responses for objects placed in the non-dominant hand. One exception was the left-handed patient JL, who was able to name objects in either hand with the same efficiency as before surgery. The other left-handed subject (SL) showed an initial decline of performance accuracy for the non-dominant hand. However, three years later, he was equally proficient with either hand whereas the right-handed patients have shown little or only modest improvements over the ensuing years. The ambidextrous patient (YE) was able to correctly identify 80% of the stimuli at his first post-operative session. It remains to be seen whether subsequent testing reveals improvements similar to that observed in the left-handed patients.

Table 4. Percentages of correct responses obtained on the intermanual matching task.

Groups	Handedness	Age at Testing	Pre-op	Post-op
Young group				
KMS	Left	4;9	-	
		6;0		90
SD	Right	6;10	-	
		7;10		83
		8;10		95
		11;11		100
PL	Right	7;6	-	
		15;6		100
YL	Right	10;8	-	
		11;0		100
		11;7		95
Older group				
SL	Left	13;6	100	
		4;11		79
		16;4		90
MB	Right	15;7	100	
		16;7		72
		23;2		75
JL	Left	16;3	100	
		19;2		77
		19;9		100
YE	Ambidextrous	18;10	100	
		20;6		100
		21;10		95
FP	Right	19;11	100	
		20;5		50
		20;11		60
MP	Right	21;2	95	
		22;7		65
		23;8		85

[1]Ages are given in years and months.

2. Intermanual Matching Task. Table 4 reports the results of the intermanual matching task. Inspection of this table reveals once more that none of the subjects tested before surgery had difficulties in cross-matching the objects (mean performance for older group: 98%). However, their ability to compare the stimuli was considerably reduced after surgery (20-50%). The exception was the ambidextrous patient who obtained perfect scores at his first postoperative testing. Again the greatest recovery of intermanual matching was observed in the two left-handed patients, whereas the right-handers in the older group have shown little recovery in the long term.

Although the younger subjects were not tested before surgery, their cross-matching abilities seemed to be well preserved after the callosotomy. In fact, one of the children (YL) obtained perfect scores at his first testing session, four months after the operation. The other children displayed also a high level of accuracy (83-90%) at initial testing and with the passage of time they too, have achieved a perfect performance.

3. Intra- and intermanual Localization of Touch. The results of the localization tasks are reported in Table 4. All patients tested before surgery had perfect or near perfect scores in the intramanual condition (mean accuracy for intramanual localization: 99%) but most of them made some mistakes when the task required a transfer of the stimulation to the other hand (mean accuracy for intermanual localization: 90%). However, their accuracy for crossed responses drastically declined after the callosotomy,

Table 5. Percentages of correct responses obtained on the intra- and intermanual localization task.

Groups	Handedness	Age at Testing[1]	Intra Pre-op	manual Post-op	Inter Pre-op	manual Post-op
Young group						
SD	Right	6;10	-		-	
		7;10	-	71	-	96*
		8;10	-	92	-	88
		11;11	-	100	-	83
PL	Right	7;6				
		15;6	-	100	-	88
YL	Right	10;8				
		11;0	-	92	-	83*
		11;7	-	100	-	not testable
Older group						
SL	Left	13;6	100		84	
		14;7		83		67
		16;4		96		55
MB	Right	15;7	100		100	
		15;11		100		59
		17;3		98		50
		23;2		88		80
JL	Left	16;3	-		-	
		18;1		100		58
		19;9		98		67
FP	Right	19;11	96		82	
		20;5		100		21
		20;11		100		21
MP	Right	21;2	100		95	
		22;4		100		13
		23;7		100		25
		23;8		100		29

[1]Ages are given in years and months.

and this decline seemed to be inversely related to the age of the patients at the time of the surgery (Table 5). Although a significant recovery was observed in one patient (MB), most of the patients have shown little or no improvement of their cross-localizing ability over time.

Once more, the two younger callosotomized patients were unable to sustain the level of attention necessary to perform this demanding task before surgery and they continued to experience difficulties after the operation. Because of these difficulties, the two patients SD and YL were tested with a simpler version of the task during their initial post-operative testing session. The modified task consisted of pointing with the index finger of the stimulated hand to the correct finger on the non-stimulated hand (see Lassonde et al. 1986). Both were able to perform this simpler version of the cross-localization task at a high level of accuracy (83-92%). In fact, the performance of the intermanual task by SD was superior to his intramanual performance. One year later, when the attention span of this patient had increased he was able to do the task in its standard form with little loss of accuracy. However, unlike on the matching and naming tasks, his performance in the crossed condition has not further improved over the years. The same holds true for PL who, nine years after the surgery, has not achieved perfect transfer of touch.

Experiment II

1-Tactile Matching Task. The left part of Figure 1A summarizes the results of the intra- and intermanual matching tests. Because of the myelopathy that accompanies the Andermann syndrome, only three patients (HB, MG, LG) were able to participate in the tactile experiment which required a manual response.

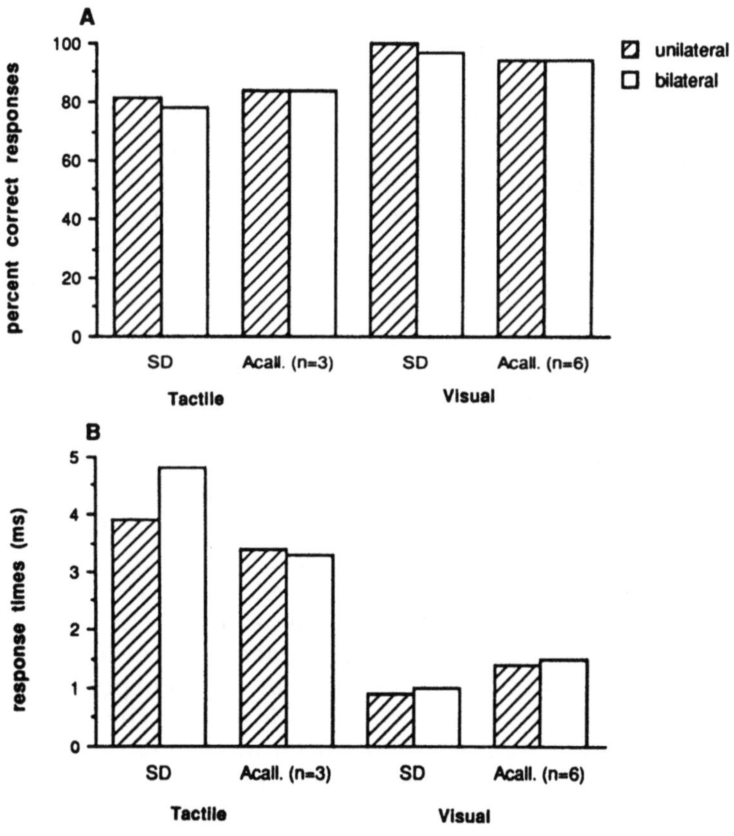

Figure 1. Results of unilateral and bilateral comparisons of tactile and visual stimuli obtained from the young callosotomized patient SD and acallosal adults. A- correct responses, B- response times.

It is evident from Figure 1A, that the callosotomized patient and the acallosal subjects performed the tasks at a high level of accuracy (78-84%) and that they were equally accurate whether uni- or bilateral integration was involved. It can also be noted that the performance of the callosotomized patient SD was comparable to that of the acallosal subjects although his intermanual performance was slightly inferior to that of the acallosal group. This discrepancy was somewhat more pronounced when response times were considered (Figure 1B). Once more the acallosals showed no important differences between intra- and intermanual responses whereas the response times of SD were longer, especially in the intermanual condition.

2. Visual Matching Task. The opposite pattern of results could be observed in the visual modality. Although all patients obtained scores in the 90 % range for both uni- and bilateral comparisons, the young callosotomized subject excelled in both conditions and outperformed the acallosal adults (Figure 1A). The same applied to response time measures: no important differences between unilateral and bilateral processing of the visual material were noted (Figure 1B). However, the responses of the callosotomized patient were considerably faster than those of the acallosals.

DISCUSSION

The results reported in the present study clearly show that the cerebral organization (or reorganization) that allows for interhemispheric transfer following early section of the corpus callosum can be observed for both the visual and the tactile modalities. The findings we have gathered during the last decade raise questions regarding the specific compensatory mechanisms that may be operating in young split-brain patients.

The possibility that some callosal fibers were spared must, of course, be considered. Only one of the young callsotomized patients underwent an MRI examination which confirmed the completeness of the section. This is the patient who has been studied most extensively by our group and whose performance has consistently shown that he is able to compensate for the callosal section.

One might argue that the differences between the younger and older groups may be related to extra-callosal damage in the latter. However, it should be pointed out that both our young and older patients had co-existing anomalies of the central nervous system. Furthermore, this did not affect their performance <u>before</u> the callosotomy. In the light of this evidence it would rather appear that the preservation of interhemispheric communication following early section of the corpus callosum is attributable to cerebral reorganization. However, the nature of the mechanisms of compensation is still a matter of speculation.

Bilateral speech has often been proposed to account for the capacity of acallosal subjects to name objects held in either hand (see Jeeves 1986, for a review). Because of the reorganization that can occur in the epileptic brain, the same mechanism(s) may be operating in callosotomized patients, especially those whose initial focus was located in the left hemisphere (e.g., Rasmussen and Milner 1977). For instance, the excellent performance of the younger subjects on the task requiring verbal identification of objects held in either hand could be accounted for by presuming that each hemisphere had direct access to the speech mechanism. In fact, this seems to be the case in our older left-handed patients, since these patients showed a greater recovery on this task than the right-handers in the older group. However, this interpretation still does not explain the excellent stereognostic performance of three of our four younger patients (SD, YL, PL). Not only were they right-handed but their initial epileptic foci were located in the right hemisphere which makes transfer of language functions highly unlikely. Furthermore, bilateral representation of speech could hardly account for the preservation of cross-localization in these patients since this task does lend itself easily to verbal coding. It is more probable that these subjects were using alternative pathways of interhemispheric communication which would not require abnormal representation of speech. Again we have to ask which pathways would be likely candidates for such transmission in the absence of the corpus callosum.

The compensation observed in visual studies, including ours (Lassonde et al. 1988), has often been attributed to an increased use and/or anatomical reinforcement of residual commissures, either the anterior commissure (Doty and Negrao 1973; Risse et al. 1978) or the intertectal commissure (e.g., Ptito et al. 1981; Sprague et al. 1970). In the tactile modality, on the other hand, an enhanced proficiency of the ipsilateral connections of the spinothalamic pathway has so far been the most plausible explanation for this type of cross-integration (e.g., Dennis 1976; Laget et al. 1977; Lassonde et al. 1986). The parallel use of ipsi- and contralateral connections would allow each hemisphere to receive sensory input from both hands, thus bypassing the interhemispheric route. However, comparison with the results of the visual experiment indicate that intrahemispheric integration of information is more limited than commissural transmission.

The differences in the quality of interhemispheric communication for the two sensory modalities observed in SD clearly point to different compensatory mechanisms for vision and touch. On the visual tasks, SD's performance was as good as that observed in the acallosal subjects who are known to be devoid of disconnection deficits. However, on the tactile tasks, SD's performance was poorer than that of the acallosals, especially in the condition requiring bimanual integration. The dissociation between the two modalities was even more apparent when response times were compared. While SD responded as quickly as the acallosals on the visual tasks, his responses were considerably longer for the tactile discriminations and more so in the interhemispheric condition.

The fact that compensation should be limited in the tactile modality is not surprising since it probably involves secondary pathways or mechanisms that were not programmed to accomplish this type of processing. A more fundamental question, however, arises as to why the adult brain should not benefit from these compensatory mechanisms to the same degree as the immature brain. The answer to this question may be found in the concept of cerebral plasticity.

It is generally assumed that cerebral plasticity is linked to the extent of functional maturity of the lesioned structure at the time of the insult (e.g., Goldman and Galkin 1978; Goldman-Rakic 1980, 1981). By deduction, plasticity should be drastically reduced once the corpus callosum has reached the final stage of maturity, presumably around the age of ten, when the myelination cycle of this commissure is completed (Yakovlev and Lecours 1967). This explanation, however, does not seem to hold since YL, a patient operated on at the age of 11, could still perform the transfer tasks at his first postoperative testing session, four months after surgery. This finding suggests that the extra-callosal pathways that must have been used by this patient are still operational even though the maturation cycle of the lesioned structure is completed, or alternatively, that the maturation of the callosum extends into puberty.

Another explanation consists in postulating that cerebral plasticity is linked to the phenomenon of synaptogenesis. There is evidence that many cortical connections undergo a stage of exuberance before synaptic stabilization occurs (for reviews see Changeux 1983; Changeux and Danchin 1976; Rakic et al. 1986). According to electron microscopic studies of frontal and occipital cortices in man, synaptogenesis starts during mid-gestation and continues during the first two years of life to decline gradually until adolescence when neuronal connections approach adult profiles (Huttenlocher et al. 1982). The stage of decreasing synaptic density is thought, in turn, to coincide with the emergence of greater precision and complexity of nervous functions.

This lack of structural specificity in the immature brain may account for cerebral plasticity in general, as suggested previously (e.g., Huttenlocher 1969; Changeux and Danchin 1976), and for the rapid recovery of function seen in our young callosotomized patients in particular. It is conceivable that congenital absence or early section of transcallosal projections may lead to the development, alteration and/or selective reinforcement of connections that would normally be eliminated during the critical period of axon retraction and synaptic elimination. The persistence of these connections (ipsilateral and/or commissural) could then explain the excellent preservation of interhemispheric communication that was seen in the patient who underwent callosotomy at age 11, that is after the presumed completion of callosal myelination.

Regardless of the exact nature of the mechanisms underlying neural compensation, the present series of experiments lead us to conclude that cerebral plasticity does not apply only to loss of cortical tissue but may also be seen following postnatal lesion of white matter. It is understood, however, that plasticity is not unlimited and that developmental processes follow time tables that may vary for different cerebral regions. Thus, once the critical period has passed, neuronal connectivity would become more static, which might explain the limited compensation seen in the callosotomized adults in this and other studies reported in the literature.

ACKNOWLEDGMENT

This study was supported by a grant from the National Science and Engineering of Canada (NSERCC) awarded to Dr. Maryse Lassonde.

REFERENCES

Bogen, J.E., 1985, The callosal syndrome, *in:* "Clinical Neuropsychology," K.M. Heilman and E. Valenstein, eds., Oxford University Press, New York.

Bogen, J.E. and Vogel, P.J., 1975, Neurological status in the long term following cerebral commissurotomy, *in:* "Les Syndromes de Disconnexion Calleuse chez l'Homme," F. Michel and B. Schottt, eds., Hôpital Neurologique, Lyon.

Changeux, J.P., 1983, "L'Homme Neuronal," Fayard, Paris.

Changeux, J.P. and Danchin, A., 1976, Selective stabilization of developing synapses as a mechanism for the specification of neuronal networks, *Nature* 264:705.

Dennis, M., 1976, Impaired sensory and motor differentiation with corpus callosum agenesis. A lack of inhibition during ontogeny? *Neuropsychologia* 14:455.

Doty, R.W. and Negrao, N, 1973, Forebrain commissures and vision, *in:* "Handbook of Sensory Physiology," Vol. VII/3B, R. Jung, ed., Springer, Berlin.

Gazzaniga, M.S., 1970, "The Bisected Brain," Appleton Century-Crofts, New York.

Gazzaniga, M.S., Bogen, J.E., and Sperry, R.W., 1963, Laterality effects in somesthesis following cerebral commissurotomy in man, *Neuropsychologia* 1:209.

Geffen, G, Nilsson, J., Quinn, K., and Teng, E.L., 1985, The effect of lesions of the corpus callosum on finger localization, *Neuropsychologia* 23:497.

Geoffroy, G., Lassonde, M., Delisle, F., and Décarie, M., 1983, Corpus callosotomy for control of intractable epilepsy in children, *Neurology* 33:891.

Geoffroy, G., Lassonde, M., Sauerwein, H., and Décarie, M., 1986, Effectiveness of corpus callosotomy for control of intractable epilepsy in children, *in:* "Two Hemispheres - One Brain," F. Lepore, M. Ptito, and H.H. Jaspers, eds., Alan Liss, New York.

Geoffroy, G., Sauerwein, H., Lassonde, M., and Décarie, M., 1983, Inclusion and exclusion criteria for corpus callosotomy for control of intractable epilepsy in children, *J. Epilepsy* 3 (Suppl):196.

Goldman P.S. and Galkin, T.W., 1978, Prenatal removal of frontal association cortex in the fetal rhesus monkey; anatomical consequences in postnatal life, *Brain Res.* 152:451.

Goldman-Rakic, P.S., 1980, Morphological consequences of prenatal injury to the primate brain, *Prog. Brain Res.* 53:1.

Goldman-Rakic, P.S., 1981, Development and plasticity of primate frontal association cortex, *in:* "Organization of the Cerebral Cortex," F.O. Schmidt, ed., MIT Press, Cambridge, Massachusetts.

Goldstein, M.N. and Joynt, R.J., 1969, The long-term effects of callosal sectioning: report of a case, *Arch. Neurol.* 20:196.

Goldstein, M.N., Joynt, R.J., and Hartley, R.B. 1975, The long-term effects of callosal sectioning: report of a second case, *Arch. Neurol.* 32:52.

Huttenlocher, P.R., 1969, Synaptic density in human frontal cortex. Developmental changes and effects of aging, *Brain Res.* 163:195.

Huttenlocher, P.R., De Courten, C., Garey, L.J., and Van der Loos, H., 1982, Synaptogenesis in human visual cortex - evidence for synapse elimination during normal development, *Neurosci. Let.* 33:247.

Jeeves. M.A., 1986, Callosal agenesis: neuronal and developmental adaptations, *in:* "Two Hemispheres - One Brain," F. Lepore, M. Ptito, and H.H. Jaspers, eds., Alan Liss, New York.

Laget, P., D'Allest, A.M., Fihey, R., and Lortholary, O., 1977, L'intérêt des potentiels évoqués somesthésiques homolatéraux dans les agénésies du corps calleux, *Rev. EEG Neurophysiol. Clin.* 7:498.

Lassonde, M., Sauerwein, H., Geoffroy, G., and Décarie, M., 1986, Effects of early and late transection of the corpus callosum in children. A study of tactile and tactuomotor transfer and integration, *Brain* 109:953.

Lassonde, M., Sauerwein, H., McCabe, N., Laurencelle, L., and Geoffroy, G., 1988, Extent and limits of cerebral adjustment to early section or congenital absence of the corpus callosum, *Behav. Brain Res.* 30:165.

Lassonde, M., Sauerwein, H., Geoffroy, G., and Décarie, M., 1990, Long-term neuropsychological effects of corpus callosotomy in children, *J. Epilepsy* 3 (Suppl):279.

Lassonde, M., Sauerwein, H., Chicoine, A.J., and Geoffroy, G., 1990, Absence of disconnection syndrome in callosal agenesis and early callosotomy: Brain reorganization or lack of structural specificity?, *Neuropsychologia* 29:481.

Ptito, M. and Lepore, F., 1983, Interocular transfer in cats with early callosal transection, *Nature* 301:513.

Ptito, M., Lepore, F., Lassonde, M., Miceli, D., and Guillemot, J.P., 1981, Le rôle du corps calleux dans le transfert interhémisphérique de l'information visuelle, *Rev. Can. Biol.* 40:61.

Rakic, P., Bourgeois, J.P., Eckenhoff, M.F., Zecevic, N., and Goldmann-Rakic, P.S., 1986, Concurrent overproduction of synapses in diverse regions of the primate cerebral cortex, *Science* 232:232.

Rasmussen, T. and Milner, B., 1977, The role of early left-brain injury in determining laterality of cerebral speech function, *Ann. N.Y. Acad. Sci.* 299:355.

Risse, G.L., Ledoux, J., Springer, S.P., Wilson, D.H., and Gazzaniga, M.S., 1978, The anterior commisssure in man: functional variation in a multisensory system, *Neuropsychologia* 16:23.

Sperry, R.W., Gazzaniga, M.S., and Bogen, J.E., 1969, Interhemispheric relationships: the neocortical commissures; syndrome of hemispheric disconnection, *in:* "Handbook of Clinical Neurology," Vol. 4, P.J. Vinken and G.W. Bruyn, eds., Elsevier, Amsterdam.

Sprague, J.M., Berlucchi, G., and DiBernardino, A. 1970, The superior colliculus and pretectum in visually guided behavior and visual discrimination in the cat, *Brain Behav. Evol.* 3:285.

Yakovlev, P.I. and Lecours, A.R., 1967, The myelogenic cycles of regional maturation of the brain, *in:* "Regional Development of the Brain Early in Life," A. Minowski, ed., Blackwell, Oxford.

NEUROLOGICAL EFFECTS OF CALLOSOTOMY

Alexander G. Reeves, M.D.[1] and Gail Risse, Ph.D.[2]

[1]Section of Neurology
Dartmouth-Hitchcock Medical Center
One Medical Center Drive
Lebanon, New Hampshire 03756

[2]Department of Neurology
Hennepin County Medical Center
Minneapolis, Minnesota

HISTORICAL BACKGROUND

Trescher and Ford reported in 1937 on alexia in the left visual field of a patient who had had section of the posterior callosum by Dandy and opined that "special methods of examination are required to demonstrate the essential symptoms." In spite of this caveat, until Sperry and his colleagues (1953, 1961) had undertaken their studies on the split-brain monkey, the right questions were not asked. Dandy appeared to be ignorant of Trescher and Ford's observations on his own patient stating that "No symptoms follow its division. This simple experiment at once disposes of the extravagant claims to function of the corpus callosum" (Dandy 1936).

Several years later, Akelaitis studied the callosotomy patients of Van Wagenen (Akelaitis 1941, 1944). Although he noted a difficulty with bimanual cooperation between the right and left hands, he was loath to attribute the change to section of the callosum. Instead, he postulated it to be the result of lateralized operative or preoperative brain damage. Traditional neuropsychological testing therefore supported Dandy's anecdotal statement.

Callosotomy as a treatment for intractable generalized seizures lay fallow from 1940 when Van Wagenen's series was interrupted by World War II, until Joseph Bogen and co-workers (Bogen and Vogel 1963; Bogen et al. 1965) rekindled the procedure in the early 1960s. One of Sperry's young colleagues, Michael Gazzaniga, was assigned to this new project (Bogen and Gazzaniga 1965). A series of studies of the bisected human

Epilepsy and the Corpus Callosum II, Edited by A.G. Reeves
and D.W. Roberts, Plenum Press, New York, 1995

callosum ensued; this series continues today and has so far elucidated a great wealth of information and speculation on the functions of the forebrain commissures and the right and left cerebral hemispheres (Gazzaniga 1970, 1985; Gazzaniga and LeDoux 1978; Seymour, Reuter-Lorenz, and Gazzaniga 1994; see also chapter by_Tramo et al. in this text.)

The neuropsychological consequences of transection of the corpus callosum -- and by necessity, the underlying hippocampal commissure -- have now been well described and are uniform enough to allow some predictability. Interhemispheric sensory disconnection syndromes and the behavioral consequences that follow posterior and complete callosotomy have been easily demonstrated at the bedside (Reeves 1991) and studied extensively (Gazzaniga 1970; Tramo 1995). Even so, the brains of patients with intractable seizures are generally and variably defective, which must be kept in mind when attempting to extrapolate findings to normal brain functions.

NEUROLOGICAL OBSERVATIONS

Standard neuropsychological and psychosocial assessments have shown little or no negative effects when compared with premorbid levels of activity (Goodman et al. 1985; Akelaitis 1941, 1944; Ferrell et al. 1984; Ritter et al.1989 abstract; Oepen et al. 1985; Mamelak 1993). Most patients show improvements, at least in part, on the basis of decreased seizures and anticonvulsant loads (Goodman et al. 1985; Ferrell et al. 1984; Ritter et al. 1989; Lezak 1982). Some negative effects have been recorded in a small number of patients and will be elaborated below. These have rarely outweighed the benefits of improved seizure control (Spencer et al. 1987). In this chapter, we will discuss the neuropsychological effects of anterior section (anterior half to two-thirds), posterior section (posterior half plus the hippocampal commissure), and complete section of the callosum, plus hippocampal commissure. Unless otherwise stated, the model subject will be left hemisphere-dominant for verbal language and right-handed.

Anterior Section

An acute and easily observable syndrome is characteristic in the hours and days following anterior callosal section (Ross et al. 1984). In our experience, this syndrome occurs (at least in part) in most patients. The major finding is a decrease in the spontaneity of speech; it may be so severe that the patient is completely mute, or it may be as mild as a slowness in initiating speech (Ross et al. 1984; Bogen 1976). Typically, once begun, speech, although it may continue to be slow, is not dysphasic (Ross et al. 1984). Other components of the syndrome can include variable degrees of paresis of the nondominant leg, a disinhibited forced grasping reflex in the nondominant hand, and urgency incontinence (Ross et al. 1984). Although the syndrome usually lasts from days to weeks, a small number of patients have been disabled for months or longer (Ross et al. 1984; Rayport et al. 1984; Sussman et al. 1984).

It has been suggested that either operative traction on the nondominant frontal parasagittal region or acute parasagittal disconnection, or both, are responsible for the syndrome (Ross et al. 1984). In support of traction injury are (a) the presence of contralateral paresis of the leg, presumably the result of compression of the parasagittal portion of motor strip which represents the leg, and (b) the contralateral forced grasp response, a finding which has been related to premotor, supplementary motor, and adjacent cingulate cortex lesions (Twitchell 1965).

Speech arrest follows unilateral stimulation of the nondominant supplementary motor area which lies anterior to the primary motor cortex (Penfield and Roberts 1966). A role in the generation of speech and movement in general has been postulated for the supplementary motor area of the dominant and nondominant hemispheres (Ross et al. 1986; Gelmers 1983; Damasio and Geschwind 1984; Brust et al. 1982; Caplan and Zervas 1978; Alexander and Schmitt 1980). Whether traction on the supplementary motor area alone or a combination of traction plus diaschisis from disconnecting the two callosally interconnected supplementary motor areas (Pandya and Rosene 1985) is the cause of the mutism (Ross et al. 1984; Macfarlene et al. 1991) is a moot point at present. Most important is the observation that the disability is usually temporary. Positron emission tomography studies in baboons following anterior callosal section support the diaschisis hypothesis in that a transient bilateral depression of cortical metabolism occurs which is predominant in the anterior parts of the brain. Metabolic recovery progressively occurs and reaches normal levels over approximately three months (Yamaguchi et al. 1990).

No clear-cut interhemispheric disconnection problem has been established following anterior two-thirds section of the corpus callosum, however, when the sectioning is carried further back toward the splenium and into the splenium, difficulties with somatosensory, auditory, and finally, visual transfer are noted (Risse et al. 1989). In contradistinction, other cases in limited numbers have been reported suggesting that limited sectioning of the corpus callosum or effects of naturally occurring midline lesions involving the callosum have interfered with some functions while not affecting others (Geschwind and Kaplan 1962; Geffen 1980; Bentin et al. 1984; Volpe et al. 1982). Neighborhood involvement of cortical structures in these patients may be the best explanation for the various and variable findings. Most of these studies did not adequately verify the extent of the lesions involved (Gordon et al. 1971; Volpe et al. 1982).

The lack of finding a consistent or persistent disconnection syndrome with anterior section of the corpus callosum is in keeping with anatomic studies describing the projection areas of fibers in the posterior callosum from the somatosensory, auditory, and visual cortices (Pandya and Seltzer 1986; Pandya et al. 1971; Pandya and Rosene 1985).

One center has reported several patients who developed dysphasic difficulties following anterior, posterior, and complete callosotomy, and these deficits have been attributed to preoperative bilateral or mixed cerebral dominance for verbal language (Spencer et al. 1988) and an implied callosally mediated hemispheric interdependence for these functions. This will be discussed further below.

Posterior Section

Interhemispheric sensory disconnection occurs following section of the posterior half of the callosum (Gazzaniga 1978; Gazzaniga and LeDoux 1978). The dominant and nondominant hemispheres are no longer able to internally transmit somesthetic or visual information to each other. Visual and tactile scannings of both halves of the environment nevertheless allow both sides of the brain to have equal access to both sides of the world. The potential disability from posterior callosal section is therefore greatly diminished.

Although there is no obvious behavioral indication of disconnection, the effects of posterior half section are readily evident at the bedside. Rapid presentation of visual stimuli (tachistoscopic) in the nondominant visual field is not reported by the patient, whose language-dominant hemisphere has no access to the information and no verbal capacity in the nondominant hemisphere to report what has been seen.

The splenium is the major route for interhemispheric visual transfer (Pandya and Rosene 1985; Risse et al 1989). If the section is carried rostrally to the callosal midpoint, encompassing parietal interconnections, tactile separation occurs. Objects placed in the nondominant hand of the blindfolded subject cannot be identified by the dominant hemisphere. The dominant hemisphere reports that it has not seen or felt anything, but it can be demonstrated that there are no nondominant hemisphere elementary visual or tactile deficits. Objects presented tachistoscopically to the nondominant visual field can be correctly identified behind a screen by the nondominant hand. Similarly, objects palpated by the hand can be re-identified from a group of objects by the same hand. What is missing is the ability to say what has been seen and felt on the left.

Auditory stimuli reach both ears almost equally from both sides of the environment. Internally, the pathways to the auditory cortex are bilateral from both cochlear systems, so each ear has access to both hemispheres. Nevertheless, with dichotic listening tasks, auditory disconnection can be demonstrated when the posterior temporal lobes are separated by posterior callosal sectioning (Musiek et al. 1985; Alexander and Warren 1988; Kaga et al. 1990; Gazzaniga et al. 1985). In our experience, when a verbal command is given to perform a task with the nondominant hand, no problem is encountered. The dominant hemisphere, which is alone capable of understanding the command, presumably controls the nondominant motor cortex through the non-severed anterior callosum which interconnects the two frontal motor systems. If the anterior callosum is then split, completing hemispheric disconnection (with the exception of the anterior commissure), the patient (his dominant hemisphere) no longer correctly performs a command to carry out a task with the nondominant hand (Gazzaniga 1970, 1985; Gazzaniga and LeDoux 1978). He may respond in some fashion with the nondominant hand, possibly either by (a) a perseveration of a response by the dominant hand which has been observed by the nondominant hemisphere or (b) mimicking a movement made by the examiner. It is presumed that the nondominant hemisphere is able to recognize the prosody (music) of a command and therefore attempts to accommodate the examiner by using nonverbal cues.

Complete Section

Most split-brain procedures are now carried out in two stages to decrease the acute postoperative morbidity of a prolonged surgical procedure (Wilson, Reeves, and Gazzaniga 1982) and to avoid the complete split-brain syndrome if possible. One-stage complete callosotomy is not, to our knowledge, practiced today. If inadequate seizure control is achieved by anterior section [the usual preliminary procedure in most centers (Spencer et al. 1987)] the callosotomy is completed. The most recent multicenter data indicate anterior section to be approximately 50% as effective for control of seizures when compared to complete section (Spencer et al. 1987).

The classic split-brain syndrome follows complete callosotomy (Gazzaniga 1970, 1985; Gazzaniga and LeDoux 1978). The sensory disconnection seen with posterior section is present, and now the dominant hemisphere no longer has access to the motor cortex of the nondominant hemisphere for the execution of skilled distal movements (Volpe 1985). Each hemisphere, however, is able to control some proximal movements of the ipsilateral limb through ipsilateral uncrossed motor pathways (Volpe 1985). The nondominant (left) hand no longer will respond correctly to verbal commands for skilled movements. Nevertheless, some incorrect behavior of the nondominant hand will usually be seen because (a) the dominant (left) hemisphere can initiate some proximal movements of the ipsilateral nondominant upper limb, and (b) the nondominant hemisphere can perceive these movements of the left arm and can also perceive the prosody of a command. If the examiner inadvertently first asks the subject to carry out a command with the dominant-hemisphere hand and is then asked to use the nondominant-hemisphere hand the same way, more often than not the task is completed correctly. The nondominant hemisphere, perceiving the prosody of a command and the proximal movement of the nondominant arm, copies what it has just seen the dominant arm do, leaving the novice examiner with the impression that the nondominant hemisphere has verbal language capability. If the examiner begins with a command to carry out a task with the nondominant hand or follows the dominant-hand task with a different command for the nondominant hand, the disconnection is revealed.

Thus, the nondominant hand no longer responds appropriately to verbal commands. A number of our patients, when asked what is wrong with their nondominant hand, have responded with considerable insight that the hand will not cooperate, implying that their dominant-hemisphere self is not in control of their nondominant-hemisphere self.

Furthermore, the nondominant hand is frequently noted in the postoperative period to act in an antagonistic manner towards the dominant hand. Basic manual tasks such as eating, opening and closing doors, turning faucets on and off, and washing hands may be difficult because the nondominant hand may terminate an action which the dominant hemisphere has initiated. For example, the dominant hand, frequently used when initiating a manual activity, may turn on the hot water. Both hands reach for the water to begin the well-learned bimanual task of washing. The nondominant hand recoils from what appears to be too hot water and then turns off the faucet. The dominant hand turns the faucet on again, appearing to tolerate the heat better. Ultimately, the nondominant hand tolerates the heat or the dominant hand decreases the heat.

Cooperation is established for previously well-learned bimanual tasks, presumably by use of external cues in most patients over a short period of time (weeks to months). New bimanual tasks are learned with difficulty, a problem which may need to be considered with younger patients. However, recent data suggests that younger patients are, in fact, more able to accommodate hemispheric separation -- at least with somesthetic-directed functions (Lassonde et al. 1986; see also chapter by Lassonde in this text). This may be on the basis of greater plasticity of sensory integration allowed by greater use of ipsilaterally projecting sensory pathways in the developing brain (Lassonde et al. 1986; see also chapter by Lassonde in this text). Furthermore, in time it is apparent that some patients are able to control some hand-finger movements with ipsilateral motor pathways (Ferguson, Rayport, and Corrie 1985). Further support for the greater compensatory plasticity of the young brain is observed in the experimental model using the callosotomized newborn rat which shows good anatomical compensation for loss of the major commissure in that synaptic density remains unchanged from normal in the developing lesioned animals (Zecevic et al. 1989).

Interhemispheric antagonism has remained a chronic problem in a small number of patients. Persistent and disabling difficulties with daily hygiene, dressing, and shopping (Ferguson, Rayport, and Corrie 1985), and in one case (Reeves and Roberts, *unpublished observations*) in almost all daily activities, have occurred. The latter patient has such great antagonism between his two hemispheres that his left and right arms abuse each other, potentially causing physical injury. The dominant (left) hemisphere has verbally expressed great hostility towards the left arm and leg, considering them another hostile person. This antagonistic behavior has created a paralyzing disability which far outweighs the benefits of the excellent seizure control afforded by callosotomy in this patient. No retrospective clues are available to allow prediction of a predisposition towards such a disabling behavioral result in other patients. Fortunately, it is most unusual, and developing cooperation between hemispheres is the rule.

One of our patients (Reeves and Roberts, *unpublished observations*) developed excellent cooperation within days of complete callosal section. When asked to carry out a task with her nondominant-hemisphere-controlled left hand (e.g., show me three fingers with your left hand), she quickly learned to place the left hand into her right hand, which would then arrange the fingers appropriately. The nondominant hemisphere is this patient willingly followed the lead of the dominant hemisphere using external cues just as it had done before surgery using internal callosally transmitted cues.

In other tasks (such as picture completions and Kohs block pattern reproduction), the dominant hemisphere, frequently less capable with visuospatial problems, has allowed the more capable nondominant hemisphere to take charge and complete the task (Gazzaniga and DeDoux 1978).

One left-handed patient with left-hemisphere dominance for verbal language is no longer able to write more than random letters or digits with his left hand following callosotomy. He has subsequently learned to write capably with his right hand (Gur et al. 1984). Similarly, a patient in our series, preoperatively right-handed but with right-hemisphere verbal language dominance, became left-handed for writing following callosotomy (Reeves and Roberts, *unpublished observations*). In both of these instances, it is presumed that the dominant hemisphere was exercising callosally mediated contralateral control of the motor system for writing. Although they had writing difficulties, neither of these two patients had other evident dysphasic difficulties. In both, one 20 and the other 11 years old, compensation was ultimately successful and writing was learned with the contralateral hand. Right-hemisphere dominance with right-handedness is rare, but it should be expected that some left-handed patients will need to transfer writing skills to their right hand after complete callosotomy (Sass et al. 1990).

A number of other, more benign behavioral examples of hemispheric independence have been observed.

A 14-year-old right-handed boy who lived by the seashore had many of the usual food aversions seen in youngsters. In particular, he had a strong aversion for white bait, a favorite of his family, which is a finger food dish of small fish, fried and eaten uneviscerated. His aversion was visual and imaginative. He had never tasted the dish. Following complete callosotomy, he was seen to be taking the fish with his left hand, putting them in his mouth and eating them. His hamburger was untouched and when his mother exclaimed over the new acceptance, he responded (with his dominant left hemisphere) that he was puzzled too, but that the fish were, in fact, good. Thereafter, he also ate the fish with his right hand. Presumably, his right hemisphere was naive to the aversion and was simply copying the rest of the family, who were evidently enjoying the dish.

Another right-handed, left-hemisphere-dominant patient, while discussing her newly controlled seizures with her physician on the phone, exclaimed that she could not continue to talk because her mother was coming into the room, and her left hand apparently did not want her to continue the conversation. It was trying to hang up the phone. When asked if this was a problem, she said no and that she would call back later, and then her left hand hung up the phone.

MEMORY DEFICITS

Memory dysfunction has been noted in a minority of patients following partial and complete callosotomy (Ferguson, Rayport, and Corrie 1985; LeDoux et al. 1974; Zaidel and Sperry 1974). A deficit in learning new material has been noted which may result from disconnecting two damaged and interdependent hippocampal formations by severing the callosum and hippocampal commissure (Ferguson, Rayport, and Corrie 1985; Novelly and Lifrak 1985). A review of the pertinent literature concerning lesions of the corpus callosum and recent memory was unable to find evidence for any clear cut recent memory dysfunction in the absence of extra callosal damage (Clark and Geffen 1989). Zaidel and Sperry (1974) and later Dimond (1976) noted deficits in attentional capacity following complete callosotomy which could account for the observed memory difficulties. The conclusions drawn from the series of ten patients studied by Zaidel and Sperry (1974) were criticized by LeDoux et al. (1974) and Clark and Geffen (1989) for not having preoperative data in patients with multiple preoperative reasons for memory and attentional difficulties. These included extra callosal brain lesions, seizures, and high-dose anticonvulsant medications. A patient studied pre- and postoperatively by LeDoux et al. (1974) showed no postoperative defects but, in fact, showed improved memory functions. Clark and Geffen (1989) in their review pointed out that some of the controversy may have arisen because many patients were tested for memory shortly after surgery at which time, deficits were found although later testing months to a year or more after surgery have shown improvement back to preoperative levels or improvement beyond preoperative levels. When pre-operative and postoperative testing were carried out by Petrucci et al. (1987), a minimal decrease in the memory quotient score was noted in only one of five patients. Two of the five showed improvement postoperatively. A detailed pre-operative and postoperative study of memory functions of four patients with anterior callosotomy, sparing the posterior callosum and hippocampal commissure is reported in this text by Tramo et al. There was no change or improvement in the parameters tested. Two patients were studied who had the hippocampal commissure sectioned in the process of sectioning all or part of the posterior callosum. A mild but significant deficit of recall was noted in both (Phelps et al. 1991). It is apparent that persistent memory difficulties do occur in some patients (see also Ferguson, Rayport, and Corrie 1985) and may be dependant upon section of the hippocampal commissure and/or parts of the posterior callosum, but as a rule these deficits are not of great enough significance to outweigh the psychological benefits of improved seizure control and a decreased anticonvulsant load (Gazzaniga 1985; Ferrel, Culver, and Tucker 1984).

LATERALIZED CEREBRAL DEFICITS

A small number of patients have emerged from surgery with persistent lateralized hemispheric deficits not seen prior to callosotomy. The centers reporting these findings have postulated the severance of successful, callosally mediated compensation for ancient lateralized cortical lesions (Novelly and Lifrak 1985; Campbell, Bogen and Smith 1981). Although the sample from which conclusions can be drawn is small, it is suggested that the presence of mixed bilateral hemispheric dominance for verbal language functions may sometimes lead to dysphasic deficits following callosotomy. In particular, it was suggested that difficulties may be more likely to arise if the hemisphere controlling handedness is not the language-dominant hemisphere (Spencer et al. 1988; Sass et al. 1990). Although transfer of writing capacity occurred following callosotomy in the two patients reported above who had crossed hemisphere control of writing, no dysphasic difficulties were noted (Gur et al.1984; Reeves and Roberts, *unpublished observations*).

Preoperative carotid amytal testing (The WADA test) (Wada and Rasmussen 1960) is, for the time being, recommended when feasible for patients undergoing callosotomy to determine the presence of unusual combinations of language dominance, which might predict postoperative difficulties (Spencer et al. 1987).

CONCLUSIONS

Dandy (1936) and Akelaitis (1941, 1944) were not prepared to ask the right questions. If they had simply flashed cards manually at the bedside, tested graphesthesia and stereognosis, and asked their patients to carry out tasks with their nondominant hand, they would have discovered the significance of the corpus callosum for interhemispheric communication in the human brain. Their observation that little overt negative behavioral change occurs following callosotomy has been largely corroborated and continues to justify callosotomy as a therapy for appropriate patients with intractable epilepsy. The positive effects on behavior which follow decreased seizures, along with the decreased need for anticonvulsants, outweigh (with few exceptions) the observable, but inconsistent, chronic negative effects. The split-brain syndrome is easily demonstrated and well accommodated by most patients. It continues to be an extraordinary resource for studies of the highest functions of the brain.

REFERENCES

Akelaitis, A.J.E., 1941, Studies on corpus callosum: higher visual function in each hemisphere's field following complete section of the corpus callosum, *Arch. Neurol. Psychiatry* 45:786-796.

Akelaitis, A.J.E., 1944, A study of gnosis, praxis and language following section of the corpus callosum and anterior commissure, *J. Neurosurg.* 1:94-102.

Alexander, M.P. and Schmitt, M.A., 1980, The aphasia syndrome of stroke in the left anterior cerebral artery territory, *Arch. Neurol.* 37:97-100.

Alexander, M.P. and Warren, R.L., 1988, Localization of callosal auditory pathway: a case study, *Neurology* 38:802-804.

Bentin, S., Sohar, A., and Moscovitch, M., 1984, Intermanual information transfer in patients with lesions in the trunk of the corpus callosum, *Neuropsychologia* 22:601-611.

Bogen, J.E., 1976, Linguistic performance on the short term following cerebral commissurotomy, *in:* "Studies in Neurolinguistics," H.A. Whitaker, ed., pp. 193-224.

Bogen, J.E. and Vogel, P.J., 1963, Treatment of generalized seizures by cerebral commissurotomy, *Surg. Forum* 14:431-433.

Bogen, J.E., Fisher, E.D., and Vogel, P.J., 1965, Cerebral commissurotomy: a second case report, *JAMA* 194:1328-1329.

Bogen, J.E. and Gazzaniga, M.S., 1965, Cerebral commissurotomy in man: minor hemisphere dominance for certain visuospatial functions, *J. Neurosurg.* 23:394-399.

Brust, J.C.M., Plank, C., Burke, A., et al., 1982, Language disorder in a right-hander after occlusion of the right anterior cerebral artery, *Neurology* 32:492-497.

Campbell, A.L., Jr., Bogen, J.E., and Smith A., 1981, Disorganization and reorganization of cognitive and sensorimotor functions in cerebral commissurotomy: compensatory roles of the forebrain commissures and cerebral hemispheres in man, *Brain* 104:493-551.

Caplan, L.R. and Zervas, N.T., 1978, Speech arrest in a dextral with right mesial frontal astrocytoma, *Arch. Neurol.* 35:252-253.

Clark, C.R. and Geffen, G.M., 1989, Corpus callosum surgery and recent memory, *Brain* 112:165-175.

Damasio, A.R. and Geschwind, N., 1984, The neural basis of language, *Annu. Rev. Neurosci.* 7:127-147.

Dandy, W.E., 1936, Operative experiences in cases of pineal tumors, *Arch. Surg.* 33:19-46.
Dimond, S.J., 1976, Depletion of attentional capacity after total commissurotomy in man, *Brain* 99:347-356.
Ferguson, S.M., Rayport, M., and Corrie, W.S., 1985, Neuropsychiatric observations on behavioral consequences of corpus callosum section for seizure control, *in:* "Epilepsy and the Corpus Callosum," A.G. Reeves, ed., Plenum Press, New York, pp. 501-514.
Ferrell, R.B., Culver, C.M., and Tucker, G.J., 1984, Psychosocial and cognitive function after commissurotomy for intractable seizures, *J. Neurosurg.* 58:374-380.
Gazzaniga, M.S., 1970, "The Bisected Brain," Appleton-Century-Crofts, New York.
Gazzaniga, M.S., 1985, Some contributions of split-brain studies to the study of human cognition, *in:* "Epilepsy and the Corpus Callosum," A.G. Reeves, ed., Plenum Press, New York, pp. 341-348.
Gazzaniga, M.S. and DeDoux, J.E., 1978, "The Integrated Mind," Plenum Press, New York.
Gazzaniga, M.S., Holtzman, J.D., Deck, M., and Lee, B., 1985, MRI assessment of human callosal surgery with neuropsychological correlates, *Neurology* 35:1763-1766.
Geschwind, N. and Kaplan, E., 1962, A human cerebral disconnection syndrome, *Neurology* 12:675-685.
Geffen, G., 1980, Phonological fusion after partial section of the corpus callosum, *Neuropsychologia* 18:613-620.
Gelmers, H.J., 1983, Non-paralytic motor disturbances and speech disorders: the role of the spplementary motor area, *JNNP* 46:1052-1054.
Gur, E.E., Gur, R.C., Sussman, N.M., O'Connor, M.J., and Vey, M.M., 1984, Hemispheric control of the writing hand: the effect of callosotomy in a left-hander, *Neurology* 34:904-908.
Kaga et al.,1990, Speech perceptions and auditory P300 potentials after section of the posterior half of the truncus of the corpus callosum, *Brain Topography* 3:175-181.
Lassonde, M., Sauerwain, H., Geoffrey, G., and Decarie, M., 1986, Effects of early and late transection of the corpus callosum in children, *Brain* 109:953-967.
LeDoux, J.E., Rissse, G.L., Springer, S.P., et al., 1974, Cognition and commissurotomy, *Brain* 100:87-104.
Lezak, M.D., 1982, "Neuropsychological Assessment, 2nd Edition," Oxford University Press, New York.
Mamelak, A.N., Baarbaro, N.M., Walker, J.A., and Laxer, K.D.,1993, Corpus callosotomy: a quantitative study of the extent of resection, seizure control, and neuropsychological outcome,*J. Neurosurg.* 79:688-95.
Musiek, F.E., Reeves, A.G., and Baron, J.A., 1985, Release from central auditory competition in the split-brain patients, Neurology 35:983-987.
Myers, R.E. and Sperry, R.W., 1953, Intraocular transfeer of a visual form discrimination habit in cats after section of the optic chiasm and corpus callosum, *Anat. Rec.* 115:351-352.
Nakasu, Y., Isozumi, T., Hioka, H., Handa, J., 1991, Mechanism of mutism following the transcallosal approach to the ventricles, Acta Neurochir. 110:146-153.
Novelly, R.A., and Lifrak, M.D., 1985, Forebrain commissurotomy reinstates effects of preexisting hemisphere lesions: an examination of the hypothesis, *in:* "Epilepsy and the Corpus Callosum," A.G. Reeves, ed., Plenum Press, New York, pp. 467-500.
Oepen, G., Schultz-Weiling, R., Zimmermann, P., et al., 1985, Long-term effects of partial callosal lesions, *Acta Neurochir.* 77:22-28.
Pandya, D.N. and Seltzer, B., 1986, The topography of commissural fibers, *in:* "Two Hemispheres: One Brain: Functions of the Corpus Callosum," F. Lepore, M. Ptito, H.H. Jasper, eds., Alan R. Liss, Inc., New York, pp. 48-73.
Pandya, D.N., Karol, E.A., and Heillronn, D., 1971, The topographical distribution of interhemispheric projections in the corpus callosum of the rhesus monkey, *Brain Res.* 32:31-43.
Pandya, D.N. and Rosene, D.L., 1985, Some observations on trajectories and topography of commissural fibers, *in:* "Epilepsy and the Corpus Callosum," A.G. Reeves, ed., Plenum Publishing Corp., New York, pp. 329-337..
Penfield, W. and Roberts, L., 1966, "Speech and Brain Mechanisms," Atheneum, New York, pp. 110 137.
Petrucci, R.J., Buchheit, W.A., Woodruff, G.C., et al., 1987, Transcallosal parafornicial approach for third ventricle tumors: neuropsychological consequences, *Neurosurgery* 20:457-464.

Phelps, E.A., Hirst, W., and Gazzaniga, M.S., 1991, Deficits in recall following partial and complete commissurotomy, *Cerebral Cortex* 1:492-498.

Rayport, M., Ferguson, S.M., and Corrie, W.S., 1984, Mutism after corpus callosum section for intractable seizure control, *Epilepsia* 25:663.

Reeves, A.G., 1991, Behavioral changes following corpus callosotomy, *in:* "Advances in Neurology, Vol. 55, D. Smith, D. Treiman, and M. Trimble, eds., Raven Press, Ltd., New York, pp. 293-300.

Risse, G.L., et al., 1989, Interhemispheric transfer in patients with incomplete section of the corpus callosum, *Arch. Neurol.* 46:437-443.

Ritter, F.J., Gates, J.R., Maxwell, R.E., et al., Behavioral Changes in Children Following Corpus Callosotomy, (Abstract), 1989, presented at the Epilepsy and Behavior Meeting, New Orleans, LA, Feb. 16-18,

Ross, M., Damasio, H., and Eslinger, P., 1986, The role of the supplementary motor area (SMA) and anterior cingulate (AC) in the generation of movement, *Neurology* 36(Suppl. 1):346.

Ross, M.K., Reeves, A.G., and Roberts, D.W., 1984, Post-commissurotomy mutism, *Ann. Neurol.* 16:114.

Sass, K.J., Novelly, R.A., Spencer, D.D., et al., 1990, Postcallosotomy language impairments in patients with crossed cerebral dominance, *J. Neurosurg.* 72:85-90.

Seymour, S.E., Reuter-Lorenz, P.A., and Gazzaniga, M.S., 1994, *Brain* 117:105-115.

Spencer, S., Gates, J.R., Reeves, A.G., et al., 1987, Corpus callosum section, *in:* "Surgical Treatment of the Epilepsies," Raven Press, New York, pp. 425-444.

Spencer, D.D., Spencer, S.S., Sass, K.J., et al., 1988, Corpus callosum section for epilepsy. III. Neurologic and neuropsychological effects, *Neurology* 38:25.

Sperry, R.W., 1961, Cerebral organization and behavior, *Science* 133:1749-1757.

Sussman, N.M., Gur, R.E., Gur, R.E., et al., 1983, Mutism as a consequence of callosotomy, *J. Neurosurg.* 59:514-519.

Trescher, J.H. and Ford, F.R., 1937, Colloid cyst of the third ventricle. Report of a case: operative removal with section of posterior half of the corpus callosum, *Arch. Neurol. Psychiatry* 37:959-973.

Twitchell, T.E., 1965, The automatic grasping responses of infants, *Neuropsychologia* 3:247-259.

Volpe, B.T., 1985, Observation of motor control in patients with partial and complete callosal section: implications for current theories of apraxia, *in:* "Epilepsy and the Corpus Callosum," A.G. Reeves, ed., Plenum Press, New York, pp. 381-391.

Volpe, B.T., Sidtis, J.J., Holtzman, J.D., et al., 1982, Cortical mechanisms involved in praxis: observations following partial and complete section of the corpus callosum, *Neurology* 32:645-650.

Wada, J. and Rasmussen, T., 1960, Intracarotid injection of sodium amytal for the lateralization of speech dominance: experimental and clinical observations, *J. Neurosurg.* 17:266-282.

Wilson, D.H., Reeves, A.G., and Gazzaniga, M.S., 1982, "Central" commissurotomy for intractable generalized epilepsy: series two, *Neurology* 32:687-697.

Yamaguchi, T., Kunimoto, M., Pappata, et al., 1990, Effects of anterior collosum section on cortical glucose utilization in baboons, *Brain* 113:937-951.

Zaidel, E. and Sperry, R.W., 1974, Memory impairment after commissurotomy in man, *Brain 97:263-272.*

Zecevic, N., Mojsilovic, J., Novakovic, B., and Rakic, Lj., 1989, Brain plasticity after corpus callosum transection in the newborn rat, *Metabolic Brain Disease* 4:25-31.

CALLOSOTOMY OR HEMISPHERECTOMY IN THE TREATMENT OF PATIENTS WITH INTRACTABLE SEIZURES AND HEMIPARESIS?

Frederick Andermann, M.D., F.R.C.P., Andre Olivier, M.D., Ph.D., F.R.C.S.(C), and Jean-Guy Villemure, M.D.,F.R.C.S.(C)

Montreal Neurological Institute and Hospital
3801 University Street
Montreal PQ H3A 2B4, Canada

Some years after hemispherectomy was shown to be effective in the treatment of intractable seizures in patients with hemiparesis, complications due to the development of cerebral hemosiderosis were being recognized.[5] Consequently, interest in this procedure and its utilization declined. At the same time the morbidity following central commissurotomy declined due to refinements in technique and restriction of the procedure to the callosum itself. The late Donald Wilson and his colleagues proposed disconnection of the cerebral hemispheres as an alternative to hemispherectomy for the control of intractable seizures.[2] One of Wilson's most striking successes was seen in a patient with infantile hemiplegia. The boy was living on a mattress on the floor because of recurrent drop attacks. Following callosotomy his seizures completely ceased.

Another recurrent argument for preferring callosotomy to hemispherectomy in the treatment of such patients is the widely held concept that parts of the damaged hemisphere might still subserve some useful function and that these functions might be preserved when a callosotomy rather than hemispherectomy is carried out.

Rasmussen,[1] and later Rasmussen and Villemure,[3] have shown that cerebral hemosiderosis can be prevented by providing some support to the opposite hemisphere after hemispherectomy. This is achieved by leaving the frontal and occipital poles in place, disconnected from the rest of the brain but with the blood supply intact. The disconnected tissue remains viable as can be seen from continuing electrical activity generated in these areas. Indeed sharp waves and spikes continue to originate in these regions although of course these discharges are not of clinical significance and they do not result in clinical manifestations.[4] This anatomically

Epilepsy and the Corpus Callosum II, Edited by A.G. Reeves
and D.W. Roberts, Plenum Press, New York, 1995

subtotal but functionally complete hemispherectomy has resulted in complete cessation of seizures in 75-80% of patients.[5] A follow-up of up to 14 years in some cases has not shown the major complication of cerebral hemosiderosis which appears usually within four years after complete hemispherectomy.

We have become aware of the necessity for detailed electrographic investigation and study of candidates for this form of treatment preoperatively. When there is no independent epileptogenic discharge from the normal or more normal side, one may expect a complete remission of attacks following operation. The presence of secondary generalized epileptic abnormality, usually in the form of irregular spike and wave, sometimes better formed over the more normal side, is not a contraindication and does not enter into this equation. Independent focal epileptogenic abnormalities however, may result in the continued occurrence of some seizures, usually a fraction of the original number. These attacks originate on the contralateral side. In two of our thirty patients functional hemispherectomy has not led to significant improvement in the seizure tendency. Both had fairly severe retardation, thus providing evidence of contralateral involvement without motor disability originating on that side. Their seizures most likely originated in both hemispheres independently.

In addition, following functional hemispherectomy, there is usually a considerable improvement in behavior and in social adaptation as originally described by Krynauw. There is also an increase in IQ, usually of the order of nine points. These changes are probably due to the removal of interference caused by the epileptogenic abnormality in the function of the "good" hemisphere. In addition, it is then possible to reduce, and in most patients, eventually to discontinue the antiepileptic medication leading to further improvement in intellectual functioning.[2] There is no evidence from postoperative neurological and neuropsychological studies that preservation of part of the damaged hemisphere leads to a higher level of functioning.

Rasmussen in his studies on subtotal hemispherectomy[1] has shown that preserving part of the damaged hemisphere significantly reduces the likelihood of stopping the attacks. Despite this, attempts have repeatedly been made to preserve the integrity of the visual field when it is present in these patients. These attempts often fail and in our experience the patients who have had subtotal resection of this type continue to have some attacks, often a fraction of their original number but requiring continuation of antiepileptic medication. The functional benefit of preservation of the visual field in the presence of a maximal hemiparesis has not been proven by postoperative evaluation.

Individuals in the fourth decade may require a shunt, even after functional hemispherectomy since absorption of CSF may not be as efficient in that age group compared to younger individuals.

In a review of the experience from various centres where callosotomy is carried out, presented in the First Palm Desert Symposium, Gates[6] has shown that only 5% of patients who undergo callosotomy have a complete and lasting cessation of attacks. In our experience with over 100 patients treated by callosotomy, complete cessation of attacks, as originally described by Wilson, was not achieved. The improvement may be very pronounced, especially after a maximal or as complete as possible resection, but even in patients who are seizure free postoperatively some attacks tend to recur later.(7) However, seizure tendency may respond much better than preoperatively to antiepileptic medication and we have found clobazam, a benzodiazepine with important antiepileptic properties, particularly useful in that way.

A major difference between our patient group and those reviewed by Gates may be due to the fact that we have not considered for callosotomy those patients who because of their preoperative neurological deficits were unlikely to have new or additional neurological impairment following functional hemispherectomy. These patients had no useful finger movement, walked with a hemiparetic gait and in most instances had a hemianopic field defect. We and Freeman and his group at Johns Hopkins have also considered functional hemispherectomy in patients with lesser motor disability but in whom the likelihood of developing maximal hemiparesis was great. These were patients with chronic encephalitis and epilepsy or Rasmussen's syndrome and patients with severe Sturge-Weber who developed increasing postictal neurological deficits and an eventual increase in the ongoing or permanent neurological abnormality. It is however, unwise to consider functional hemispherectomy in patients who have a lesser neurological defect and who are able to use their fingers or their hand well unless they have a clearly progressive disorder. In the long run, the increased postoperative neurological deficit may not be well tolerated by the family and by the patient despite the reduction and most often cessation of the seizures.

The possibility of staging the procedures, e.g., attempting a callosotomy first, with a view to carrying out a hemispherectomy later has been considered in some centres.[7,8,9] This, perhaps because of the hope that the damaged hemisphere may still contribute positively to the functioning of the individual, despite the available evidence that in patients with hemispheric damage of this type there is, if anything, an increase rather than a decrease in neurological functioning. Callosotomy remains a palliative procedure and despite Wilson's striking success with the patient he initially described, cessation of seizures should not be expected or predicted. Staging of surgical procedures for the treatment of epilepsy is best considered in patients where a small operation also holds out the possibility of complete cessation of the attacks. It should be an option particularly when a more extensive resection may be expected to lead to additional deficits which have to be weighed against the benefits. In patients whose neurological deficit is unlikely to be increased by functional hemispherectomy, a callosotomy should not be attempted first since it may lead to improvement but not to cessation of the attacks, which functional hemispherectomy brings about in a large percentage of patients.

When a widespread epileptogenic area leading to a secondary generalized epileptic disorder is present the team is also faced with the dilemma of section or resection. Removal of the epileptogenic area may be associated with increased and unacceptable neurological deficits. The results of callosotomy in such patients, in our experience, have been superior to those obtained in patients without evidence for lateralization or for a hemispheric epileptogenic abnormality. On the other hand, the development of secondary generalization is not necessarily a contraindication to resective surgery. In children with focal abnormalities and focal seizures who eventually develop hypsarhythmia and infantile spasms, which presuppose a generalized epileptic process, removal of the lesion may lead to cessation of infantile spasms even though these apparently represent the result of a generalized epileptic process.[10] The main problem in presurgical evaluation is to obtain a good electroclinical correlation between the clinical manifestations and electrical abnormalities. Only then is it possible to plan a rational strategy.

With increasing experience, the indications for the various surgical procedures used for the treatment of epilepsy are continuously being reassessed and refined. Specific indications for one or the other procedure are emerging. Surgery is unlikely to provide the ideal solution to some of these severe epileptic problems in the long run. However, until improved pharmacological approaches replace it, patients should not be deprived of the improvement in the quality of their lives which surgical treatment can bring about.

REFERENCES

1. T. B. Rasmussen, Hemispherectomy for seizures revisited, *Can. J. Neurol. Sci.* 10:71-78 (1983)
2. D.W. Wilson, C. Culver, M. Waddington, and M. Gazzaniga, Disconnection of the cerebral hemispheres: an alternative to hemispherectomy for the control of intractable seizures, *Neurology* 25:1149-1153 (1975).
3. T. B. Rasmussen and J.G. Villemure, Cerebral hemispherectomy for seizures and hemiplegia, *Clev. Clin. J. Med.* 56(1):S62-8; Discussion - S79-83 (1989).
4. S. Smith, F. Andermann, J.G. Villemure, T.B. Rasmussen, and L.F. Quesney, Functional hemispherectomy II: EEG findings, spiking from isolated brain, and prediction of outcome, *Neurol.* 41(11):1790-1794 (1991).
5. P. Tinuper, F. Andermann, J.G. Villemure, T.B. Rasmussen, and L.F. Quesney, Functional hemispherectomy for treatment of epilepsy associated with hemiplegia: rationale,indications, results and comparison with callosotomy, *Ann. Neurol.* 24(1):27-34 (1988).
6. J. Gates. Verbal communication at International Symposium on Surgical Treatment of the Epilepsies, Palm Desert, California, February, 1986.
7. H. Oguni, A. Olivier, F. Andermann, and J. Comair. Anterior callosotomy in the treatment of medically intractable epilepsies: a study of 43 patients with a mean followup of 39 months, *Ann. Neurol. 30(3):357-64 (1991).*
8. D.D. Spencer and S.S. Spencer, Corpus callosotomy in the treatment of medically intractable secondarily generalized seizures of children, *Clev. Clin. J. Med.*, 56(Supp.1):S69-78; Discussion, S79-83 (1989).
9. K.J. Sass, D.D. Spencer, S.S. Spencer, R.A. Novelly, P.D. Williamson, and R.H. Mattson, Corpus callosotomy for epilepsy. II: neurological and neuropsychological outcome, *Neurology* 38(1):24-8 (1988).
10. H.T. Chugani, W.D. Shields, D.A. Shewmon, D.M. Olson, M.E. Phelps, and W.J. Peacock, Infantile spasms: I: PET identified focal cortical dysgenesis in cryptogenic cases for surgical treatment, *Ann. Neurol.* 27(4):406-13 (1990).

MULTIPLE SUBPIAL TRANSECTION: A PHYSIOLOGICAL APPROACH TO EPILEPSY SURGERY

Frank Morrell, M.D.,[1] Michael Smith, M.D.,[1]
and Walter Whisler, M.D., Ph.D.[2]

[1]Department of Neurological Sciences
[2]Department of Neurosurgery
Rush Medical Center
1753 West Congress Parkway
Chicago, Illinois 60612

A discussion of multiple subpial transection (MST) in a text devoted to corpus callosotomy and its consequences requires some explanation. Callosotomy is the prototype of *disconnection* procedures in which the main goal is the prevention of spread of epileptic discharge from one hemisphere to the other. Elimination of this main route of interhemispheric communication reduces the likelihood that the two sides of the brain will be concurrently engulfed in paroxysmal activity. The occurrence of those particularly threatening symptoms (loss of consciousness, sudden falling or massive spasms) that depend on bihemispheric discharge is thereby prevented. What callosotomy does not do is interfere in any way with the generation of the primary epileptogenic disturbance.

On the contrary, MST has, as its principal therapeutic target the disruption of the capacity of the cortex to generate epileptiform modes of activity. The procedure achieves this goal by interrupting the tangential intracortical fibers which are necessary for sufficient synchronization to produce a paroxysmal "spike". Thus, MST inactivates the epileptic focus itself and removes the epileptogenic potential at its source. Accordingly, it prevents the focal, as well as the secondarily generalized seizures. Although, technically, MST may be said to be a disconnection procedure, the aim of the disconnection is quite different from that of callosotomy. Like callosotomy, however, MST does not involve the removal of cerebral tissue.

Ablation of cerebral tissue has the inevitable consequence that all functions subserved by that tissue are lost. Therefore, ablation of an epileptogenic focus that happens to lie in speech cortex or primary motor and sensory regions is not a viable option for patients with medically intractable seizures. In order to explicitly address this issue, we developed and have utilized a new surgical approach we call multiple subpial transection. The technique involves the selective interruption of the short, horizontal, intracortical fibers at 5 mm intervals along those gyri exhibiting electrocorticographically recorded epileptiform activity. Critical features of the procedure are the strict sparing of the main vertically oriented incoming and outgoing neural elements, minimal trauma to the cortical blood supply and preservation of an intact pial investment over the crown of the gyrus where it abuts the dura.

The rationale behind the introduction of this approach depended on three kinds of experimental observation. These were: (1) evidence from our own laboratory indicating that there was a critical mass of side-to-side connectivity that was necessary for the generation of an epileptic spike and that an important and distinctive mode of spread of epileptic activity, the Jacksonian march, involved tangential intracortical fibers (Morrell and Hanbery 1969; Wilder and Morrell 1970; Tharp 1971; Morrell 1985; see also Reichenthal and Hocherman 1977; Luders et al. 1981). The second set of observations (2) were those of Sperry and his colleagues in the mid-'50's (Sperry and Miner 1955; Sperry et al. 1955). They demonstrated that dicing the cortex from above did not interfere with learned behavior thought to depend on the cortex so treated. The third set of findings (3) were those of Mountcastle (1967), Hubel and Wiesel (1962), Asanuma and Sakata (1967) and Morrell (1967) demonstrating that the main functional attributes of cerebral cortical tissue depended upon vertically oriented cortical columns. We reasoned that a surgical method which disrupted horizontal or tangential fibers, but did not interfere with the basic columnar organization or the axons feeding into or out of the cortical columns might eliminate the capacity for epileptogenic discharge without impairing the principal physiological action of cerebral cortex. An extensive series of experiments with the rhesus monkey provided direct support for both elements of this hypothesis, i.e., for the dependence of epileptiform discharge on tangential connections and of normal function largely on the vertically oriented elements. Detailed evidence was obtained for both the primary motor cortex and for the temporal lobe. Finally, the procedure was extended to human cases of focal epilepsy. As is indicated below, our experience, thus far in man has also supported the basic hypothesis.

In this paper we review the outcome in 45 patients with a minimum two years of follow-up (range 2-25 years). Outcome evaluation must include at least two distinct features, namely, (1) the functional consequence of transection, and (2) the result with respect to seizure control. In our consideration of the functional outcome it is important to stress that these operations are carried out under local anesthesia in all patients except those under the age of nine (the Landau-Kleffner cases). Functional mapping by direct electrical stimulation (as well as the recording of somatosensory evoked potentials in the case of the postcentral gyrus) is the basis of our designation of eloquent cortex and the delineation of its boundaries (Morrell et al. 1989).

FUNCTIONAL OUTCOME

Five patients have undergone transection of Broca's area. None became mute or complained of any difficulty in communication. All could understand speech and could read or write. There was a mild decrease in verbal fluency when these patients were compared to age-matched normals.

Twenty patients had transection of Wernicke's area and the supramarginal and angular gyri. All continued to speak and to understand complex commands, all could read, write, name objects, repeat phrases and comprehend complex sentences. There was no evidence of apraxia. As with the Broca's area patients, quantitative testing disclosed some diminution in verbal fluency. It is important to stress that no patient, either the anterior or posterior language cortex groups, had a verbal deficit greater than that present preoperatively. The diminution in verbal fluency was only evident when the patient with speech area transection was compared with age-matched normals. Thus, all patients considered themselves, and were considered by their relatives and peers, improved in language function following surgery.

Furthermore, four children with Landau-Kleffner syndrome (acquired epileptic aphasia) who had been essentially mute for at least two years, underwent MST of the entire perisylvian speech cortex. All of these children are now speaking and understanding speech (Morrell et al. 1989, 1990). These latter cases demonstrate not only that the operative procedure leaves normal function intact, but that, in addition, by eliminating epileptogenic capacity, can actually restore physiological action of a cortical region that had become incapable of carrying out its normal role.

Twenty-three patients had transection of the pre-central gyrus. All continue to use the contralateral extremity for writing, eating and most other motor activities with at least the same degree of skill used preoperatively. (Some patients had a degree of hemiparesis and atrophy before operation).

One patient has undergone extensive quantitative evaluation of motor control parameters before and after transection. She has shown dramatic improvement over her preoperative status in both spatial and temporal control of joint movement at assessments made six months and one year following precentral transection.

Multiple subpial intracortical transection was carried out in the post central gyrus in 19 cases. No patient exhibited a loss of primary tactile modalities. In most (see below), there was no loss of two-point discrimination, stereognosis, barognosis or tactile reading. In most patients there was a moderate impairment of fine finger movement in the contralateral hand, definitely more pronounced than was ever seen with precentral transection. Some degree of neglect of the contralateral hand was seen in four cases. In two of them, the neglect disappeared entirely in six months, in one, it is detectable on careful examination (two years post-op), but not noted by the patient or his parents, and in one, the symptom has persisted for four years.

MORBIDITY

In six of the 45 patients, the operation was associated with a small, deep venous hemorrhage that caused injury to the vertically oriented elements and, therefore, produced transient loss of function. In one patient there was a small infarct and in another subcortical hemorrhage -- both resulting in a minor, but permanent deficit. In both cases the lesion occurred at some distance from the transection site and should be considered a complication of the surgery, but not specifically of transection.

OUTCOME WITH RESPECT TO SEIZURE CONTROL

Of the 45 patients whose outcome is summarized in this report, twenty-five patients (55%) are seizure-free, five (11%) have rare seizures, seven (16%) have a 90% reduction in seizure frequency and eight (18%) have had no worthwhile improvement. Thus, 82% had a worthwhile result. Since most patients had some resection as well as transection, however, one cannot attribute the outcome solely to the new procedure. Thirteen patients had transection only. Of the latter group, seven patients (53%) are seizure-free, one (7%) has rare seizures, two (15%) have 90% reduction and three (23%) are unimproved, i.e., 75% have had a worthwhile outcome. Mortality has been nil, and no patient is more impaired than before operation (including the two with minor permanent deficits). It is also worth emphasizing that MST was originally reserved for those patients with the most desperate and therapeutically recalcitrant of the epilepsies. Therefore, a third of our series proved to have progressive disease (Rasmussen's encephalitis, 9/45; tumor, 5/45; subacute sclerosing panencephalitis, 1/45). For such patients, the operation can not be curative; new epileptogenic foci develop in brain regions other than those operated on. But it is of considerable interest that when seizures eventually did recur in these cases, they arose in non-transected cortical areas - never in the transected zone.

SOME IMPORTANT TECHNICAL ISSUES

1. Elimination of Electrical Abnormality

It is important to emphasize the fact that the success of MST truly depends on the complete elimination of the electrocorticographically defined epileptogenic discharge. To the extent that the procedure fails to eliminate spiking so too will it fail to control the targeted seizures. Unfortunately, the converse, i.e., successful elimination of epileptiform discharge, is not invariably associated with therapeutic success - as is well known from experience with other electrocorticographic monitoring. Nevertheless, incomplete abolition of the epileptic discharge is definitely a poor prognostic sign. One of the causes of residual spikes in transected cortex is making the transections too far apart, i.e., having more than the 5 mm. interval between strokes (Morrell et al. 1989). A contiguous area larger than $25mm^2$ is capable of sufficient synchrony to generate paroxysmal discharge and of mediating transcortical spread.

A second major cause of residual spikes is an extensive epileptogenic source in the depths of a sulcus. Such deep sources are rare in the common traumatic lesions and glial or meningocerebral scars. Sulcal sources are more commonly found in the viral encephalitides or infiltrating tumors where the prognosis is limited in any case. Sometimes it is feasible to reach the deep sulci if the area to be transected is adjacent to a gyrus that can be safely removed. Such a removal can be designed so as to expose and make accessible to transection the deep sulcal wall of an eloquent epileptogenic gyrus. The mesial aspect of the pre or post central gyrus often has to be dealt with in just this fashion.

2. Use of Subpial Transection to Extend the Scope and Effectiveness of Standard Resective Surgery.

As alluded to above, many of our patients have had combination procedures in which a resection has been combined with subpial transection. The most common circumstance in which this combined approach has been taken is that of dominant temporal lobe lesions in which the epileptogenic discharge extends into Wernicke's area. In such cases we will often perform a relatively small anterior temporal resection followed by subpial transection of the epileptogenic zone in the posterior, superior and middle temporal gyrus. We believe that in this way we have often converted what would have been inadequate therapeutic effort into a successful one. In a similar fashion, an epileptogenic region of lateral frontal cortex may be treated with partial excision followed by subpial transection of adjacent but clearly electrically abnormal Broca's area or the precentral gyrus.

3. Bleeding

Subarachnoid bleeding always occurs since the capillary bed is invariably torn. Such venous oozing does not present any clinical hazard and has not led to any late development of irritative lesions in our experience. As noted above, however, occasionally a larger vein is inadvertently encroached upon causing a deep venous hemorrhage that may damage vertically oriented elements and thus produce clinical symptoms. Even these are usually transient but in three patients mild but permanent deficits have occurred. In general, however, the procedure is remarkably safe since all trauma occurs in the outer grey matter far away from major vessels and from the major afferent and transcortical pathways.

Thus MST permits an extension of surgical treatment to brain regions heretofore considered inoperable and may even be entertained as a first step in the surgical therapy of any accessible cortex in which the nature of the pathology does not mandate excision.

REFERENCES

Asanuma, H. and Sakata, H., 1967, Functional organization of a cortical efferent system examined with focal depth stimulation in cats, *J. Neurophysiol.* 30:35-54.

Hubel, D.H. and Wiesel, T.N., 1962, Receptive fields, binocular interaction and functional architecture in the cat's visual cortex,*J. Physiol.* 160:106-154.

Luders, H., Bustamente, L., Zablow, L., et al., 1981, The independence of closely spaced discrete experimental spike foci, *Neurology* 31:846-851.

Morrell, F., 1967, Electrical signs of sensory coding, *in:* "The Neurosciences, a Study Program," G.C. Quarton, T. Melnechuk, and F.O. Schmitt, eds., Rockefeller Univ. Press, New York, pp.452-469.

Morrell, F., 1985, Callosal mechanisms in epileptogenesis: Identification of two distinct kinds of spread of epileptic activity, *in:* "Epilepsy and the Corpus Callosum," A.G. Reeves, ed., Plenum Press, New York, pp. 99-130.

Morrell, F. and Hanbery, J.W., 1969, A new surgical technique for the treatment of focal cortical epilepsy, *Electroencephalogr. Clin. Neurophysiol.* 26:120.

Morrell, F., Cooper, M., Ali, A., Smith, M.C., Pierre-Louis, S.J.C., and Whisler, W.W., 1990, Landau-Kleffner syndrome: Blood flow, metabolism and electrophysiologic studies, *Epilepsia* 31:672.

Morrell, F., Whisler, W.W., and Bleck, T.P., 1989, Multiple subpial transection: A New approach to the surgical treatment of focal epilepsy, *J. Neurosurg.* 70:231-239.

Morrell, F., Whisler, W.W., Smith, M.C., Pierre-Louis, S.J.C., Brocken, C., Ali, A., Cooper,, M., and Andrews, R.V., 1989, Landau-Kleffner syndrome: Treatment with multiple subpial transection, *Epilepsia* 30:693.

Mountcastle, V.B., 1957, Modality and topographic properties of single neurons of cat's somatic sensory cortex, *J. Neurophysiol.* 20:408-434.

Reichenthal, E. and Hocherman, S., 1977, The critical cortical area for develoment of penicillin-induced epilepsy, *Electroencephalog. Clin. Neurophysiol.* 42:248-251.

Sperry, R.W. and Miner, N., 1955, Pattern perception following insertion of mica plates into visual cortex, *J. Comp. Physiol. Psychol.* 8:463-469.

Sperry, R.W., Miner, N., and Myers, R.E., 1955, Visual pattern perception following subpial slicing and tantalum wire implantations in visual cortex, *J. Comp. Physiol. Psychol.* 48:50-58.

Tharp, B.R., 1971, The penicillin focus: A study of field characteristics using cross-correlation analysis, *Electroencephalog. Clin. Neurophysiol.* 31:45-55.

Wilder, B.J. and Morrell, F., 1967, Cellular behavior in secondary epileptic lesions, *Neurology* 17:1193-1204.

IV

NEUROPSYCHOLOGY

HEMISPHERIC SPECIALIZATION AND INTERHEMISPHERIC INTEGRATION: INSIGHTS FROM EXPERIMENTS WITH COMMISSUROTOMY PATIENTS

M.J. Tramo, M.D.,[1] K. Baynes, Ph.D.,[2] R. Fendrich, Ph.D.,[2]
G.R. Mangun, Ph.D.,[2] E.A. Phelps, Ph.D.,[3]
P.A. Reuter-Lorenz, Ph.D.,[4] and M.S. Gazzaniga, Ph.D.[2]

[1]Department of Neurobiology, Harvard Medical School;
Neurology Service, Massachusetts General Hospital

[2]Center for Neuroscience
University of California at Davis

[3]Department of Psychology
Yale University

[4]Department of Psychology
University of Michigan

INTRODUCTION

During the decade following the first edition of this book, methodological advances in a number of related disciplines have greatly enriched our fund of knowledge concerning the role of the corpus callosum in higher brain functions. At the microanatomical level, new tracer techniques have provided a detailed picture of interhemispheric connectivity patterns among functionally specialized cortical areas in anthropoids (for review see Pandya and Seltzer 1986). At the gross anatomical level, magnetic resonance imaging (MRI) has provided multiplanar, high-resolution images of the living human brain by which to: 1) analyze callosal morphometry; 2) delineate precisely the extent of transection in commissurotomy patients; and 3) correlate *in vivo* the extent of transection with functional effects (e.g., Gazzaniga et al. 1985, 1989; Oppenheim et al. 1989). At the behavioral level, the facility with which microcomputers can be used to generate stimuli, manipulate

Epilepsy and the Corpus Callosum II, Edited by A.G. Reeves
and D.W. Roberts, Plenum Press, New York, 1995

specific stimulus features, monitor a variety of response modes, and measure response latencies has enhanced the fine-grained assessment of cognitive, perceptual, and motor functions in neurological patients. Meanwhile, increasing numbers of collaborations among cognitive psychologists and clinical neuroscientists have provided the impetus to combine these methods in pursuit of brain-behavior correlations.

Several goals motivate experiments with commissurotomy patients. One is to determine the degree to which the hemispheres differ in their capacity to perform a particular function. This goal has received a great deal of attention since the early years of split-brain research. The split-brain paradigm differs from the classical lesion approach to studies of hemispheric specialization in that each hemisphere is structurally intact and can therefore be examined as a whole. Consequently, functional asymmetries can be assessed without drawing inferences about the function of a damaged part. Moreover, the contribution of intrahemispheric disconnection to the observed deficit is not at issue. While caution is warranted in the face of long-standing, medically-intractable epilepsy, the sensitivity of MRI to previously occult epileptogenic lesions allays at least some of the concern that the hemispheres of commissurotomy patients harbor structural abnormalities. A second goal, which is also pursued in studies of patients with intrahemispheric lesions, is to establish the dissociability of functionally specialized subsystems in order to reconstruct the modular organization of the brain. For example, lesion localization studies in neurological patients have long since established that the left hemisphere is dominant for language in the vast majority of humans. However, language itself can be broken down into a number of subcomponents, e.g., phonology, syntax, and semantics. Furthermore, language functions can be accessed via separate sensory modalities, e.g., the auditory modality in all neurologically normal humans and the visual one in literate individuals. Is the left hemisphere dominant to the same degree for each functional subcomponent and each sensory modality? If the right hemisphere of a split-brain patient demonstrates competence in one subcomponent of language, will it demonstrate competence in all? Answers to these questions provide insights into mechanisms of interhemispheric integration, a third topic of interest in split-brain research. For example, if the subcomponents of a particular function are differentially represented in the left and right hemisphere, it is likely that overt manifestations of that function as a whole involve interhemispheric integration. A fourth goal is the elucidation of cortical-subcortical interactions. For example, if cooperation between the hemispheres is observed despite complete commissurotomy, it is likely that the task at hand entails the participation of paracallosal pathways and/or subcortical structures. The nature of cortical-subcortical interactions has been of particular interest in studies of visuomotor integration. A fifth goal is to determine the functional specificity of the corpus callosum along its rostrocaudal extent. This can be achieved by studying patients who have undergone partial commissurotomy and patients in whom a small part of callosum has been inadvertently spared during surgery. MRI has contributed greatly to this last endeavor.

In the present chapter, we review experiments with ten commissurotomy patients that were recently carried out in our laboratory.

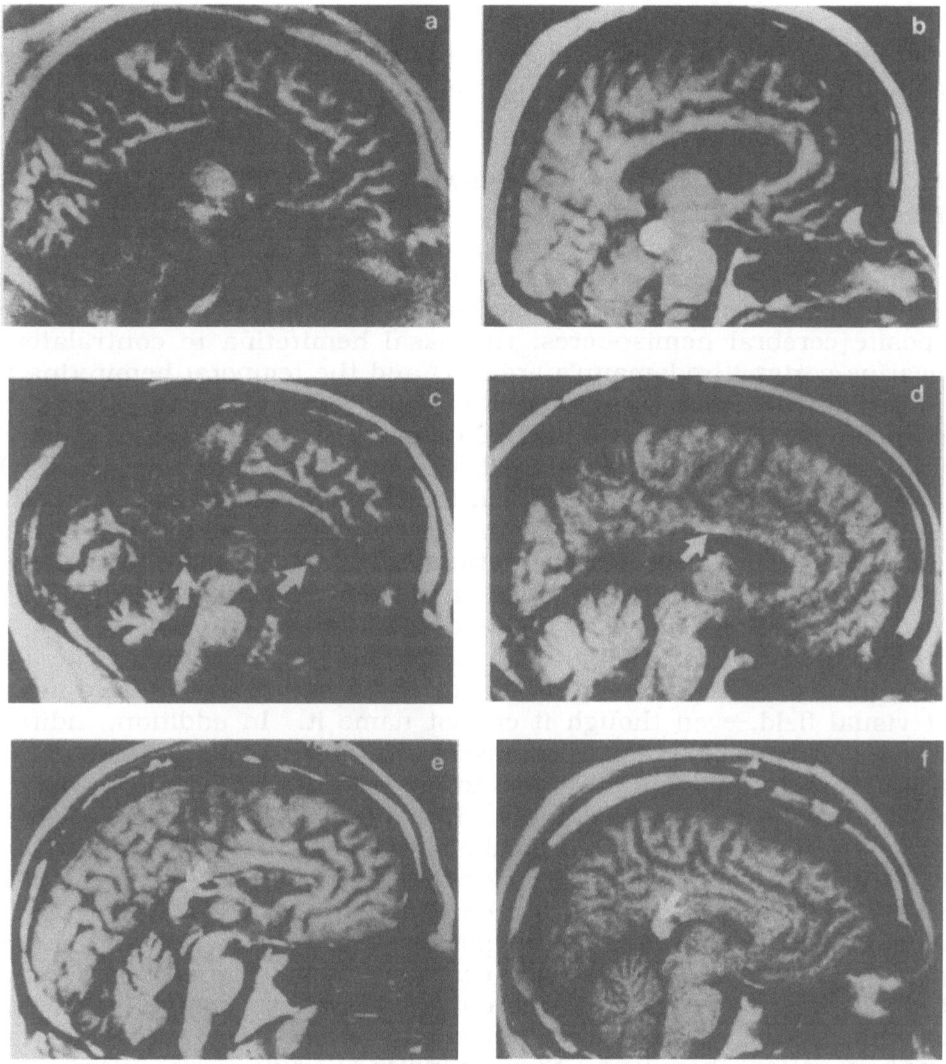

Figure 1. T1-weighted mid-sagittal MRI scans of six of the ten commissurotomy patients who participated in the experiments discussed in this chapter. **a.** J.W. The callosum and hippocampal commissure are transected. **b.** D.R. The callosum, except for a small remnant at the inferior rostrum, and the hippocampal commissure are transected. The circinate, hyperintense signal in the region of the quadrigeminal cistern probably represents a lipoma. **c.** V.P. The callosum, except for small remnants at the extremes of the splenium and rostrum (arrows), and the hippocampal commissure are transected. **d.** E.B. The splenium, approximately half of the posterior body of the callosum, and the hippocampal commissure are transected. The arrow demarcates the anterior border of the lesion. **e.** G.M. The rostrum, genu, and most of the body of the callosum are transected; the splenium (arrow) and hippocampal commissure are spared. **f.** S.C. The rostrum, genu, and most of the body are transected; the splenium (arrow) and hippocampal commissure are spared. In all cases, the anterior commissure was spared. Operative notes as well as MRI scans were used to estimate the extent of commissurotomy. (Reprinted from Phelps, et al. 1991, Deficits in recall following partial and complete commissurotomy, *Cerebral Cortex* 1:494, with permission from Oxford University Press.)

Mid-sagittal MRIs of six of these patients are illustrated in Figure 1. The neurological histories of nine have been previously described (Gazzaniga and Sperry 1967; Tramo and Bharucha 1991; Phelps et al. 1991; Baynes et al. 1992). Separate sections of the chapter are devoted to one or more experiments germane to major topics of interest in cognitive neuroscience: vision, audition, attention, language, and memory.

INTERHEMISPHERIC INTEGRATION OF VISUAL INFORMATION: BEHAVIORAL EXPERIMENTS

In the mammalian visual system, the nasal and temporal hemiretinas of each eye project to opposite lateral geniculate bodies and hence to opposite cerebral hemispheres: the nasal hemiretina to contralateral calcarine cortex (Brodamann's area 17) and the temporal hemiretina to ipsilateral calcarine cortex. Thus, the left hemisphere receives information from the right visual field and the right hemisphere receives information from the left visual field. Since the principal route by which visual information transfers between the hemispheres in humans is the splenium, patients with posterior or complete transections of the callosum cannot judge whether stimuli flashed tachistoscopically to opposite visual fields are the same or different (Gazzaniga et al. 1965). This is one of the hallmarks of the callosal disconnection syndrome.

However, this visual isolation may not be absolute. The left hemisphere often reports that it has detected a stimulus presented in the left visual field, even though it cannot name it. In addition, indirect measures of visual transfer, which assess the ability of one hemisphere to utilize visual information presented to the other, have sometimes yielded positive results. For example, Holtzman (1984) observed that commissurotomy patients can combine information from the two visual hemifields when they make saccades to a visual target. Sergent (1987) reported that some commissurotomy patients can make cross-field judgments concerning the relative positions of stimuli. Trevarthen and Sperry (1973) reported that some commissurotomy patients performed above chance on cross-field matching tasks (e.g., matching of line orientation) when stimuli were sufficiently large and presented in the periphery of each hemifield. Since transfer of crude visual information has been reported in patients in whom both the callosum and anterior commissure have been sectioned, it has been proposed that paracallosal transfer is mediated by subcortical pathways, such as the commissure of the superior colliculus (Holtzman 1984; Myers and Sperry 1985; Sergent 1990; see also Seymouth et al. 1994 for an alternative view).

We recently examined the interhemispheric integration of information about orientation, form, luminance, and position in patient J.W. In an attempt to preserve stimulus discriminability despite any degradation of stimulus features that might occur during paracallosal transfer, we utilized four types of elementary stimuli that were defined largely by low spatial frequencies (i.e., stimulus identification did not require spatially detailed information; Table 1). A minimum of 64 trials were run with each type of stimulus using a two-alternative forced-choice paradigm. In all trials, stimulus pairs were simultaneously presented for 200 ms, one to each visual hemifield with its medial edge 2° from the point of central fixation. J.W. indicated whether the stimulus in the right visual field and the stimulus in the left visual field were the "same" or "different" by pressing the corresponding key on a computer keyboard. The side of the response hand was counterbalanced across trials.

Table 1. Stimuli used to examine interhemispheric integration of visual information.

Stimulus	Attribute To Be Matched
Low frequency sine wave grating patches, horizontal vs. vertical (0.5 and 1 cpd, 4° and 2° diameter gabors, 66% center contrast)	orientation
4° x 4° figures drawn with 1°-thick black lines (e.g. "x" vs. "+")	form
2° white (80 cd/m^2) vs. black (<1 cd/m^2) square patches against a grey (30 cd/m^2)background	luminance
2° square white patches offset vertically +/- 3°	relative position

Performance was at chance for stimuli varying with respect to orientation, form, and luminance. The side of the response hand did not significantly affect accuracy. When both stimuli were presented in the same visual hemifield, each hemisphere was able to perform same-different discriminations at near-perfect levels. These results indicate that information about orientation, form, and luminance cannot be transferred from one hemisphere to the other following callosal transection.

For the relative displacement task, J.W. was correct on 78 of 128 trials (61%). This level of accuracy was marginally better than chance. This finding suggests limited paracallosal transfer of visual information concerning the relative position of objects.

We subsequently repeated the orientation task with patient V.P. As with J.W., her performance was at chance.

We questioned whether the lack of evidence for coarse visual transfer for orientation, form, and luminance might be related to the combined effects of: 1) short stimulus duration; and 2) the poor signal-to-noise ratio caused by degradation of stimulus features during paracallosal transfer. We reasoned that longer stimulus durations might be required to provide the disconnected hemisphere with enough information to make a correct judgment. However, stimulus durations less than 200 ms - i.e., shorter than the latency of saccadic eye movements - are necessary to ensure that commissurotomy patients cannot move their eyes in time to avail the stimulus to both hemispheres. To overcome this problem, we employed a computer-controlled Purkinje image eyetracker and mirror image-stabilizer system. The stabilizer moves images being viewed in tandem with the subject's eye movements. The system thus holds image positions fixed on the desired part of the retina so that any attempt by the subject to saccade toward a target does not affect the side or site of cortical stimulation. Consequently, we were able to present stimuli over much longer durations. Upon re-examining the transfer of information about line orientation, we presented patches of square wave gratings for 5s each; the patches were oriented horizontally or vertically and were 2° in diameter with fundamental spatial frequencies of 1, 2, 4, or 8 cycles per degree (cpd). Again, the medial edges of the patches were 2° from the vertical meridian. J.W. performed 48 or more trials at each spatial frequency. As with short stimulus durations, performance at matching orientations between the visual fields was at chance. For the four spatial frequencies employed, J.W.'s accuracy rates were 38%, 52%, 54%, and 42%, respectively. Therefore, the previously observed failure of J.W. to transfer information about orientation from one hemisphere to the other cannot be attributed to short stimulus duration. This conclusion probably applies to the results obtained with V.P. and to those obtained using the other types of stimuli; we are presently collecting data to test these predictions.

These experiments provide little evidence for the transfer of crude visual information between the disconnected hemispheres, at least for stimuli occupying between 2-6° of the visual field lateral to the point of central fixation along the vertical meridian. Given the borderline significance of our findings concerning relative position, more data are needed concerning this type of visual information. If some visual impression is conveyed paracallosally between the hemispheres, it is very informationally impoverished. Myers and Sperry (1985) have observed that when there is communication between the disconnected hemispheres, the information transferred appears to be "largely connotative, contextual or orientational [spatial] in nature." Such information may be descriptive but not genuinely pictorial. In accord with this supposition, the current data obtained using stimuli with low spatial frequencies indicate that the idea of a dim and/or blurry image does not really apply to the kinds of visual information that might be transferred paracallosally.

INTERHEMISPHERIC INTEGRATION OF VISUAL INFORMATION: ELECTROPHYSIOLOGICAL EXPERIMENTS

Scalp-recorded event-related potentials (ERPs) that reflect the processing of visual information by the occipital cortex provide an objective means of assessing interhemispheric visual transfer in commissurotomy patients. The P1 component (90-130 ms latency) and N1 component (160-190 ms latency) of visual ERPs can be evoked by hemifield stimulation and recorded from scalp electrodes located over both the contralateral and ipsilateral occipital cortex (for review see Regan 1990). For example, stimulation of the left visual hemifield evokes ERPs from the right occipital cortex and, after a delay of 5-15 ms, from the left occipital cortex (Rugg et al. 1984; Mangun and Hillyard 1988; Saron and Davidson 1989). One hypothesis that might account for this latency difference between contralateral and ipsilateral responses is that neural impulses generating ERPs at ipsilateral scalp sites are propagated over longer distances than those generating ERPs over contralateral scalp sites. Thus, the shorter latency P1 and N1 components might be generated directly via the retino-geniculo-calcarine pathway, while the longer latency P1 and N1 components might be generated via the retino-geniculo-calcarine pathway plus callosal projections connecting contralateral visual cortex to ipsilateral visual cortex. Alternatively, the latency differences between contralateral and ipsilateral P1 and N1 components might indicate that subcortical projections are involved, or that neural generators in the contralateral visual cortex contribute to the responses recorded over both hemispheres (so-called "paradoxical lateralization").

In order to evaluate these alternative hypotheses, we recently measured visual ERPs in complete commissurotomy patients L.B. and J.W. (Mangun et al. 1991). In both patients, the latencies and amplitudes of the contralateral P1 and N1 components were within normal limits, but the ipsilateral P1 and N1 components were absent, indicating that the integrity of the corpus callosum is essential to the generation of these ipsilateral ERPs. These results are consistent with the previous findings of Rugg et al. (1984), who showed that the ipsilateral N1 component is absent in patients with agenesis of the corpus callosum.

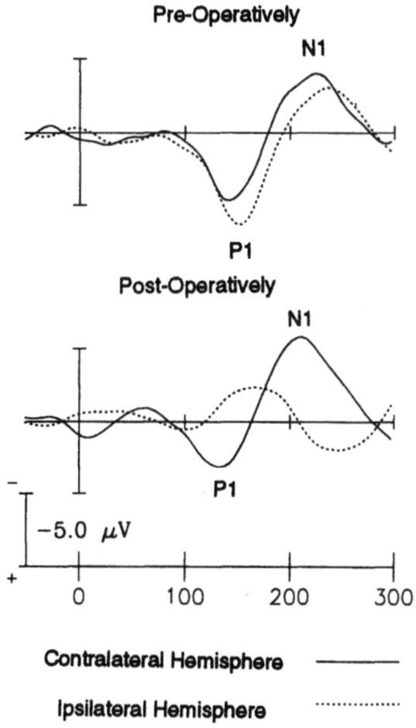

Figure 2. Visual ERPs evoked in patient L.A. by stimuli presented in the left visual field before (top) and after (bottom) transection of the splenium and posterior body of the callosum. Waveforms recorded from lateral occipital scalp sites contralateral (solid traces) and ipsilateral (dotted traces) to the stimulated hemifield are superimposed on one another. Pre-operatively, L.A. displayed a normal pattern for the P1 and N1 components: contralateral peaks with ipsilateral peaks of slightly longer latency (10-15 ms). Post-operatively, L.A. still produced contralateral occipital scalp P1 and N1 components, but there was an absence of these ERPs at ipsilateral scalp sites.

In a follow-up study, we measured visual ERPs in patient L.A. before and after the posterior completion of his two-stage callosotomy. While L.A. fixated a central point, we presented vertically oriented bars (2.0^{o} x 1.0^{o}) one at a time in the two hemifields. Stimuli were 100 ms duration and were separated by 200-500 ms; the location in which the stimulus was presented and the interstimulus interval were varied randomly. In one out of six stimulus presentations, the bar was blue; in the others it was red. The order of stimulus presentations was randomized. L.A. was instructed to press a button with his left hand when a red bar appeared in the left visual hemifield and another button with his right hand. When a red bar appeared in the right hemifield. Eye movements were monitored by

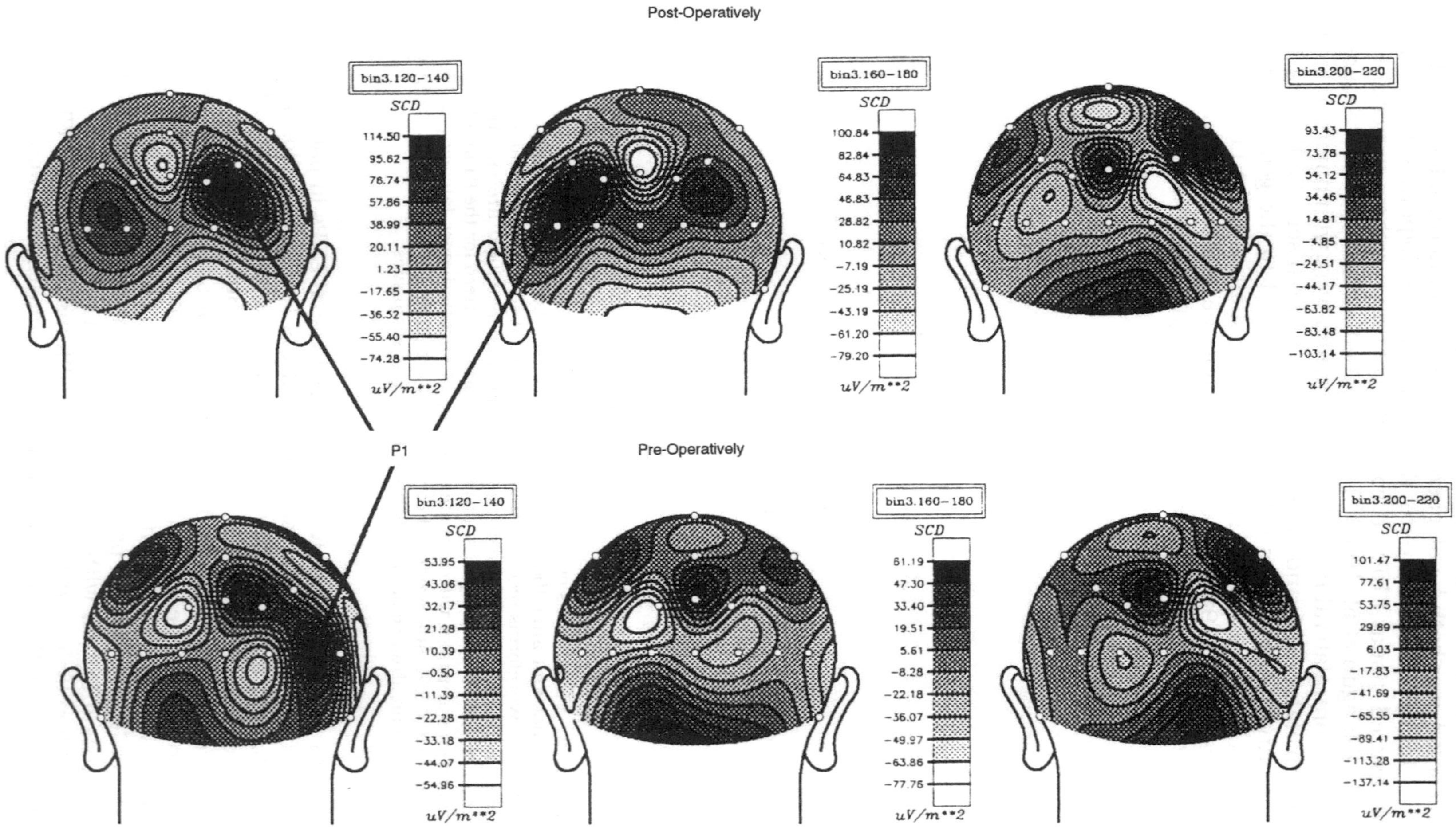

Post-Operatively
bin3.120-140
SCD
114.50
95.62
76.74
57.86
38.99
20.11
1.23
-17.65
-36.52
-55.40
-74.28
uV/m**2
bin3.160-180
SCD
100.84
82.84
64.83
46.83
28.82
10.82
-7.19
-25.19
-43.19
-61.20
-79.20
uV/m**2
bin3.200-220
SCD
93.43
73.78
54.12
34.46
14.81
-4.85
-24.51
-44.17
-63.82
-83.48
-103.14
uV/m**2
P1
Pre-Operatively
bin3.120-140
SCD
53.95
43.06
32.17
21.28
10.39
-0.50
-11.39
-22.28
-33.18
-44.07
-54.96
uV/m**2
bin3.160-180
SCD
61.19
47.30
33.40
19.51
5.61
-8.28
-22.18
-36.07
-49.97
-63.86
-77.76
uV/m**2
bin3.200-220
SCD
101.47
77.61
53.75
29.89
6.03
-17.83
-41.69
-65.55
-89.41
-113.28
-137.14
uV/m**2

electro-oculography and by direct inspection of a magnified video image.

Figure 2 shows ERPs evoked by left-field stimuli before and after posterior completion of his callosotomy. Following transection of the anterior two-thirds of the callosum, left visual field stimuli elicited P1 and N1 components over the contralateral occipital scalp and, with a slight delay, over the ipsilateral scalp. Following posterior extension of the transection, the contralateral P1 and N1 components remained intact, but the ipsilateral P1 and N1 components were abolished.

Figure 3 shows topographic maps that display the scalp distributions of current source densities over successive points in time after the onset of the visual stimulus. Before transection of the splenium and posterior body, the pattern of cortical activation was similar to that reported for normal subjects: a contralateral current source maximum was followed by an ipsilateral occipital scalp current source maximum. After posterior transection, this secondary ipsilateral maximum was absent.

Our findings in L.A., L.B., and J.W. strongly support the hypothesis that the posterior third of the corpus callosum mediates activation of visual cortex ipsilateral to the stimulated hemifield. Moreover, the observation that ipsilateral visual ERPs were preserved in L.A. following anterior callosotomy argues against the possibility that the results in L.B. and J.W. were merely the consequence of long-standing epilepsy and/or anticonvulsant treatment.

CORTICAL REPRESENTATION OF THE VISUAL FIELD: THE VERTICAL MERIDIAN

In cats and monkeys, some neurons lying along the border of areas 17 and 18 project to the corresponding region in the contralateral hemisphere via the corpus callosum (for reviews see Berlucchi 1972; Hubel 1988; Payne and Siwek 1991). In addition, some of the cells in this region have receptive fields that overlap the vertical meridian of the visual field. If the optic chiasm is cut, these neurons maintain binocular receptive fields that overlap the vertical meridian, such that stimulation of the left eye defines the right side of the receptive field and stimulation of the right

Figure 3. Topographic current density maps of scalp activity evoked in L.A. by left field stimuli. Each map was interpolated using a spherical spline algorithm and represents current flow perpendicular to the skull and scalp. Current sources (dark) and sinks (light) related to the P1 and N1 components are illustrated. *Top:* Before transection of the splenium and posterior body, there is a current source over the contralateral (right) occipital scalp at 120-140 ms latency (left panel). This is followed by current source over the ipsilateral (left) occipital scalp at 160-180 ms (middle panel). Finally, these contralateral and ipsilateral P1 sources are followed at 200-220 ms by a source-sink pair reflecting the N1 component over contralateral (right) scalp (right panel). *Bottom:* After transection of the splenium and posterior body, there is still a contralateral occipital current source at 120-140 ms (left panel); however, the ipsilateral occipital current source fails to appear by 160-180 ms (middle panel) or even 200-220 ms (right panel). These current density topographies preferentially favor nearby brain electrical activity and provide a reference-independent view of scalp-recorded ERPs.

eye the left (Berlucchi and Rizzolatti 1968). Direct evidence that almost all splenial fibers are excited by stimuli presented within 4° of the vertical meridian has been obtained in the cat via microelectrode recordings (Berlucchi et al. 1967; Hubel and Wiesel 1967). Thus, the callosum provides a cohesive representation of the two visual hemifields that is securely joined along the vertical meridian. An additional mechanism exists in cats and monkeys by which the vertical meridian of the visual field could be represented in both hemispheres: there is a narrow zone at the *retinal* vertical midline where ganglion cells projecting to the left and right lateral geniculate bodies intermingle (Stone 1966; Bunt and Minckler 1977; Leventhal et al. 1988). Such cells could allow for a representation of the vertical meridian in each hemisphere that is independent of callosally-mediated integration.

Similar mechanisms could account for the phenomenon of macular sparing in patients with hemianopia secondary to lesions of the visual cortex. Unfortunately, histological and electrophysiological data that would support this hypothesis are unavailable in humans. Moreover, psychophysical investigations of normal human subjects have consistently failed to find evidence of a dual representation of the vertical meridian (Harvey 1978; Haun 1978; Lines and Milner 1983). Furthermore, overlap in the blood supply to caudal occipital cortex from the middle and/or posterior cerebral arteries might account for at least some cases of macular sparing following stroke.

If the cortical representation of the vertical meridian were bilateral in humans, and if it did not arise solely on the basis of callosal projections between the left and right visual cortices, it would be possible for commissurotomy patients to match stimuli presented at the vertical meridian with stimuli presented in either visual hemifield. We recently carried out such an experiment with patient V.P. (Fendrich and Gazzaniga 1989). One simple geometric form (the "reference" stimulus, e.g., a square or triangle) that subtended 30' of visual arc was presented 2.5° from the vertical meridian for 1200 ms; 1000 ms after the onset of the latter stimulus, a second form (the "target" stimulus) that subtended 15' was presented for 200 ms in the opposite hemifield only 15', 30', or 1° from the vertical meridian. We monitored eye movements with a Purkinje eyetracker and eliminated all trials in which the eyes deviated from fixation by more than 30' during presentation of the reference or more than 7' during presentation of the target. V.P. indicated whether the reference and the target were the "same" or "different" by pressing the corresponding key on a computer keyboard. The side of the response hand was counterbalanced across trials.

The data are summarized in Table 2. V.P.'s performance was not significantly better than chance on between-field comparisons. In contrast, accuracy was near perfect when the reference and the target were presented within the same hemifield. The side of the response hand had no effect on response accuracy. These results argue against a bilateral cortical representation of the vertical meridian, or at least a bilateral cortical representation that is independent of callosally mediated interhemispheric integration.

However, we noticed that in none of the six between-field conditions did V.P.'s accuracy fall below 50%. Furthermore, when the reference

Table 2. Within-field and between-field comparisons of form at the vertical meridian.

	Reference in Right Visual Field			
Visual Field of Target	Right	Left	Left	Left
Distance from Vertical Meridian	15'	15'	30'	1°
Accuracy (N≥96)	0.96	0.56	0.58	0.60
	Reference in Left Visual Field			
Visual Field of Target	Left	Right	Right	Right
Distance from Vertical Meridian	15'	15'	30'	1°
Accuracy (N≥96)	0.85	0.65	0.56	0.50

Table 3. Between-field comparisons of orientation at the vertical meridian.

Spatial Frequency (cpd)	1	2	4	8
Response Accuracy (N=120)	0.62	0.80	0.69	0.63

was 15' within the left visual field, accuracy on between-field comparisons 64.6%) was only about 20% below that on within-field comparisons (85.4%). These observations raised the possibility that visual information presented *at* the vertical meridian was available to both hemispheres but that the quality of the information was not adequate for V.P. to make the required comparison between different geometric forms. More recently, we have obtained data in patient J.W. that fit with this possibility. Using the computer-controlled Purkinje image eyetracker and mirror image-stabilizer system to lateralize stimuli for long durations, we presented pairs of square wave gratings on opposite sides of the midline for 5s. We also ran trials using the customary stimulus duration, 200 ms. Each grating was 2° diameter with its medial edge 1° lateral to the intersection of the vertical and horizontal meridians. Spatial frequencies were systematically varied at 1, 2, 4, and 8 cpd. One hundred twenty trials were carried out at each stimulus duration and spatial frequency. J.W.'s left hemisphere stated whether the orientations of the gratings (horizontal vs. vertical) were the "same" or "different."

When the stimulus pairs were presented for 200 ms, J.W. performed no better than chance (42% at 4 cpd; 52% at 2 cpd). However, when the stimulus pairs were presented for 5s, performance was significantly above chance (Table 3).

These results indicate that in the absence of callosal transfer, each hemisphere has access to degraded visual information from a region that extends about 1° across the vertical meridian. The animal data cited above suggest this may be due to a zone at the retinal vertical midline that gives rise to uncrossed nasal and crossed temporal projections. We are currently investigating whether this zone of nasotemporal overlap: 1) is uniform along the vertical meridian; 2) is symmetrical with respect to the horizontal meridian; and 3) expands or contracts in the region subserving central vision.

VISUAL ATTENTION

Since visual information lateralized within the left and right hemifields is relayed to opposite hemispheres by the retino-geniculo-calcarine pathways, shifting visual attention from one hemifield to the other across the vertical meridian requires: 1) interhemispheric integration via the corpus callosum; 2) subcortical integration of retino-tectal input; and/or 3) cortical-subcortical integration. In order to assess the role of the callosum in lateralized shifts of visual attention, we recently carried out a series of experiments with patients S.C., J.W., and V.P. (Reuter-Lorenz and Fendrich 1990). We employed Posner's spatial pre-cueing task, which had been previously used to study attentional control in normal subjects (e.g., Posner 1980) and patients with focal unilateral and bilateral hemispheric lesions (e.g., Posner et al. 1984; Tramo et al. 1989).

Each patient fixated a central dot on a computer screen that displayed four empty squares. The squares were aligned symmetrically along the horizontal axis, two on each side of the fixation point. After a warning tone, one of the squares brightened for 200 ms. After an interval of 500 ms, a target ("X") appeared in one of the squares. In 70% of trials, the "X" appeared in the square that had been brightened. Since the target appeared in the spatial location to which the patient had been cued, we term these "valid" trials. In 16% of trials, the "X" appeared in one of the other three squares; we term these "invalid" trials. There are thus two types of invalid trials: one in which the "X" appears in the same hemifield as the cue ("within-field/invalid trials"), and one in which the "X" appears in the hemifield contralateral to the cue ("between-field/invalid trials"). In the remaining 14% of trials, no target appeared; these "catch" trials were included to discourage the patients from responding to the onset of the cue. Each patient was asked to press a button as quickly as possible when the target "X" appeared anywhere on the screen. Reaction time and response accuracy were recorded on-line via computer. Eye movements were monitored via direct inspection of a magnified video image. The side of the response hand was counterbalanced across blocks of trials.

For normal subjects, reaction times are significantly shorter when the target appears at a cued location than when it appears at a non-cued location (Posner 1980). Since the eyes remain fixated on the central point throughout each trial, the difference in reaction times between valid and invalid trials reflects the effect of attention on target detection. We refer to this difference as the "cost" of attending to the wrong location. This cost is typically in the range of 50-100 ms. To compare intrahemispheric versus interhemispheric mechanisms of attentional control, one can measure the costs associated with the two different types of invalid trials: within-field versus between-field, respectively. In within-field/invalid trials, only one hemisphere is stimulated; consequently, it is unlikely that interhemispheric integration is required by the task. In between-field/invalid trials, the cue and target appear on opposite sides of the vertical meridian; consequently, it is possible that interhemispheric integration is required by the task. Thus, if the callosum mediates the interhemispheric integration of visual attention, commissurotomy patients should show much greater costs on across-field/invalid trials than on within-field/invalid trials.

Figure 4 presents the results of our experiments in terms of the costs associated with within-field/invalid trials and between-field/invalid trials. Costs are calculated by taking the difference between the mean reaction times for each invalid condition minus the mean reaction time for the valid condition. On within-field/invalid tri-als, J.W. and V.P. showed costs that are within the range of the controls, indicating that intrahemispheric mechanisms of attentional control were operating at normal levels. However, on between-field/invalid trials, costs were significantly greater for J.W. and V.P. than for the controls. This result indicates that rapid shifts of attention between the two hemifields relies on the integrity of the corpus callosum. S.C., in whom the anterior two-thirds of the callosum was transected, showed normal performance on across-field/invalid trials as well as within-field/invalid trials. S.C.'s performance suggests that the posterior third of the corpus callosum plays a critical role in the efficient allocation of attention between the left and right visual fields.

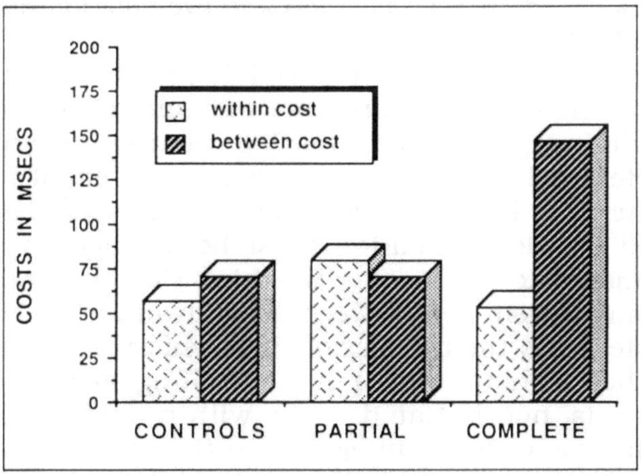

Figure 4. Costs for within-field/invalid trials and between-field/invalid trials for normal subjects and for patients S.C., J.W., and V.P. (Reprinted from Reuter-Lorenz and Fendrich, 1990, Orienting Attention Across..., *J. Cog. Neurosci.* 2:233 with permission from MIT Press.)

We questioned whether the cost difference between within-field/invalid trials and between-field/invalid trials might be related to: 1) differences in the spatial magnitude of the attentional shift (on average greater for the between-field/invalid condition); and/or 2) differences related to the spatial organization of the hemifields with respect to either meridian (vertical as well as horizontal). Therefore, in a follow-up experiment with J.W., we measured the costs of shifting visual attention within versus between the upper and lower hemifields. No difference was found. Thus, the previously observed within-field versus between-field difference is specific to shifts of attention across the vertical meridian.

Information about the eye position, head position, and body position of the observer as well as the visual location of the stimulus is utilized by the cortex to direct visual attention in space (for reviews see Heilman et al. 1985, Andersen 1989). For example, an egocentric representation based on body-centered coordinates contributes to left hemispatial neglect in patients with right parietal lobe lesions. We thus investigated whether shifts of attention across the vertical meridian by J.W. were based on body-centered coordinates (Reuter-Lorenz and Fendrich 1990). In our original experiment, the left side of the display was positioned in the left visual hemifield *and* to the left of the body, likewise for the right. We decoupled these representations of space by turning the screen approximately 90° and having J.W. tilt his head in the same direction. Thus, the target locations remained aligned with the horizontal meridian of the retina and head but they were aligned more or less vertically with respect to the body. With this arrangement, the magnitude of the deficit should have been reduced if the deficit previously observed on between-field trials were due to impairment in shifting between representations of the left and right hemibody. It was not: the pattern of costs was the same as those previously observed when target locations were aligned horizontally with respect to body coordinates. These results suggest that the costs of across-field shifts of attention in our commissurotomy patients were due to an impairment in shifting attention across the left and right *visual fields*. The possible contributions of spatial representations based on head position and eye position have not yet been assessed (see also Ladavas et al. 1993).

Posner and colleagues have proposed that performance on the spatial-precueing task incorporates three mechanisms of attentional control: engagement, disengagement, and moving attention (Posner et al. 1984; Posner and Petersen 1990). The engage component enables attention to be allocated to the cued location. On invalid trials, when the target occurs at a non-cued location, attention must be disengaged from the cued location and moved to the target. We think that the problem arises for commissurotomy patients on between-field/invalid trials because splenial transection interrupts the transfer of visual information about the target to the cued hemisphere, which is controlling attention. Therefore, disengagement is not initiated. On within-field/invalid trials, the hemisphere that controls attention is also the one that directly receives sensory input about the target; in this case, all three mechanisms proceed normally. This interpretation fits well with the finding that between-field shifting is preserved following anterior callosotomy: visual information about the target is transferred from the stimulated hemisphere to the cued hemisphere via the splenium and disengagement occurs normally.

Commissurotomy patients perform differently on the spatial pre-cueing task than patients with left hemispatial neglect due to right parietal lobe damage. The latter often manifest a selective deficit in disengagement that impairs their ability to shift attention to a visual target presented to the left of the cue, regardless of whether the target falls within the left or right hemifield (Posner et al. 1987). In contrast, our commissurotomy patients showed a deficit that was bidirectional but specific to shifts across the vertical meridian of the visual field. This suggests that parietal damage and hemispheric disconnection can interfere with different aspects of the attention shifting process.

VISUOMOTOR INTEGRATION AND GAZE CONTROL

Given the present evidence that the corpus callosum is essential for certain aspects of attentional control, how do commissurotomy patients meet the visuomotor demands of daily life? One possibility is that each hemisphere controls ipsiversive as well as contraversive saccades. While such a notion challenges the traditional view that each hemisphere controls only contraversive saccades, evidence supporting the possibility of bidirectional gaze control has been obtained via microelectrode recordings of cerebral cortex in awake-behaving anthropoids (e.g., Bizzi and Schiller 1970; Bruce and Goldberg 1984).

In order to assess whether each hemisphere exerts bidirectional gaze control in humans, we recently examined the ability of J.W.'s hemispheres to control saccadic eye movements in the horizontal plane (Hughes et al. 1992). The experiment involved two tasks that utilized the same stimulus conditions but different response requirements. First, we needed to establish that the visual information used to elicit a saccade was restricted to a single hemisphere. A red, green, or yellow light was flashed in one or the other visual hemifield for 150 ms. J.W. was asked to name the color. Since J.W.'s right hemisphere was not able to speak, we expected response accuracy to be at chance for left visual field stimuli if color information could not be transferred paracallosally. Indeed, naming accuracy was at chance (35%) for left visual field stimuli and flawless (100%) for right visual field stimuli. These results provided strong evidence that J.W. could not transfer visual information about color from the right hemisphere to the left hemisphere; combined with the above data concerning the lack of interhemispheric visual transfer in this complete commissurotomy patient, it was also likely that color information could not be transferred from the left hemisphere to the right hemisphere. Thus, these color cues could be used to selectively probe the saccadic abilities of the separated hemispheres.

In the second task, J.W. was instructed to saccade towards a target that was positioned either lateral to a light cue (contraversive condition) or medial to a light cue (ipsiversive condition). The direction of the saccade was indicated by the color of the light (e.g., red = contraversive; yellow = ipsiversive; blue = no eye movement). We found that each hemisphere performed well above chance when cued to make ipsiversive as well as contraversive saccades, and there was no difference between the hemispheres with respect to overall accuracy. These results indicate that each hemisphere exerts bidirectional gaze control.

An interesting feature of J.W.'s performance on this task was his tendency to overshoot the final destination when making ipsiversive saccades. Normal observers typically undershoot on this sort of saccade task. Since overshoot was not observed when J.W. made contraversive saccades, it is unlikely that anticonvulsant medications caused the overshoot on ipsiversive saccades. Instead, we believe that J.W.'s overshoot served to bring the target destination into the visual field of the hemisphere that generated the saccade. An example will illustrate this point. If a yellow light (ipsiversive cue) were presented to J.W.'s left visual field, his right hemisphere would correctly generate a rightward saccade towards a target. Since the eyes overshoot, the target would fall within the left visual field. Thus, overshoot on ipsiversive saccades would enable the hemisphere that generated the saccade to view the target and adjust the final eye position.

Our findings corroborate previous reports of bidirectional gaze control in hemispherectomy patients (Troost et al. 1972; Sharpe et al. 1978). In commissurotomy patients, ipsiversive control of saccadic eye movements, combined with saccadic overshoot, may constitute a means by which each disconnected hemisphere can gain access to visual information from all parts of the visual world, thereby compensating, at least in part, for the effects of commissurotomy on visual perception and attention.

HEMISPHERIC SPECIALIZATION IN AUDITORY PERCEPTION AND COGNITION

Considerably fewer split-brain experiments have been carried out on auditory functions than on visual functions. One reason for this discrepancy is that it is not possible to lateralize acoustic stimuli to one or the other hemisphere with the same certainty or facility. The corpus callosum is only one of several commissures connecting the left and right central auditory pathways. Acoustic stimuli presented to one ear are already represented on both sides of the brain at the level of the superior olivary complex, and there are additional connections between the two sides more rostrally in the pons and midbrain (for review see Webster 1992).

Most studies of auditory functions in commissurotomy patients have been carried out using the dichotic listening technique, in which different acoustic stimuli are presented simultaneously to the left and right ears (Broadbent 1956). This technique takes advantage of the following: 1) ascending auditory projections are predominantly contralateral; and 2) cortical auditory evoked potentials elicited in the hemisphere contralateral to the stimulated ear have a shorter latency than those elicited in the ipsilateral hemisphere (for reviews see Sidtis 1982; Efron 1985). Presumably, in the face of competing dichotic stimuli, each hemisphere is activated predominantly by the stimulus presented to the contralateral ear. Most normal, right-handed subjects thus recognize a slightly but significantly higher proportion of speech sounds presented to the right ear than the left. Patients who have undergone complete or posterior commissurotomy show even greater right ear superiority, which results from both an increase in right ear scores and a decrease in left ear scores (Milner et al. 1968; Sparks and Geschwind 1968; Gazzaniga et al. 1975; Springer and Gazzaniga 1975; Musiek et al. 1984, 1985, 1989; Sidtis 1984). Behavioral-MRI correlations in a recent case of focal callosal hemorrhage following rupture of a post-traumatic pericallosal artery aneurysm indicate that interhemispheric auditory projections pass through the posterior body of the callosum (Tramo et al. 1991). This organization would be homologous to that defined by histological methods in non-human primates (for review see Pandya and Seltzer 1986).

Among the commissurotomy patients discussed in this chapter, J.W. and V.P. were previously reported to manifest the typical pattern of performance when spoken or written responses were used to assess the recognition of dichotic consonant-vowel syllables; however, when a non-verbal response mode was used, V.P.'s right hemisphere as well as her left was able to perform at near-perfect levels (Sidtis et al. 1981). In addition, Sidtis (1984) demonstrated a left ear advantage for pitch recognition using dichotic harmonic complex tones, i.e., tones composed of a sine wave and one or more harmonics.

Several investigators have questioned the reliability with which the dichotic method can be used to probe hemispheric specialization for auditory functions (e.g., Pizzamiglio et al. 1974; Blumstein et al. 1975; Teng 1981). Considerable intra-subject and inter-subject variability has been observed in normals, and a smaller proportion of right-handers show a significant right ear advantage than would be expected on the basis of clinico-anatomical correlations in patients with left hemisphere lesions. For verbal stimuli that contain little in the way of rapid acoustic transients (e.g., vowels), laterality effects are weaker and intra- and inter-subject variability greater. One dichotic experiment uncovered an illusory phenomenon: given a sequence of dichotic tones alternating in pitch, the listener usually lateralized the higher tone to the right ear, even when it was presented to the left ear (Deutsch 1974). There are also questions concerning whether ear differences reflect laterality effects involving: 1) cortical and/or subcortical processes; and 2) auditory attention and/or auditory perception.

As a result of the uncertainties surrounding the interpretation of results obtained with the dichotic listening technique, and in view of our interest in non-verbal auditory processing, we chose a different approach in our recent split-brain experiments. Rather than attempt to lateralize acoustic stimuli, we lateralized the response choices by presenting them visually with a tachistoscope. Thus, acoustic stimuli were presented to both hemispheres but the response choices were available to only one hemisphere. Any resultant laterality effects would thus be attributable to hemispheric asymmetries involving auditory perception and/or auditory-visual integration. We have used this approach to study timbre discrimination and recognition (Tramo and Gazzaniga 1989), virtual pitch perception (Bharucha, Tramo, and Zatorre, unpublished data), tonal consonance perception, (Tramo and Bharucha 1991) and the priming of tonal consonance perception by harmonic context (Tramo and Bharucha 1991) in J.W., V.P., and D.R. We present the results of the latter two experiments here.

The perception of tonal consonance relies on the capacity of the auditory system to: 1) resolve the frequency components of simultaneous tones; and 2) extract their harmonic relationships (for reviews see Terhardt 1984; DeWitt and Crowder 1987; Greenwood 1991). In general, listeners judge simultaneous tones as dissonant when: 1) the tones are very close in frequency, such that their frequencies cannot be resolved; and 2) the frequencies of the tones do not form simple integer ratios (e.g., 2:1, 3:2, 4:3), such that their pitches do not blend or fuse into an integrated whole.

In order to study tonal consonance perception in commissurotomy patients V.P. and J.W., we adapted the paradigm of Bharucha and Stoeckig (1986, 1987). The listener is asked to judge whether a combination of simultaneous tones sounds "in-tune" or "out-of-tune." In one condition, the tones were harmonically related to one another (harmonic condition); in the second condition, some of the tones were harmonically related and some were not (quasi-harmonic condition). The simultaneous tones comprised a musical chord, the major triad, the basis of harmonic practice in classical Western music (Apel 1972). The three notes of a major triad belong to the same harmonic series; the lowest note of the triad (the root) forms a frequency ratio of 4:5 with the second highest note (the major third) and a frequency ratio of 2:3 with the third highest note (the perfect fifth). In the quasi-harmonic condition, the fifth was not perfect - it was mistuned a fraction of a semitone. We digitally synthesized 12 in-tune and 12 out-of-tune major triads that differed with respect to their roots (one for each note in the chromatic scale, C, C#, D, D#, E, etc.). Each was 2s in duration.

Our procedure is schematized in Figure 5. The stimulus was played over a loudspeaker placed in front of the patient; its loudness was well above hearing threshold and adjusted to be comfortable and customary for the subject. At the onset of the stimulus, response choices were flashed to one or the other hemisphere (duration = 150 ms; location = left or right visual field, 1.5^{o} away from the point of fixation along the horizontal meridian). In one quadrant, a line drawing of a smiling face and the word "IN" were flashed; simultaneously in the other quadrant, a grimacing face and the word "OUT" were flashed. Although the right hemispheres of V.P. and J.W. are able to read single words, we used non-verbal as well as verbal response choices to avoid biasing the experiment in favor of the left hemisphere. The subjects task was to point to the response choice that matched the sound using the ipsilateral hand; thus, the hemisphere that saw the response choices controlled the response hand. One block of 96 trials was run in pseudorandom order with each patient, 48 trials for each hemisphere.

The results are summarized in Figure 6. For both V.P. and J.W., overall response accuracy was significantly better than chance when response choices were presented to the right hemisphere. However, when response choices were presented to the left hemisphere, J.W.'s performance was not significantly better than chance and V.P.'s performance was marginally better than chance. Furthermore, V.P.'s performance was significantly better when response choices were presented to the right hemisphere than when they were presented to the left hemisphere. These results indicate that, for both patients, the right hemisphere is superior to the left with respect to tonal consonance perception. In J.W.'s case, the left hemisphere appears to be incompetent.

We then set out to examine the effects of harmonic context on tonal consonance perception in V.P. and J.W. Bharucha and Stoeckig (1986, 1987) had demonstrated in normal subjects (both non-musicians and musicians) that the perceived consonance of a musical chord (the target) is influenced by its harmonic relationship to a preceding chord (the prime). The priming of "in-tune" versus "out-of-tune" judgments by harmonic context in the tonal consonance task is analogous to the priming of "word" versus "non-word" judgments by semantic context in lexical decision tasks (e.g., Milberg and Blumstein, 1981). For example, when a subject is asked to decide if a target letter-string (e.g., "nurse") constitutes a word, his/her response accuracy is higher when the target is preceded by a semantically related word (e.g., "doctor") than when it is preceded by an unrelated word (e.g., "paper"). Similarly, when a subject is asked to decide if a target chord (e.g., Gmaj) is in-tune, his/her response accuracy is higher when the target is preceded by a harmonically related chord (e.g., Fmaj, which is also in the key of Cmaj) than by an unrelated chord (e.g., Dbmaj). The cognitive processes mediating the effects of harmonic context on tonal consonance perception are specific to the auditory modality and operate exclusively outside the verbal domain. Our experiment thus provided an opportunity to study hemispheric mechanisms subserving auditory priming in the non-verbal domain.

As in the tonal consonance task described above, there were two target conditions, harmonic and quasi-harmonic, and each patient was instructed to judge whether the target sounded in-tune or out-of-tune. The present task differed in that a prime chord was played for 3s immediately before the target chord. In half the trials, the prime chord was harmonically related to the target chord; in the other half of trials, it was not. Harmonic relatedness was defined on the basis of previous data

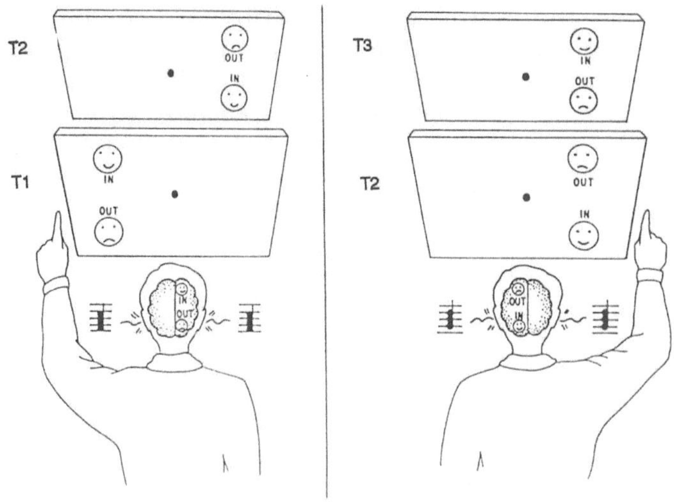

Figure 5. Paradigm used to study tonal consonance perception and auditory priming in commissurotomy patients V.P. and J.W. The sound stimulus, a musical chord, was played in free field over a loudspeaker. In the example shown at the left, the chord is an Fmaj triad, at the right, a Gmaj triad. At the onset of the stimulus, response choices were flashed for 150 ms to one or the other hemisphere. In the example shown at the left, the response choices on the first trial (T1) were flashed in the left visual field. For each choice, both verbal and non-verbal referents of the "in-tune" and "out-of-tune" alternatives were provided. The patient pointed to his/her response with the hand ipsilateral to the visual stimulus. Thus, the hemisphere that saw the response choices controlled the pointing. The side and quadrant of the response choices were counterbalanced over successive trials (T2..). (Reprinted from Tramo and Bharucha, 1991, Musical Priming by the right..., *Neuropsychologia* 29(4):3115 with kind permission from Pergamon Press, Ltd., Oxford, UK.)

in normal subjects, which are consistent with the claims of formal music theory and the transition probabilities that characterize harmonic progression in Western music (for reviews see Bharucha 1987 and Krumhansl 1990).

For both V.P. and J.W., overall response accuracy was significantly better than chance when response choices were presented to the right hemisphere. However, when response choices were presented to the left hemisphere, response accuracy was not significantly better than chance for both patients. Furthermore, performance was significantly better on right hemisphere trials than on left hemisphere trials for both V.P. and J.W. Finally, only the right hemisphere trials showed the normal interaction between harmonic relatedness and chord intonation for both patients. Since V.P.'s left hemisphere performed above chance when only targets were presented (see above), and since we had previously observed priming despite impaired tonal consonance perception in a patient with bilateral lesions of auditory cortex (Tramo et al. 1990), our results indicate that associative auditory functions mediating harmony perception are lateralized within the right hemispheres of V.P. and J.W.

While our findings are consistent with the view that musical functions are mediated by the right hemisphere, we emphasize that the foregoing experiments pertain to specific aspects of music perception and cognition. For example, our previous results concerning timbre recognition in J.W., V.P., and D.R. indicate that both hemispheres are able to recognize musical instrument sounds. Moreover, the findings in numerous cases in the neurological literature are consistent with the interpretation that the sum of auditory perceptual and cognitive functions mediating musical experience are distributed within both hemispheres. The degree to which the hemispheres are specialized to carry out non-verbal auditory functions that are relevant to prephonemic processing of speech sounds, voice perception, environmental sound perception, and music perception is a topic of ongoing research.

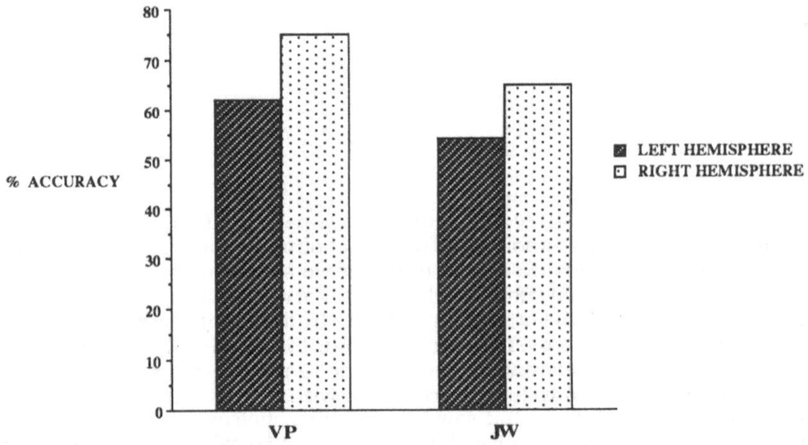

Figure 6. Performance of V.P. and J.W. on a task that required them to determine whether harmonic and quasi-harmonic combinations of simultaneous tones sounded "in-tune" or "out-of-tune."

LANGUAGE FUNCTIONS OF THE RIGHT HEMISPHERE

Since the original report of patient W.J. by Gazzaniga, Bogen, and Sperry in 1962, it has been clear that the corpus callosum avails the sensations and perceptions of the right hemisphere to left hemisphere structures mediating speech production. Following a brief period of mutism caused by the combined effects of surgical retraction, edema, diaschisis, and disconnection, conversational language appears to be normal (see also Reeves and Risse chapter in this text); however, commissurotomy patients fail to name items placed out-of-view in their left hand or presented tachistoscopically in the left visual field (for reviews see Gazzaniga 1970; Bogen 1985). Providing these patients with a non-verbal means of response, such as pointing to a matching item with the left hand, allows them to demonstrate that stimuli presented to the right hemisphere have been apprehended, as evidenced above in our experiments on vision and audition. Thus, information about the world can be processed by the right hemisphere but cannot be transferred to the left hemisphere for oral report. Using non-verbal response modes, it has been possible to measure the performance of the right hemisphere on a variety of tasks that tap into three of the principal functional subcomponents of language: semantics, phonology, and syntax.

With respect to semantic functions, the right hemispheres of two of the first three California patients, L.B. and N.G., can match written and spoken words to pictures (Zaidel 1976, 1982). For both patients, the right hemisphere comprehends more spoken words than written ones. The right hemispheres of four additional patients in the California series were subsequently found to be able to derive the meanings of spoken and written words (R.Y., A.A., N.W., and C.C.; for review see Zaidel 1990). The same abilities have also been demonstrated by the right hemispheres of four commissurotomy patients from the Dartmouth (P.S., J.W., D.R.) and Ohio (V.P.) series (Gazzaniga et al. 1977, 1979, 1982, 1984; Sidtis et al. 1981; Baynes et al. 1992). Unlike L.B. and N.G., the right hemispheres of J.W. and D.R. possess auditory and visual lexicons that are equally well developed. The right hemisphere of V.P. but not J.W. has manifested electrophysiological evidence of semantic priming in a lexical decision task (Kutas et al. 1988).

The ability to decode speech sounds in order to derive their meaning indicates that the right hemispheres of the ten commissurotomy patients cited above are competent with respect to certain aspects of phonological processing. For example, since the right hemispheres of L.B., N.G., P.S., J.W., V.P., and D.R. were able to match pictures to auditory words, all demonstrated competence for phonological decoding at the *single word* level. However, as noted in the previous section on auditory perception, the right hemispheres of J.W. and V.P. differed in their capacity to discriminate consonant-vowel *syllables*. Moreover, the right hemispheres of only two patients (P.S. and V.P.) can *generate speech*.

With respect to syntax, fewer observations are available. In one experiment, the right hemispheres of J.W. and V.P. were able to determine whether sentences presented auditorily contain grammatical violations; however, in another experiment, neither was able to use syntactic information to derive the meaning of structurally ambiguous sentences or of sentences constructed in the passive voice (Gazzaniga et al. 1984; Baynes and Gazzaniga 1988). The right hemispheres of L.B. and N.B. have demonstrated some right hemisphere sensitivity to word class and simple morphosyntactic information (e.g., tense markings; Zaidel 1977, 1978). However, the memory requirements of tasks using grammatically complex sentences may confound a straightforward interpretation of the results as they relate to syntactic functions per se.

The results concerning right hemisphere language functions in commissurotomy patients have been interpreted differently by different investigators (for reviews see Gazzaniga 1983; Zaidel 1983; Baynes 1990). On the one hand, Gazzaniga and colleagues have stressed the variability and plasticity found in their patient pool. On the other hand, Zaidel has concluded that the characteristics of right hemisphere language are more or less uniform among his split-brain patients. These characteristics include: 1) better comprehension than expression; 2) better auditory-verbal comprehension than visual-verbal comprehension; 3) better semantic than syntactic and phonological functions; and 5) good ability to use nonverbal cues (e.g., facial expressions, voice intonation, body gestures) to facilitate language comprehension. At present, the number of psycholinguistic experiments with commissurotomy patients is too small to sort out the possible effects of variables such as the side(s) and site(s) of electroencephalographic abnormalities, seizure control, age, gender, and family history of sinistrality. As the number of commissurotomy patients increases in the years to come, fine-grained, prospective analyses of right hemisphere language functions could provide information useful to the rehabilitation of left hemisphere stroke patients with aphasia.

LANGUAGE FUNCTIONS: READING AND LEXICAL ACCESS

Many experiments with normal readers have analyzed how the mental representation of a written word (sometimes called the visual word form) is accessed. Intuitively, one might suppose that we work our way through a word, examining and recognizing each letter, then rapidly assembling the letters into a word. If letter-by-letter reading were the mechanism by which visual word forms are recognized, then the more letters a word has, the longer it would take to read the word. In fact, data from normal subjects reveal that length is not the most important variable in word recognition. Moreover, it takes longer to recognize a letter when it is presented alone than when it is part of a word - that is, knowing the word helps to perceive the letter. This counter-intuitive phenomenon is known as the word superiority effect. It is not possible to account for the word superiority effect as a result of sophisticated guessing strategies or response bias. Neither word shape nor envelope explain the phenomenon, since the effect is resistant to changes in type case, font, and format (for review see Pollatsek and Rayner 1989). The word superiority effect is generally viewed as evidence that: 1) the brain stores an abstract cognitive representation of the visual word form; and 2) perceptual processing of letters is facilitated in "top-down" fashion through the activation of this representation.

We recently studied the word superiority effect in patient J.W. (Reuter-Lorenz and Baynes 1992) using the task developed by Reicher (1969) and modified by Bub et al. (1989) for studies of patients with pure alexia (also known as alexia without agraphia, a syndrome that is usually caused by a lesion of the left occipital lobe and splenium). A clear effect was observed for words presented in the left as well as right visual field, indicating that visual word forms can be accessed by the right hemisphere as well as the left hemisphere.

Two additional results of interest emerged. First, in order to achieve 75% accuracy on the task, substantially longer exposure durations were required by the right hemisphere than by the left: 216 ms versus 46 ms, respectively. This suggests that right hemisphere performance with short stimulus durations is compromised by degradation of stimulus features in the course of perceptual processing rather than by the absence of cognitive representations of word form. Second, left hemisphere errors occurred more frequently for targets in the middle of letter strings than for targets at the beginning or end of the strings; in contrast, right hemisphere errors increased with the distance of the target from the beginning of the letter string. This hemispheric difference in error patterns suggests that letters presented to the left hemisphere are processed in parallel, while letters presented to the right hemisphere are processed in serial.

In general, normal subjects show word length effects when letters are presented in the left visual field; in addition, word length effects have been observed under unusual conditions of right visual field stimulation (e.g., with vertical presentations; Young and Ellis 1985; Bub and Lewine 1985; Ellis et al. 1988; Eviatar and Zaidel 1991). Bub and Lewine (1988) have interpreted these observations as evidence that callosal transfer of visual information about words from the right to the left hemisphere requires sequential processing; differences occurring early in the perceptual processing of word forms would thus be responsible for the visual field differences observed in J.W. In contrast, Young and Ellis (1985) do not believe complex interactions characteristic of the word superiority effect can be accounted for by a callosal transfer hypothesis; instead, they propose that the right hemisphere uses a mode of lexical access that requires a temporary graphemic store only when the words are in unusual formats. Our findings demonstrate word length effects within the right hemisphere itself and thus support hypotheses that do not require callosal transfer in order to obtain such effects. Furthermore, our results suggest that the inherent limitations of right hemisphere language functions restrict its ability to use parallel modes of lexical access. The resulting delay in lexical access may diminish the amount of lexical information that can be used rapidly to assemble the mental framework for language expression and comprehension.

MEMORY FUNCTIONS: RECALL AND RECOGNITION

To some extent, the terms "commissurotomy" and "callosotomy" have been used interchangeably in the neuropsychological literature to describe the lesions of split-brain patients. However, distinctions can be made. For example, the former term would be more appropriate when the anterior commissure as well as the callosum is sectioned. Even when the anterior commissure is left intact, complete callosotomies and posterior callosotomies always or almost always include transection of the hippocampal commissure, a thin sheet of crossing efferent projections from the hippocampi that typically pass just beneath the posterior body of the callosum. Given the well-established role of the hippocampus in memory functions (for reviews see Victor and Agamanolis 1990; Zola-Morgan and Squire 1993), it seems reasonable to hypothesize that disruption of interhemispheric integration at this level leads to memory impairment.

Memory functions have been reported as both impaired (Zaidel and Sperry 1974) and intact (LeDoux et al. 1977; Sass et al. 1988) in split-brain patients. In their review of the literature, Clark and Geffen (1989) suggest that the presence or absence of memory deficits depends on the extent and/or site of transection: patients with complete or posterior commissurotomies manifest a mild memory deficit, whereas those with anterior transections do not. We present original data from our own studies that support the following arguments: 1) transections that include the hippocampal commissure cause a mild memory deficit; 2) the deficit involves recall but not recognition; and 3) this pattern of deficit may be attributed to hemispheric specialization for different functional subcomponents of memory (Phelps et al. 1991; Phelps and Gazzaniga 1992).

Auditory-verbal recall and visual-nonverbal recall were examined pre-operatively and post-operatively in six commissurotomy patients using the Logical Memory and Picture Reproduction subtests of the Wechsler Memory Scale (Wechsler 1945). In four post-operative cases, S.C, G.M., J.J., and E.S., the hippocampal commissure was intact (these patients underwent anterior callosotomies); in two cases, the hippocampal commissure was transected in the course of posterior (E.B.) and near-complete (D.R.) callosotomies (Fig. 1). All patients were maintained on anticonvulsant drugs at the times of testing.

In the Logical Memory subtest, each patient was asked to recall each of two short stories immediately after it was read aloud by the examiner. The two stories combined contained a total of 46 facts, and the scoring was based on the number recalled. In the Picture Reproduction subtest, the patient was asked to reproduce each of four abstract line drawings immediately after it was presented on a card. The four drawings combined contained 14 features, and the scoring was based on the number reproduced. Following Russell (1975), each patient was also asked to recall the stories and reproduce the drawings 30 minutes after initial presentation; thus, retention as well as registration and retrieval was assessed.

The results are summarized in Table 4. A positive value indicates that the post-operative score was higher than the pre-operative score. At the time of pre-operative testing, all patient scores were within one standard deviation of the mean of fifty normal subjects of similar age, except for D.R., whose score was more than two standard deviations above the mean (for review of normal data, see Lezak 1983).

On immediate story recall, three of the four patients with intact hippocampal commissures (HC-intact) showed either no post-operative decline or improvement in performance; in contrast, both patients with hippocampal commissurotomies (HC-cut) showed a post-operative decline that was greater than one standard deviation below the mean of the HC-intact group.

On immediate picture reproduction, three of the four HC-intact patients showed either no post-operative decline or improvement; in contrast, both HC-cut patients showed a marked post-operative decline that was greater than two standard deviations below the mean of the HC-intact group.

On delayed story recall, two of the four HC-intact patients showed post-operative improvement, and both showed improvement in the proportion of information that could be retained during the delay. For E.B., the only HC-cut patient in whom delayed recall was tested, there was a post-operative decline that: 1) was greater than that of all four HC-intact patients; and 2) was greater than one standard deviation below the mean of the HC-intact group. E.B. showed only a mild post-operative decline in the proportion of information that could be retained during the

TABLE 4. Pre-operative and post-operative scores on the Logical Memory and Picture Reproduction Subtests of the Weschler Memory Scale in patients with and without transection of the hippocampal commissure (HC). Each value represents the post-operative score minus the pre-operative score. For each testing interval, the proportion retained (R) is the score after the 30 min. delay (30') divided by the score upon immediate recall (0'). NT = not tested. NA = not appropriate.

	Story Recall			Drawing Reproduction		
	0'	30'	R	0'	30'	R
HC INTACT						
S.C.	+0.50	+0.50	+0.03	+1.0	- 1.0	- 0.14
G.M.	+0.25	- 1.25	- 0.18	+3.0	0.0	- 0.73
E.S.	- 2.25	- 1.25	- 0.08	- 1.0	- 2.0	- 0.30
J.J.	0.00	+0.75	+0.06	0.0	+4.0	+0.57
Mean	- 0.38	- 0.31	- 0.04	+0.8	+0.2	- 0.15
SD	±1.10	±0.94	±0.09	±1.5	±2.3	±0.47
HC CUT						
E.B.	- 1.75	- 1.50	- 0.04	- 3.0	- 3.0	- 0.33
D.R.	- 9.00	NT	NT	- 6.0	NT	NT
Mean	- 5.38	NA	NA	- 4.5	NA	NA
SD	±3.62	NA	NA	±1.5	NA	NA

delay; E.B.'s R value matched the mean of the HC-intact group. Thus, the difference in delayed recall between E.B. and the HC-intact patients was not due to a difference in retention but to a difference in the amount of information that had been initially encoded.

On delayed picture reproduction, two of the HC-intact patients showed either no post-operative decline or post-operative improvement. In contrast, E.B. showed a severe drop in post-operative performance that was greater than all four HC-intact patients and more than one standard deviation below their mean. One HC-intact patient showed post-operative improvement in the proportion of registered information that could be retrieved after the 30 minute delay; E.B. showed a drop that was greater than three of the HC-intact patients but within one standard deviation of the mean of the four HC-intact patients. Again, this suggests that a difference in encoding rather than a difference in retention was the principal basis of E.B.'s poorer performance.

In order to compare the pre-operative and post-operative performance in HC-intact versus HC-cut patients, separate sign tests were carried out. Overall, the difference in pre- and post-operative performance was marginally significant for the HC-cut patients and non-significant for the HC-intact patients.

Verbal memory was tested further in patients S.C. and G.M. using a task similar to that developed by Klatzky (1980) to study recognition as well as recall in normal subjects. Four lists of twenty words each were presented pre-operatively and four different lists were presented post-operatively. Each word was printed on a card, contained one or two syllables, and had an average frequency of 20 occurrences per million

(Kucera and Francis 1967). The list was presented at a rate of 5s per card. Recall was tested immediately after presentation and after delays of 30s, 2 min. and 2 hrs. with intervening distractors. Following attempts at delayed recall, each patient was given a two-alternative forced choice recognition task in which he had to decide which of two words printed on a card had been presented.

Both S.C. and G.W. showed post-operative improvement in immediate recall and in delayed recall at all three intervals. The mean and standard deviation for the number of words recalled across the four test intervals was 4.25 ± 3.28 pre-operatively; post-operatively, they scored 8.00 ± 3.42. This improvement in visual-verbal recall following anterior callosotomy was consistent with that observed for auditory-verbal recall (Table 4) and argued against the possibility that these patients harbored verbal memory deficits that were too subtle to detect using the Logical Memory subtest. Because both pre-operative and post-operative scores on the recognition task were at ceiling, it was not possible to discern a post-operative improvement in recognition.

We also examined word list recall and recognition in patients J.W. and V.P., in whom the hippocampal commissure had been transected several years previously. The experimental design was similar to that described above for S.C. and G.W., except that recall and recognition were tested once in each experimental session, ten mins.. after list presentation. The performance of J.W. and V.P. was compared with that of two neurologically normal subjects of similar age and education.

A marked impairment in recall was found for J.W. and V.P. After the ten min. delay, their mean and standard deviation were 1.4 ± 1.3; the normal controls scored 5.4 ± 1.8. On some of the word lists, J.W. and V.P. did not recall a single word; the fewest number of words recalled by either normal control on any list was three. Furthermore, J.W. and V.P. recalled fewer words at one hr. than the two HC-intact patients recalled at two hrs (4.0 ± 3.0). With respect to recognition, J.W. and V.P. showed no deficit: they recognized 17.0 ± 2.5 words, while the controls recognized 17.9 ± 1.5 words.

As a whole, the foregoing results suggest that auditory-verbal memory, visual-verbal memory, and visual-nonverbal memory are more likely to be impaired following commissurotomies that include the hippocampal commissure than those that spare it. Our findings are thus consistent with the hypothesis that the interhemispheric integration of information flowing from the hippocampi is essential to normal memory. The deficit appears to involve recall but not recognition. Given the limited data available, these conclusions must remain tentative in anticipation of additional prospective studies that compare the pre-operative and post-operative performance of patients with and without transection of the hippocampal commissure and that take into account the multiplicity of clinical variables that influence performance on memory tests (e.g., focus, duration, and control of seizures; type, number, and levels of anticonvulsants).

MEMORY FUNCTIONS: INFERENCE AND INTERPRETATION

Why might disconnecting the hemispheres lead to a dissociation between recall and recognition? Experiments in normal subjects indicate that recall is more sensitive than recognition to the elaboration and organization of to-be-remembered stimuli and to multiple representations of events (Paivio 1971; Bransford and Johnson 1972). For example, a subject's ability to recall a word is better if the word represents a concrete item that can be encoded visually as well as verbally (e.g., "chair") than if the word represents an abstract concept that can only be encoded verbally (e.g., "value"); in contrast, the facilitation provided by multiple forms of representation is not as pronounced in tests of recognition. If the two hemispheres use different mechanisms to encode information for subsequent recall, disrupting interhemispheric integration could have a detrimental effect. There is evidence to support the notion that the left and right hemispheres mediate different aspects of memory functions (for review see Milner 1975). Experiments with commissurotomy patients indicate that the left hemisphere is specialized for inference and interpretation (Gazzaniga 1985); these skills play an important role in memory (Bartlett 1932; Bransford and Johnson 1972).

In order to analyze hemispheric differences in memory functions that involve inference and interpretation, we carried out the following experiment with patients J.W. and V.P. Each patient was presented with three sets of 80 line drawings. The drawings in each set showed a sequence of events (e.g., a man showering, shaving, donning a shirt, donning shoes, etc.) that were common to a particular scene (e.g., getting ready for work). Each sample drawing was presented in full view for 3s. After a 90 min. delay with intervening distractors, a comparison drawing was presented in full view for each patient to look at as long as he/she needed to decide whether or not the drawing had been presented previously. There were three comparison conditions: 1) the drawing had been presented previously; 2) the drawing had not been presented previously and was inconsistent with the event depicted in the scene (e.g., the same man fixing a television); and 3) the drawing had not been presented previously but was consistent with the event depicted in the scene (e.g., the same man donning pants). The patient was then asked to fixate a central point on a computer screen and the response choices, "YES" and "NO", were flashed simultaneously in alternate quadrants of one visual hemifield; the patient indicated his/her response by pointing with the ipsilateral hand.

For the first two conditions, both the left and right hemispheres performed significantly above chance, and their performance did not differ significantly. That is, each hemisphere was capable of: 1) recognizing the originally presented drawings; and 2) rejecting drawings that were inconsistent with the original scene. However, for the third condition, the performance of the left and right hemispheres differed dramatically: the left hemisphere was no better than chance and the right hemisphere was significantly above chance; analysis of the error pattern revealed that the left hemisphere was biased in favor of false alarms. That is, it inferred from the overall scene that it had seen the drawing when, in fact, it had not.

These results indicate that the left hemisphere uses inference and interpretation to encode and/or retrieve information. The left hemisphere is thus more likely than the right to be influenced by the expected sequence of events, even though this influence may lead it astray during task performance.

SUMMARY

Our experimental findings support the following hypotheses about hemispheric specialization and interhemispheric integration: 1) the interhemispheric transfer of visual information concerning orientation, form, and luminance is abolished following posterior commissurotomy, at least for stimuli presented ≥ 2° lateral to the vertical meridian; 2) there is limited paracallosal transfer of visual information concerning the relative position of objects; 3) despite commissurotomy, each hemisphere has access to crude visual information from that portion of the ipsilateral visual field lying at or in close proximity to the vertical meridian; 4) the posterior third of the corpus callosum mediates the generation of visual evoked potentials from occipital cortex ipsilateral to the stimulated hemifield; 5) the posterior third of the corpus callosum mediates shifts of visual attention between the hemispheres; 6) shifts of visual attention across the vertical meridian are computed on the basis of retinotopic representations of space, head position, and/or eye position, but not on the basis of body position; 7) each hemisphere has the capacity to generate both ipsiversive and contraversive saccades in response to a visual stimulus; 8) auditory structures lying within and/or projecting to the right hemisphere mediate tonal consonance perception, which relies on the capacity of the auditory system to resolve the frequency components of simultaneous tones and extract their harmonic relationships; 9) the right hemisphere is sensitive to the effects of harmonic context on tonal consonance perception and can support auditory priming in the nonverbal domain; 10) language capacities of the right hemisphere are evident in a minority of commissurotomy patients and the degree of competence varies from patient to patient; 11) in general, right hemispheres that manifest language competence possess greater semantic than syntactic and phonological capacities; 12) right hemisphere reading, when present, involves the slow, serial processing of letters instead of the rapid, parallel processing of letters normally involved in lexical access; 13) the interhemispheric integration of hippocampal output via the hippocampal commissure contributes to normal memory, particularly with respect to recall; and 14) the capacity to use inference and interpretation in mnemonic processing is lateralized within the left hemisphere.

ACKNOWLEDGMENTS

We gratefully acknowledge the contributions of our collaborators to this work: Jamshed Bharucha, Steven Hillyard, William Hirst, Howard Hughes, Marta Kutas, Elizabetta Ladavas, William Loftus, Frank Musiek, and C. Mark Wessinger. We thank Joseph Bogen, Mark Rayport, David Roberts, Alexander Reeves, and Donald Wilson for the opportunity to examine their patients. This work was supported by N.I.H. NS17778, NS22626, DC00071, and the James S. McDonnell Foundation.

REFERENCES

Apel, W., 1972, "Harvard Dictionary of Music, 2nd Ed.," Belknap/Harvard, Cambridge.

Andersen, R.A., 1989, Visual and eye movement functions of the posterior parietal cortex, *Ann. Rev. Neurosci.* 12:377-403.

Bartlett, F.C., 1932, "Remembering: A Study in Experimental and Social Psychology," Cambridge University Press, Cambridge.

Baynes, K., 1990, Language and reading in the right hemisphere: highways or byways of the brain?, *J. Cognit. Neurosci.* 2:159-179.

Baynes, K. and Gazzaniga, M.S., 1988, Right hemisphere language: insights into normal language mechanisms? *in:* "Language, Communication and the Brain," F. Plum, ed., Raven Press, NY, pp. 117-126.

Baynes, K., Tramo, M.J., and Gazzaniga, M.S., 1992, Reading with a limited lexicon in the right hemisphere of a callosotomy patient, *Neuropsychologia* 30:187-200.

Berlucchi, G., 1972, Anatomical and physiological aspects of visual functions of corpus callosum, *Brain Res.* 37:371-392.

Berlucchi, G., Gazzaniga M.S., and Rizzolatti, G., 1967, Microelectrode analysis of transfer of visual information by the corpus callosum, *Arch Ital. Biol.* 105:583-596.

Berlucchi, G. and Rizzolatti, G., 1968, Binocularly driven neurons in visual cortex of split-chiasm cats, *Science* 159:308-310.

Bharucha, J.J., 1987, Music cognition and perceptual facilitation: a connectionist framework, *Music Perc.* 5:1-30.

Bharucha, J.J. and Stoeckig, K., 1986, Reaction time and musical expectancy: priming of chords, *J. Exp. Psychol.: Hum. Percept. Perform.* 12:403-410.

Bharucha, J.J. and Stoeckig, K., 1987, Priming of chords: Spreading activation or overlapping frequency spectra? *Percept. Psychophys.* 41:519-524.

Bizzi, E. and Schiller, P.H., 1970, Single unit activity in the frontal eye fields of un-anesthetized monkeys during head and eye movement, *Exp. Brain Res.* 10:151-158.

Blumstein, S., Goodglass, H., and Tartter, V., 1975, The reliability of ear advantage in dichotic listening, *Brain Lang.* 2:226-236.

Bogen, J.E., 1985, The callosal syndrome, *in:* "Clinical Neuropsychology, 2nd Edition," K.M. Heilman and E. Valenstein, eds., Oxford University Press, NY, pp. 295-338.

Bransford, J.P. and Johnson, M.K., 1972, Contextual prerequisites for understanding: some investigations for comprehension and recall, *J. Verbal Learn. Verbal Behav.* 11:717-726.

Broadbent, D.E., 1956, Successive responses to simultaneous stimuli, *Q. J. Exp. Psychol.* 8:145-162.

Bruce, C.G. and Goldberg, M.E., 1984, Physiology of the frontal eye fields, *Trends Neurosci.* 7:436-441.

Bub, D.N., Black, S.E., and Howell, J., 1989, Word recognition and orthographic context effects in a letter-by letter reader, *Brain Lang.* 36:357-376.

Bub, D.N. and Lewine, J., 1988, Different modes of word recognition in the left and right visual fields, *Brain Lang.* 33:161-188.

Bunt, A.H. and Minckler, D.S., 1977, Foveal sparing. New anatomical evidence for bilateral representation of the central retina, *Arch. Ophthalmol.* 95:445-1447.

Clark, C.R., and Geffen, G.M., 1989, Corpus callosum surgery and recent memory, *Brain* 112:165-175.

Deutsch, D., 1974, An auditory illusion, *Nature* 251:307-309.

DeWitt, L.A., and Crowder, R.G., 1987, Tonal fusion of consonant musical intervals: the oomph in Stumpf, *Perc. Psychophys.* 41:73-84.

Efron, R., 1985, The central auditory system and issues related to hemispheric specialization, *in:* "Assessment of Central Auditory Dysfunction. Foundation and Clinical Correlates," M.L. Pinheiro and F.E. Musiek, eds., Williams and Wilkins, Baltimore, pp. 143-154.

Ellis, A.W., Young, A.W., and Anderson, C., 1988, Modes of word recognition in the right and left cerebral hemispheres, *Brain Lang.* 35:254-273.

Eviatar, Z., and Zaidel, E., 1991, The effects of word length and emotionality on hemispheric contribution to lexical decision, *Neuropsychologia* 29:412-428.

Fendrich, R., and Gazzaniga, M.S., 1989, Evidence of foveal splitting in a commissurotomy patient, Neuropsychologia 27:273-281.
Gazzaniga, M.S., 1970, "The Bisected Brain," Appleton-Century-Crofts, NY.
Gazzaniga, M.S., 1983, Right hemisphere language following bisection: a 20-year perspective, *Amer. Psychol.* 38:525-537.
Gazzaniga, M.S., 1985, "The Social Brain," Basic Book, New York.
Gazzaniga, M.S., Bogen, J.E., and Sperry, R.W., 1962, Some functional effects of sectioning the cerebral commissures in man, *Proc. Natl. Acad. Sci. U.S.A.* 48:1765-1769.
Gazzaniga, M.S., Bogen, J.E., and Sperry, R.W., 1965, Observations on visual perception after disconnexion of the cerebral hemispheres in man, *Brain* 88:221-236.
Gazzaniga, M.S., Holtzman, J.D., Deck, M.D.F., and Lee, B.C.P., 1985, MRI assessment of human callosal surgery with neuropsychological correlates, *Neurology* 35:1763-1766.
Gazzaniga, M.S., Kutas, M., Van Pettern, C., and Fendrich, R., 1989, Human callosal function: MRI-verified neuropsychological functions, *Neurology* 39:942-946.
Gazzaniga, M.S., LeDoux, J.E., and Wilson, D.H., 1977, Language, praxis, and the right hemisphere: clues to some mechanisms of consciousness, *Neurology* 27:1144-1147.
Gazzaniga, M.S., Risse, G.L., Springer, S.P., Clark, A.B., and Wilson, D.H., 1975, Psychologic and neurologic consequences of partial and complete cerebral commissurotomy, *Neurology* 25:10-15.
Gazzaniga, M.S., Smylie, C.S., Baynes, K., McCleary, C., and Hirst, W., 1984, Profiles of right hemisphere language and speech following commissurotomy, *Brain Lang.* 22:206-220.
Gazzaniga, M.S. and Sperry, R.W., 1967, Language after section of the cerebral commissures, *Brain* 90:131-148.
Gazzaniga, M.S., Volpe, B.T., Symlie, C.S., Holtzman, J., and Wilson, D.H., 1982, Evidence for paracallosal verbal transfer after callosal section: a possible consequence of bilateral language organization, *Brain* 105:53-63.
Gazzaniga, M.S., Volpe, B.T., Symlie, C.S., Wilson, D.H., and LeDoux, J.E., 1979, Plasticity in speech organization following commissurotomy, *Brain* 102:805-815.
Gazzaniga, M.S. and Sperry, R.W. (1967) Language after section of the cerebral commissures, *Brain* 90:131-138.
Greenwood, D.D., 1991, Critical bandwidth and consonance in relation to cochlear frequency-position coordinates, *Hear. Res.* 54:164-208.
Harvey, L.O., Jr. , 1978, Single representation of the visual midline in humans, *Neuropsychologia* 16:601-610.
Haun, F., 1978, Functional dissociation of the hemispheres using foveal visual input, *Neuropsychologia* 16:725-733.
Heilman, K.M., Watson, R.T., and Valenstein, E., 1985, Neglect and related disorders, *in:* "Clinical Neuropsychology, 2nd Edition," K.M. Heilman and E. Valenstein, eds., Oxford University Press, NY, pp. 243-293.
Holtzman, J.D., 1984, Interaction between cortical and subcortical visual areas: evidence from human commissurotomy patients, Vision *Res.* 24:801-813.
Hubel, D.H., 1988, "Eye, Brain, and Vision," Sci. Amer. Library, NY.
Hubel, D.H. and Wiesel, T.N., 1967, Cortical and callosal connections concerned with the vertical meridian of visual fields in the cat,*J. Neurophysiol.* 30:1561-1573.
Hughes, H.C., Reuter-Lorenz, P.A., Fendrich, R., and Gazzaniga, M.S., 1992, Bidirectional control of saccadic eye movements by the disconnected cerebral hemispheres, *Exp. Brain Res.* 91:335-339.
Klatzky, R.L., 1980, "Human Memory, 2nd Edition," W.H. Freeman and Co., San Francisco.
Krumhansl, C.L., 1990, "Cognitive Foundations of Musical Pitch," Oxford Univ. Press, NY.
Kucera, H. and Francis, W.N., 1967, "Computational Analysis of Present Day American English," Brown Univ. Press, Providence.
Kutas, M., Hillyard, S.A., and Gazzaniga, M.S., 1988, Processing of semantic anomaly by right and left hemispheres of commissurotomy patients, *Brain* 111:553-576.
Ladavas, E., DePesce, M., Mangun, G.R., and Gazzaniga, M.S., 1993, Variations in attentional bias in the two disconnected cerebral hemispheres, *Cognit. Neuropsychol.* (in press)
LeDoux, J.E., Risse, G., Springer, S., Wilson, D.H., and Gazzaniga, M.S., 1977, Cognition and commissurotomy, *Brain* 110:87-104.

Leventhal, AG., Ault, S.J., and Vitek, DJ., 1988, The nasotemporal division of the primate retina: the neural bases of macular sparing and splitting, *Science* 240:66-67.
Lezak, M.D., 1983, "Neuropsychological Assessment," 2nd Ed., Oxford Univ. Press, N.Y.
Lines, C.R. and Milner, A.D., 1983, Nasotemporal overlap in the human retina investigated by means of simple reaction time to lateralized light flash, *Exp. Brain Res.* 50:166-172.
Mangun, G.R. and Hillyard, S.A., 1988, Spatial gradients of visual attention: behavioral and electrophysiological evidence, *EEG Clin. Neurophysiol.* 70:417-428.
Mangun, G.R., Luck, S.J., Gazzaniga, M.S., and Hillyard, S.A., 1991, Electrophysiological measures of interhemispheric transfer of visual information: studies in split-brain patients, *Soc. Neurosci. Abstr.* 17:866.
Milberg, W. and Blumstein, S.E., 1981, Lexical decision and aphasia: evidence for semantic processing, *Brain Lang.* 14:371-385.
Milner, B., 1975, "Hemispheric Specialization and Interaction," M.I.T. Press, Cambridge.
Milner, B., Taylor, L., and Sperry, R.W., 1968, Lateralized suppression of dichotically presented digits after commissural section in man, *Science* 161:184-186.
Musiek, F.E, Kibbe, K., and Baran, J.A., 1984, Neuroaudiological results from split-brain patients, *Semin. Hear.* 5:219-229.
Musiek, F.E, Kurdziel-Schwan, S., Kibbe, K.S., Gollegly, K.M., Baran, J.A, Rintelmann, W.F., 1989, The dichotic rhyme task: results in split-brain patients, *Ear Hear.* 10:33-39.
Musiek, F.E., Reeves, A.G, and Baran, J.A., 1985, Release from central auditory competition in the split-brain patient, *Neurology* 35:983-987.
Myers, J.J. and Sperry, R.W., 1985, Interhemispheric communication after section of the forebrain commissures, Cortex, 21:249-260.
Oppenheim, J.S., Skerry, J.E., Tramo, M.J., and Gazzaniga, M.S., 1989, Magnetic resonance imaging morphology of the corpus callosum in monozygotic twins, *Ann. Neurol.* 26:100-104
Pandya, D.N. and Seltzer, B., 1986, The topography of commissural fibers, *in:* "Two Hemispheres - One Brain: Functions of the Corpus Callosum," A.R. Liss, NY, pp. 47-73.
Paivio, A., 1971, "Imagery and Verbal Processes," Rineholt and Wilson, NY.
Payne, B.R. and Siwek, D.F., 1991, The visual map in the corpus callosum of the cat, *Cerebral Cortex* 1:173-188.
Phelps, E.A., Hirst, W., and Gazzaniga, M.S., 1991, Deficits in recall following partial and complete commissurotomy, *Cerebral Cortex* 1:492-498.
Phelps, E.A. and Gazzaniga, M.S., 1992, Hemispheric differences in mnemonic processing: The effect of left hemisphere interpretation, *Neuropsychologia* 30:293-297.
Pizzamiglio, L., DePascalis, C., and Vignati, A., 1974, Stability of dichotic listening test, *Cortex* 10:203-205.
Pollatsek, A. and Rayner, K., 1989, Reading, *in:* "Foundations of Cognitive Science," M.I. Posner, ed., M.I.T. Press, Cambridge, pp. 401-436.
Posner, M.I., 1980, Orienting of attention, *Quart. J. Exp. Psychol.* 32:3-25.
Posner, M.I. and Petersen, S.E., 1990, The attention system of the human brain, *Ann. Rev. Neurosci.* 13: 25-42.
Posner, M.I., Walker, J.A., Friedrich, F.F., and Rafal, R.D., 1984, Effects of parietal injury on covert orienting of attention, *J. Neurosci.* 4:863-74.
Posner, M.I., Walker, J.A., Friedrich, F.F., and Rafal, R.D., 1987, How do the parietal lobes direct covert attention?, *Neuropsychologia* 25:135-146.
Regan, D., 1990, "Human Brain Electrophysiology," Elsevier, Amsterdam.
Reicher, G.M., 1969, Perceptual recognition as a function of meaningfulness of stimulus material, *J. Exp. Psychol.* 81:275-280.
Reuter-Lorenz, P.A. and Baynes, K., 1990, A pre-lexical basis for letter-by-letter reading: a case study, *J. Cognit. Neuropsychol.* 7:1-20.
Reuter-Lorenz, P.A. and Fendrich, R., 1990, Orienting attention across the vertical meridian: evidence from callosotomy patients, *J. Cognit. Neurosci.* 2:232-238.
Reuter-Lorenz, P.A. and Fendrich, R., 1992, Oculomotor programming and covert orienting: the effects of central versus peripheral precues, *Perc. Psychophys.* 52:336-344.

Rugg, M., Lines, C., and Milner, A., 1984, Visual evoked potentials to lateralized visual stimuli and the measurement of interhemispheric transfer time, *Neuropsychologia* 22:215-225.

Rugg, M., Milner, A., and Lines, C., 1985, Visual evoked potentials to lateralized stimuli in two cases of callosal agenesis, *J. Neurol. Neurosurg. Psychiat.* 48:367-373.

Russell, E.W., 1975, A multiple scoring method for assessment of complex memory functions, *J. Consult. Clin. Psychol.* 43:800-809.

Saron, C. and Davidson, R, 1989, Visual evoked potential measures of interhemispheric transfer time in humans, *Behav. Neurosci.* 103:1115-1138.

Sass, K.J., Spencer, D.D., Spencer, S.S. Novelly, R.A., Williamson, P.D., and Mattson, R.H., 1988, Corpus callosotomy for epilepsy. II. Neurologic and neuropsychological outcome, *Neurology* 38:24-38.

Sergent, J., 1987, A new look at the human split brain, *Brain* 110:1375-1392.

Sergent, J., 1990, Furtive incursions into bicameral minds: integrative and coordinating role of subcortical structures, *Brain* 113:537-568.

Seymour, S., Reuter-Lorenz, P.A., and Gazzaniga, M.S., 1994, The disconnection syndrome: basic findings reaffirmed, *Brain* (in press)

Sharpe, J.A., Lo, A. W., and Rabinovitch, H.E., 1979, Control of the saccadic and smooth pursuit systems after cerebral hemidecortication, *Brain* 102:387-403.

Sidtis, J.J., 1982, Predicting brain organization from dichotic listening performance: cortical and subcortical functional asymmetries contribute to perceptual asymmetries, *Brain Lang.* 17:287-300.

Sidtis, J.J., 1984, Music, pitch perception, and the mechanisms of cortical hearing, *in*: "Handbook of Cognitive Neuroscience," M.S. Gazzaniga, ed., Plenum, NY, pp. 91-114.

Sidtis, J.J., Volpe, B.T., Wilson, D.H., Rayport, M., and Gazzaniga, M.S., 1981, Variability in right hemisphere language function after callosal section: evidence for a continuum of generative capacity, *J. Neurosci.* 1:429-458.

Sparks, R. and Geschwind, N., 1968, Dichotic listening in man after section of neocortical commissures, *Cortex* 4:3-16.

Springer, S.P. and Gazzaniga, M.S., 1975, Dichotic testing of partial and complete split brain subjects, *Neuropsychologia* 13:341-346.

Stone, J., 1966, The naso-temporal division of the cat's retina, *J. Comp. Neurol.* 136:585-600.

Teng, E.L., 1981, Dichotic ear difference is a poor index for the functional asymmetry between the cerebral hemispheres, *Neuropsychologia* 19:235-240.

Terhardt, E., 1984, The concept of musical consonance: a link between music and psychoacoustics, *Music Perc.* 1:276-295.

Tramo, M.J. and Bharucha, J.J., 1991, Musical priming by the right hemisphere post-callosotomy, *Neuropsychologia* 29:313-325.

Tramo, M.J., Bharucha, J.J., and Musiek, F.E., 1990, Music perception and cognition following bilateral lesions of auditory cortex, *J. Cognit. Neurosci.* 2:195-212.

Tramo, M.J. and Gazzaniga, M.S., 1989, Discrimination and recognition of complex tonal spectra by the cerebral hemispheres: differential lateralization of acoustic-discriminative and semantic-associative functions in auditory pattern perception, *Soc. Neurosci. Abstr.* 15:1060.

Tramo, M.J., Guglielmo, M.A., Reuter-Lorenz, P., and Gazzaniga, M.S., 1990, Functional dissociations in hemispatial neglect: brainprints and quantitative lesion localization, *Ann. Neurol.* 28:255.

Tramo, M.J., Musiek, F.E., and Gazzaniga, M.S., 1991, Disruption of interhemispheric integration of complex auditory information following focal hemorrhage into the posterior body of the corpus callosum, *Soc. Neurosci. Abstr.* 17:1484

Trevarthan, C.B. and Sperry, R.W., 1973, Perceptual unity of the ambient visual field in human commissurotomy patients, *Brain* 96:547-570.

Troost, B.T., Weber, R.B., and Daroff, R.B., 1972, Hemispheric control of eye movements I: Quantitative analysis of refixation saccades in a hemispherectomy patient, *Arch. Neurol.* 27:441-448.

Victor, M. and Agamanolis, D., 1990, Amnesia due to lesions confined to the hippocampus: A clinical-pathologic study, *J. Cognit. Neurosci.* 2:246-257.

Webster, D.B., 1992, An overview of mammalian auditory pathways with an emphasis on humans, *in*: "The Mammalian Auditory Pathway: Neuroanatomy," D.B. Webster, A.N. Popper, and R.R. Fay, eds., Springer-Verlag, NY, pp. 1-22.

Wechsler, D.A., 1945, A standardized memory scale for clinical use, *J. Psychol.* 19:87-95.

Young, A.W. and Ellis, A.W., 1985, Different modes of lexical access for words presented in the left and right visual hemifields, *Brain Lang.* 24:326-358.

Zaidel, D. and Sperry, R.W., 1974, Memory impairment following commissurotomy in man, *Brain* 97:263-272.

Zaidel, E., 1976, Auditory vocabulary of the right hemisphere following brain bisection or hemidecortication, *Cortex* 12:191-211.

Zaidel, E., 1977, Unilateral auditory language on the token test following cerebral commissurotomy and hemispherectomy, *Neuropsychologia* 15:1-18.

Zaidel, E., 1978, Lexical organization in the right hemisphere, *in:* "Cerebral Correlates of Conscious Experience," P.A. Buser and A. Rougeul-Buser, eds., Elsevier, Amsterdam, pp. 177-187.

Zaidel, E., 1982, Reading in the disconnected right hemisphere: an aphasiological perspective, *in:* "Dyslexia: Neuronal, Cognitive, and Linguistic Aspects," Y. Zotterman, ed., Pergamon Press, Oxford, pp. 67-91.

Zaidel, E., 1983, A response to Gazzaniga: language in the right hemisphere, an empirical perspective, *Amer. Psych.* 38:542-546.

Zaidel, E., 1990, Language functions in the two hemispheres following complete cerebral commissurotomy and hemispherectomy, *in:* "Handbook of Neuropsychology, Vol. 4," F. Boller and J. Grafman, eds., Elsevier, Amsterdam, pp. 115-150.

Zola-Morgan, S. and Squire, L.R., 1993, Neuroanatomy of memory, *Ann. Rev. Neurosci.* 16:547-564.

INDEX

MIX
Papier aus verantwortungsvollen Quellen
Paper from responsible sources
FSC® C105338

If you have any concerns about our products,
you can contact us on
ProductSafety@springernature.com

In case Publisher is established outside the EU,
the EU authorized representative is:
Springer Nature Customer Service Center GmbH
Europaplatz 3, 69115 Heidelberg, Germany

Printed by Libri Plureos GmbH
in Hamburg, Germany